Lecture Notes in Physics

Editorial Board

H. Araki, Kyoto, Japan
E. Brézin, Paris, France
J. Ehlers, Potsdam, Germany
U. Frisch, Nice, France
K. Hepp, Zürich, Switzerland
R. L. Jaffe, Cambridge, MA, USA
R. Kippenhahn, Göttingen, Germany
H. A. Weidenmüller, Heidelberg, Germany
J. Wess, München, Germany
J. Zittartz, Köln, Germany

Managing Editor

W. Beiglböck
Assisted by Mrs. Sabine Lehr
c/o Springer-Verlag, Physics Editorial Department II
Tiergartenstrasse 17, D-69121 Heidelberg, Germany

Springer
Berlin
Heidelberg
New York
Barcelona
Budapest
Hong Kong
London
Milan
Paris
Santa Clara
Singapore
Tokyo

The Editorial Policy for Proceedings

The series Lecture Notes in Physics reports new developments in physical research and teaching – quickly, informally, and at a high level. The proceedings to be considered for publication in this series should be limited to only a few areas of research, and these should be closely related to each other. The contributions should be of a high standard and should avoid lengthy redraftings of papers already published or about to be published elsewhere. As a whole, the proceedings should aim for a balanced presentation of the theme of the conference including a description of the techniques used and enough motivation for a broad readership. It should not be assumed that the published proceedings must reflect the conference in its entirety. (A listing or abstracts of papers presented at the meeting but not included in the proceedings could be added as an appendix.)

When applying for publication in the series Lecture Notes in Physics the volume's editor(s) should submit sufficient material to enable the series editors and their referees to make a fairly accurate evaluation (e.g. a complete list of speakers and titles of papers to be presented and abstracts). If, based on this information, the proceedings are (tentatively) accepted, the volume's editor(s), whose name(s) will appear on the title pages, should select the papers suitable for publication and have them refereed (as for a journal) when appropriate. As a rule discussions will not be accepted. The series editors and Springer-Verlag will normally not interfere with the detailed editing except in fairly obvious cases or on technical matters.

Final acceptance is expressed by the series editor in charge, in consultation with Springer-Verlag only after receiving the complete manuscript. It might help to send a copy of the authors' manuscripts in advance to the editor in charge to discuss possible revisions with him. As a general rule, the series editor will confirm his tentative acceptance if the final manuscript corresponds to the original concept discussed, if the quality of the contribution meets the requirements of the series, and if the final size of the manuscript does not greatly exceed the number of pages originally agreed upon. The manuscript should be forwarded to Springer-Verlag shortly after the meeting. In cases of extreme delay (more than six months after the conference) the series editors will check once more the timeliness of the papers. Therefore, the volume's editor(s) should establish strict deadlines, or collect the articles during the conference and have them revised on the spot. If a delay is unavoidable, one should encourage the authors to update their contributions if appropriate. The editors of proceedings are strongly advised to inform contributors about these points at an early stage.

The final manuscript should contain a table of contents and an informative introduction accessible also to readers not particularly familiar with the topic of the conference. The contributions should be in English. The volume's editor(s) should check the contributions for the correct use of language. At Springer-Verlag only the prefaces will be checked by a copy-editor for language and style. Grave linguistic or technical shortcomings may lead to the rejection of contributions by the series editors. A conference report should not exceed a total of 500 pages. Keeping the size within this bound should be achieved by a stricter selection of articles and not by imposing an upper limit to the length of the individual papers. Editors receive jointly 30 complimentary copies of their book. They are entitled to purchase further copies of their book at a reduced rate. As a rule no reprints of individual contributions can be supplied. No royalty is paid on Lecture Notes in Physics volumes. Commitment to publish is made by letter of interest rather than by signing a formal contract. Springer-Verlag secures the copyright for each volume.

The Production Process

The books are hardbound, and the publisher will select quality paper appropriate to the needs of the author(s). Publication time is about ten weeks. More than twenty years of experience guarantee authors the best possible service. To reach the goal of rapid publication at a low price the technique of photographic reproduction from a camera-ready manuscript was chosen. This process shifts the main responsibility for the technical quality considerably from the publisher to the authors. We therefore urge all authors and editors of proceedings to observe very carefully the essentials for the preparation of camera-ready manuscripts, which we will supply on request. This applies especially to the quality of figures and halftones submitted for publication. In addition, it might be useful to look at some of the volumes already published. As a special service, we offer free of charge LaTeX and TeX macro packages to format the text according to Springer-Verlag's quality requirements. We strongly recommend that you make use of this offer, since the result will be a book of considerably improved technical quality. To avoid mistakes and time-consuming correspondence during the production period the conference editors should request special instructions from the publisher well before the beginning of the conference. Manuscripts not meeting the technical standard of the series will have to be returned for improvement.

For further information please contact Springer-Verlag, Physics Editorial Department II, Tiergartenstrasse 17, D-69121 Heidelberg, Germany

Frieder Lenz   Harald Grießhammer
Dieter Stoll (Eds.)

Lectures on QCD

Foundations

Springer

Editors

Frieder Lenz
Harald Grießhammer
Dieter Stoll
Institut für Theoretische Physik III
Universität Erlangen-Nürnberg
Staudtstrasse 7
D-91058 Erlangen, Germany

Cataloging-in-Publication Data applied for.

Die Deutsche Bibliothek - CIP-Einheitsaufnahme

Lectures on QCD / Frieder Lenz ... (ed.). - Berlin ; Heidelberg
; New York ; Barcelona ; Budapest ; Hong Kong ; London ;
Milan ; Paris ; Santa Clara ; Singapore ; Tokyo : Springer, 1997
 (Lecture notes in physics ; 481)
 ISBN 3-540-62543-7
NE: Lenz, Frieder [Hrsg.]; GT

ISSN 0075-8450
ISBN 3-540-62543-7 Springer-Verlag Berlin Heidelberg New York

This work is subject to copyright. All rights are reserved, whether the whole or part of the material is concerned, specifically the rights of translation, reprinting, re-use of illustrations, recitation, broadcasting, reproduction on microfilms or in any other way, and storage in data banks. Duplication of this publication or parts thereof is permitted only under the provisions of the German Copyright Law of September 9, 1965, in its current version, and permission for use must always be obtained from Springer-Verlag. Violations are liable for prosecution under the German Copyright Law.

© Springer-Verlag Berlin Heidelberg 1997
Printed in Germany

The use of general descriptive names, registered names, trademarks, etc. in this publication does not imply, even in the absence of a specific statement, that such names are exempt from the relevant protective laws and regulations and therefore free for general use.

Typesetting: Camera-ready by the authors/editors
Cover design: *design & production* GmbH, Heidelberg
SPIN: 10550510   55/3144-543210 - Printed on acid-free paper

Preface

The two volume set "Lectures on QCD" provides an introductory overview of Quantum Chromodynamics, the theory of strong interactions. In a series of articles, the fundamentals of QCD are discussed and significant areas of applications are described. Emphasis is put on recent developments. The field-theoretic basis of QCD is the focus of the first volume. The topics discussed include lattice gauge theories, anomalies, finite temperature field theories, sum-rules, the Skyrme model, and supersymmetric QCD. Applications of QCD to the phenomenology of strong interactions form the subject of the second volume. There, investigations of deep inelastic lepton–nucleon scattering, of high energy hadronic reactions and studies of the quark–gluon plasma in relativistic heavy ion collisions are presented.

These articles are based on lectures delivered by internationally well known experts on the occasion of a series of workshops organised by the "Graduiertenkolleg on Strong Interaction Physics" of the Universities of Erlangen-Nürnberg and Regensburg in the years 1992–1995. The workshops were held at "Kloster Banz". Kloster Banz is a former monastery overlooking the valley of the river Main and still serves, for some days of the year, as the stage where certain canons and orthodoxies are vigorously formulated.

Inspired by the atmosphere of the site, the workshops were set up with the aim of introducing novices in the field to the basics of QCD. Accordingly, the character of the lectures was pedagogical rather than technical. With the organisation of these workshops we have attempted to establish a new form in graduate education. Graduate students of the "Graduiertenkolleg" constituted a large fraction of the audience. They have worked out these articles on QCD in collaboration with the lecturers.

Thanks are due to Jutta Geithner for technical help in the preparation of these proceedings. The support of the "Graduiertenkolleg" by the Deutsche Forschungsgemeinschaft was instrumental in this endeavor and is gratefully acknowledged.

Erlangen, January 1997   F. Lenz
                                                            H. W. Grießhammer
                                                            D. Stoll

Contents

**Fascinating Field Theory: Quantum Field Theory,
Renormalization Group, and Lattice Regularization**
P. Hasenfratz; Notes by F. Kleefeld and T. Kraus 1

**Lattice Gauge Theory
and the Structure of the Vacuum and Hadrons**
J.W. Negele; Notes by H.W. Grießhammer and D. Lehmann 36

Topological Effects on the Physics of the Standard Model
R. Jackiw; Notes by H.W. Grießhammer, O. Schnetz, G. Fischer,
and S. Simbürger 90

Semiclassical Aspects of Quantum Field Theories
L. O'Raifeartaigh; Notes by M. Engelhardt and S. Lenz 128

Anomalies in Gauge Theories
M.A. Shifman; Notes by M. Engelhardt and D. Stoll 157

QCD Sum Rules
M.A. Shifman; Notes by M. Engelhardt and D. Stoll 170

The Skyrme Model
I. Klebanov; Notes by M. Engelhardt and D. Stoll 188

Introduction to Supersymmetry
Notes by H.W. Grießhammer, D. Lehmann, and M. Seeger 217

Subject Index 272

Fascinating Field Theory: Quantum Field Theory, Renormalization Group, and Lattice Regularization*

P. Hasenfratz[1];
Notes by F. Kleefeld[2] and T. Kraus[2]

[1] Institute for Theoretical Physics, University of Bern, Siedlerstrasse 5, 3012 Bern, Switzerland
[2] Institute for Theoretical Physics, University of Erlangen–Nürnberg, Staudtstr. 7, 91058 Erlangen, Germany

1 Introduction

Quantum field theory gives a unified prescription of physical phenomena where collective behaviour plays an important role. This collective behaviour is independent of the microscopical details and can lead to mind-boggling phenomena such as dynamical mass generation, spontaneous symmetry breaking, confinement, etc. The main purpose of these lectures is to communicate our fascination about the way quantum field theories work.

In the introduction we try to explain and motivate some of the notions entering the title of these lectures and answer a few 'why questions' concerning the subject.

1.1 Why Field Theory (As Opposed to Point Mechanics)?

In most of the interesting problems in physics a large number of degrees of freedom are involved. Very often we are not interested in the individual motion of one of the degrees of freedom, but rather in the collective phenomena created by the correlated motion of a large number of variables.

Consider a pendulum, for example. Its motion is a simple problem in classical mechanics. If we consider many of them coupled by springs, modes might occur which involve the coordinated, correlated motion of all the degrees of freedom (*waves*). For these kinds of phenomena a field theoretic description is very natural.

---

* Lectures presented at the workshop "Lattice QCD and Dense Matter" organised by the Graduiertenkolleg Erlangen–Regensburg, held on October 11th–13th, 1994 in Kloster Banz, Germany

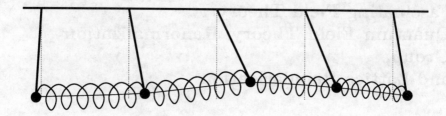

Fig. 1. *Collective phenomena.*

1.2 Why Quantum Field Theory (QFT) (Rather Than Quantum Mechanics (QM)) for the Collective Excitations?

In microphysics, the birth and death, the *creation* and *annihilation* of excitations (particles) is a basic feature, which is observed everywhere.

Take for example the absorption and emission of a photon by an atom. QM already has problems with this relatively simple example, especially with induced emission.[1] There are, however, high energy scattering events where the number of produced particles lies in the hundreds, even in the thousands. QM works with normalized wavefunctions and the associated interpretation excludes the possibility of particle creation and annihilation. We need a prescription with states of an arbitrary number of exitations and with the possibility to communicate between them. This is just what QFT provides us with.

1.3 Why Lattice Regularization?

Some of the mathematical operations which enter the definition of a QFT require a careful limiting procedure. In field theory, the variables are associated with space-time points. These variables have some kind of self-interaction, whereas the elementary interaction between different degrees of freedom is over infinitesimal distances as expressed by derivatives. Already in the classical theory, the definition of a derivative requires the temporary introduction of a finite increment (of the argument of the function considered) which disappears at the end by some limiting procedure

$$f'(x) = \lim_{a \to 0} \frac{f(x+a) - f(x)}{a}. \tag{1}$$

Similarly, the very definition of a QFT requires the temporary introduction of a defining framework called *regularization*, which disappears from the theory by a limiting process. The way to introduce and remove the regularization is a highly non-trivial problem.

Accepting the fact that we need some regularization scheme – why should we resort to *lattice* regularization? Let us mention some of the reasons.

---

[1] Strictly speaking, it does not belong to the realm of QM.

- In general, in any problems which require numerical analysis (e.g. integrals, partial differential equations, ...), the standard procedure is to introduce meshes. Also the path integral can be defined in a natural way by introducing meshes. Consider the probability amplitude for the propagation of a particle from the coordinates (x_i, t_i) to (x_f, t_f) in QM:

$$K(x_f, t_f; x_i, t_i) = \sum_{\text{all paths}} e^{iS_{\text{cl}}(x(t))}, \qquad (2)$$

where S_{cl} is the classical action and the sum extends over all paths connecting the points (x_i, t_i) and (x_f, t_f). The sum over all paths in equation (2) can be defined by introducing a space-time lattice (see Fig. 2)).

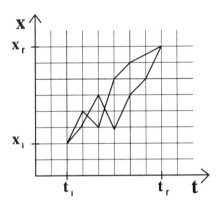

Fig. 2. *Paths on a space-time lattice.*

- Lattice regularization overcomes some of the shortcomings of other schemes. Lattice regularization is defined non-perturbatively, unlike, for example, dimensional regularization which can only be applied in the context of perturbation theory.
  Lattice regularization is very convenient for gauge systems, or for theories with constrained variables. Choosing a momentum cutoff, the constraints may become very cumbersome. Consider for example the non-linear σ model, where the spins lie on a unit sphere: $\mathbf{S}(\mathbf{x})^2 = 1$. This constraint is simple in configuration space, but complicated in momentum space. Even worse, a momentum cutoff violates gauge invariance.

1.4 Why Renormalization Group?

We begin with a slight detour, asking ourselves about the common feature of all field theories (QCD, QED, electro-weak interaction, Φ^4, ...). Their most important common property is *locality*. Locality means that the elementary interaction in the action takes place only between infinitesimally separated points. On the other hand, we are interested in finite correlation lengths of the systems described by field theories. In QCD, for example, we want to obtain a correlation length of the order fm, which is the typical scale for hadronic objects. Therefore some miraculous thing has to happen that ensures we obtain a finite correlation length at the end in the solution (in the Greens functions), despite the fact that the distance of our basic interaction in the action approaches zero. Collective behavior is needed to overcome this paradox.

Let us look at the relevant scales in momentum space (Fig. 3). We are interested in some region of physical momenta which is well below the cutoff. The predictions in this physical region are influenced by by the high momentum modes also. At first sight this seems to be a problem, since we do not know what the elementary interaction between, say, an electron and a photon at the cut-off should look like. For the field theoretical prescription to be successful we need the results of the theory in the physical region to be independent of the details of the interaction near the cutoff (i.e. independent of the details of the interaction, or of the regularization scheme). This property, called universality, is produced typically by the collective motion of a large number of degrees of freedom. This is the property which allows us to calculate the anomalous magnetic moment of the electron to many digits precision in spite of our limited knowledge on the form of the interaction at very high energies.

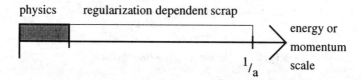

Fig. 3. *Momentum scales.*

It would be useful to define a procedure which eliminates the uninteresting degrees of freedom at high momenta, but takes into account their effect on the remaining variables exactly. This procedure is the *renormalization group transformation (RG)*. RG ideas not only provide a deep insight into the way QFTs work, on properties like universality and renormalizability of a theory, but they lead to powerful quantitative methods also [1].

2 Quantum Field Theories (QFTs) Versus Critical Phenomena in Classical Statistical Physics

2.1 Formal Relation Between QFTs and Classical Statistical Physics

Consider an n-point function in QFT in Minkowski space, with $\Phi(x)$ as the generic field

$$\langle 0 | T\left(\hat{\Phi}(x_1), \hat{\Phi}(x_2), \ldots, \hat{\Phi}(x_n)\right) | 0 \rangle. \tag{3}$$

In path integral language this expression is equal to

$$\text{eq.}(3) = \frac{\int D\Phi\, \Phi(x_1)\Phi(x_2)\ldots\Phi(x_n) \exp\left(i \int_{-\infty}^{+\infty} dt \int_{-\infty}^{+\infty} d\mathbf{x}\, \mathcal{L}(\Phi, \partial_\mu \Phi)\right)}{Z}, \tag{4}$$

where $Z = \int D\Phi \exp\left(i \int dt \int d\mathbf{x}\, \mathcal{L}\right)$ is the normalization and $x = (t, \mathbf{x}) = (t, x_1, \ldots, x_{d_s})$. For the Lagrangian $\mathcal{L}$ we might take

$$\mathcal{L}_{\text{Minkowski}} = \frac{1}{2}\partial_0 \Phi \partial_0 \Phi - \sum_{i=1}^{d_s} \frac{1}{2}\partial_i \Phi \partial_i \Phi - V(\Phi). \tag{5}$$

For many reasons it is helpful to perform an analytic continuation and go to Euclidean space,

$$t = x_0 \to -i x_d$$

$$i \int dt \int d\mathbf{x} \to \int d^d x \tag{6}$$

where $d = d_s + 1$. By Wick rotation, the relative minus sign between the time and the spatial derivatives disappears,

$$\mathcal{L}_{\text{Minkowski}} \to -\left(\sum_{i=1}^{d} \frac{1}{2}\partial_i \Phi \partial_i \Phi + V(\Phi)\right) = -\mathcal{E}(\Phi, \partial \Phi), \tag{7}$$

and we get

$$Z = \int D\Phi \exp\left(-\int d^d x\, \mathcal{E}(\Phi, \partial \Phi)\right), \tag{8}$$

$$\langle \Phi(x_1)\Phi(x_2)\ldots\Phi(x_n)\rangle = \frac{\int D\Phi\, \Phi(x_1)\Phi(x_2)\ldots\Phi(x_n) \exp\left(-\int d^d x\, \mathcal{E}\right)}{Z}. \tag{9}$$

Equations (8,9) are nothing but the partition function and correlation functions for a system in classical statistical mechanics. The QFT of fields in d_s space dimensions is transformed – by continuation to Euclidean space – into classical statistical physics of fields in $d = d_s + 1$ Euclidean dimensions. The Euclidean action plays the role of $E/k_B T$, where E is the classical energy of the d dimensional system.

2.2 Mass (Gap) m, Correlation Length ξ

The n-point (correlation) functions also determine the spectrum of the theory. Take (in Minkowski space) the operator in the Heisenberg picture

$$\hat{\Phi}(t, \mathbf{0}) = e^{i\mathcal{H}t}\hat{\Phi}_S e^{-i\mathcal{H}t}, \tag{10}$$

where $\hat{\Phi}_S$ is the field operator in the Schrödinger picture at the space point $\mathbf{0}$. Then

$$\langle 0 | \hat{\Phi}(t, \mathbf{0}) \hat{\Phi}(0, \mathbf{0}) | 0 \rangle = \sum_n e^{-i(E_n - E_0)t} \left|\langle 0 | \hat{\Phi}_S | n \rangle\right|^2, \tag{11}$$

where we have used the closure relation

$$1 = \sum_n | n \rangle \langle n | \tag{12}$$

for the energy eigenstates $| n \rangle$ of the system. If we perform the analytic continuation to Euclidean space, we obtain

$$t \to -ix_d$$

$$\langle 0 | \hat{\Phi}(t, \mathbf{0}) \hat{\Phi}(0, \mathbf{0}) | 0 \rangle \to \sum_n e^{-(E_n - E_0)x_d} |c_n|^2. \tag{13}$$

If the state $\hat{\Phi}_S | 0 \rangle$ is orthogonal to the groundstate $| 0 \rangle$ then

$$\langle 0 | \hat{\Phi}(t, \mathbf{0}) \hat{\Phi}(t, \mathbf{0}) | 0 \rangle \stackrel{x_d \to \infty}{\Longrightarrow} e^{-\Delta E x_d} |c_1|^2,$$

$$\Delta E = E_1 - E_0 \doteq m_{df}. \tag{14}$$

The dimensionful quantity m_{df} is the *mass gap* in the channel of $\hat{\Phi}$. As m_{df} determines how fast the correlator in (14) falls off as a function of x_d, the inverse of m_{df} can be identified with the correlation length ξ_{df}:

$$\xi_{df} = \frac{1}{m_{df}}. \tag{15}$$

Note again that m_{df} and ξ_{df} are dimensionful physical observables and their physical units are given by: $[m_{df}] = g$, $[\xi_{df}] = cm$.

2.3 QFTs Versus Critical Phenomena

There is a deep similarity between the problems of a QFT and those of critical phenomena. For illustration, consider a large piece of magnet. The magnet is built of atoms forming a lattice structure with a lattice distance $a \sim O(\text{Å})$. At a given generic temperature $T > T_C$, the net magnetization is zero, the distance over which the magnetic spins are correlated (correlation length) is of the order of the lattice spacing (see Fig. 4 on the next page). Let us tune the temperature towards the critical temperature where where spontaneous magnetization

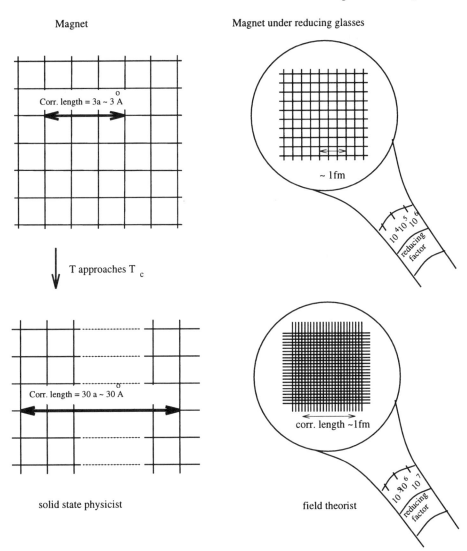

Fig. 4. *Solid State Physics versus Field Theory.*

sets in. In approaching T_C, the correlation length gets larger and larger, several hundreds of lattice units (hundreds of Ås), and beyond. A field theorist looks at this problem through his special glasses. These are reducing glasses. Through his glasses, the original lattice unit and correlation length is $O(1fm)$. This is a trivial rescaling. It is more important that the field theorist increases the reducing factor of his glasses in such a way that through his glasses, the correlation length stays the order of $1fm$ even when the temperature gets close to the Curie point (see Fig. 4). For the solid state physicist, the correlation length approaches macroscopical distances, while looking through the increasingly reducing glasses of the field theorist, the correlation length remains constant, but the lattice becomes increasingly fine grained. At the Curie point, the field theorist obtains a continuum field theory ($a \to 0$) with finite $O(1 fm)$ correlation length. His colleague has a lattice with fixed lattice unit ($a \sim 1$ Å) and macroscopical correlation length. It is clear, however, that the main difficulty – obtaining and controlling the large scale collective behavior – is a common problem for them.

3   Renormalization Group (RG) Transformation

We shall discuss RG transformations both in momentum and in configuration space. The systematic elimination of large momentum (short distance) fluctuations is done by:

– *thinning out* the variables
– *relabelling* and *rescaling* the remaining variables

The momentum space regularization with a sharpe cutoff is conceptually simple and it is well suited to explain the steps of a RG transformation. For a free field theory and in leading order of perturbation theory it leads to a technically simple problem. We start our discussion with this case. In general, however the sharpe cutoff creates difficulties. In addition, momentum space regularization is not suited for contrained and gauge systems, as we discussed before. All these problems are avoided by using a regularization in configuration space. In the second part of this section we discuss RG transformations in this context. In the following the notation $\Phi(x)$ is used for a scalar field, $\tilde{\Phi}(k)$ for its Fourier transform:

$$\Phi(x) = \int_k e^{-ikx} \tilde{\Phi}(k) \quad ,$$

using the following abbreviation for the d–dimensional integration in Euclidean momentum space:

$$\int_k = \int_{-\infty}^{\infty} \frac{d^d k}{(2\pi)^d} \quad .$$

3.1   RG Transformation in Momentum Space

We choose a cutoff in momentum space:

$$|k| \leq \Lambda^{\text{cut}} \quad .$$

The next step is to express everything in terms of *dimensionless* quantities in units of the cutoff, i.e. introduce the dimensionless momentum variable q:

$$q_\mu = \frac{k_\mu}{\Lambda^{\text{cut}}} \quad \Rightarrow \quad |q| \leq 1 \quad ,$$

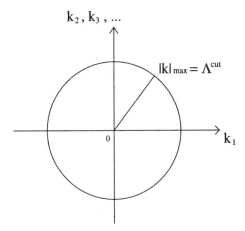

Fig. 5. *Considered range of momentum.*

and the dimensionless field variables $\varphi(q)$:

$$\varphi(q) = \tilde{\Phi}(k) \cdot \left(\Lambda^{\mathrm{cut}}\right)^{\frac{d+2}{2}} .$$

The generating functional in Euclidean space then reads

$$Z = \prod_{|q|\leq 1} \int d\varphi(q)\, e^{-S(\varphi)} . \tag{16}$$

The idea is now to integrate out in a cascading procedure the irrelevant high momentum parts of the degrees of freedom in Z to come down to relevant momentum scales which determine physics. The cascading procedure is done as follows. One splits up the fields $\varphi(q)$ into two parts, $\varphi_0(q)$, which is defined in the region $|q| \leq \frac{1}{2}$, and $\varphi_1(q)$, which is defined in the region $\frac{1}{2} \leq |q| \leq 1$:

$$\varphi(q) = \underbrace{\varphi_0(q)}_{|q|\leq \frac{1}{2}} + \underbrace{\varphi_1(q)}_{\frac{1}{2}\leq |q|\leq 1} . \tag{17}$$

Integration over $\varphi_1(q)$ (*"thinning out"*) leads to the effective action $\overline{S}(\varphi_0)$:

$$Z = \prod_{|q|\leq \frac{1}{2}} \int d\varphi_0(q) \prod_{\frac{1}{2}\leq |q|\leq 1} \int d\varphi_1(q)\, e^{-S(\varphi_0 + \varphi_1)} =$$

$$= \prod_{|q|\leq \frac{1}{2}} \int d\varphi_0(q)\, e^{-\overline{S}(\varphi_0)} .$$

Now one replaces the dimensionless momentum q by $q' = 2q$ (*"relabelling"*) where

$$0 \leq |q'| \leq 1 .$$

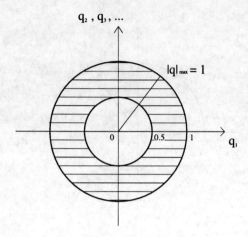

Fig. 6. Range of q (The shaded region is integrated out by one RG transformation step.).

This leads to the following change in the field variables ("*rescaling*"):

$$\varphi'(q') = \underbrace{\zeta^{-1}}_{\text{const}} \varphi_0(q) \quad . \tag{18}$$

For the generating functional one obtains

$$Z = \prod_{|q'|\leq 1} \int d\varphi'(q') e^{-S'(\varphi')} \stackrel{!}{=} \prod_{|q|\leq 1} \int d\varphi(q) e^{-S'(\varphi)} \quad . \tag{19}$$

In the second step primed fields and integration variables were just replaced by unprimed symbols. The net effect of the RG transformation between eqs.(16) and (19) is just the following change in the action:

$$S(\varphi) \xrightarrow{RG} S'(\varphi) \quad .$$

The low energy (long distance) correlations remain unchanged by the RG transformation as one can check ($|q_1|, |q_2| \ll 1$):

$$<\varphi(q_1)\varphi(q_2)>|_S \quad =$$

$$= \frac{1}{Z} \prod_{|q|\leq 1} \int d\varphi(q)\, \varphi(q_1)\, \varphi(q_2)\, e^{-S(\varphi)} \quad =$$

$$= \frac{1}{Z} \prod_{|q| \le \frac{1}{2}} \int d\varphi_0(q) \; \varphi_0(q_1) \varphi_0(q_2) \prod_{\frac{1}{2} \le |q| \le 1} \int d\varphi_1(q) \, e^{-S(\varphi_0 + \varphi_1)} =$$

$$= \frac{1}{Z} \prod_{|q| \le \frac{1}{2}} \int d\varphi_0(q) \; \varphi_0(q_1) \varphi_0(q_2) \, e^{-\overline{S}(\varphi_0)} \qquad \text{eq. (18)}$$

$$= \frac{1}{Z} \zeta^2 \prod_{|q'| \le 1} \int d\varphi'(q') \; \varphi'(\frac{q_1}{2}) \varphi'(\frac{q_2}{2}) \, e^{-S'(\varphi')} \; .$$

We thus have the following expression for the correlation function

$$\underbrace{<\varphi(q_1)\varphi(q_2)>|_S}_{\sim \delta(q_1+q_2)\,G^{(2)}(q_1)\big|_S} = \zeta^2 \underbrace{<\varphi'(2q_1)\varphi'(2q_2)>|_{S'}}_{\sim \delta(2q_1+2q_2)\,G^{(2)\prime}(2q_1)\big|_{S'}}$$

Using $\delta(2q) = 2^{-d}\delta(q)$ one obtains the relation for the two point Greens functions:

$$G^{(2)}(q)\Big|_S = \zeta^2 \, 2^{-d} \, G^{(2)\prime}(2q)\Big|_{S'} \; . \qquad (20)$$

If the propagator has a pole at $q^2 = -m^2$ (m is measured in Λ^{cut} units, $m \ll 1$), i.e. $G^{(2)}(q) \sim (q^2 + m^2)^{-1}$, then the pole will be shifted by the RG transformation to the position $m' = 2m$, as

$$G^{(2)\prime}(q) \sim \frac{1}{\left(\frac{q}{2}\right)^2 + m^2} \sim \frac{1}{q^2 + (2m)^2} \sim \frac{1}{q^2 + (m')^2} \; .$$

Although the dimensionful mass $m_{df} = m * \Lambda^{\text{cut}}$ and so the dimensionful correlation length $\xi_{df} \sim (m_{df})^{-1}$ is not changed by the RG transformation, we have the following changes for the action $S(\varphi)$, the dimensionless mass m, the dimensionless correlation length $\xi \sim m^{-1}$ and the cutoff Λ^{cut} itself:

$$S(\varphi) \xrightarrow{RG} S'(\varphi)$$

$$m \xrightarrow{RG} 2m$$

$$\xi \xrightarrow{RG} \frac{\xi}{2}$$

$$\Lambda^{\text{cut}} \xrightarrow{RG} \frac{\Lambda^{\text{cut}}}{2} \; . \qquad (21)$$

3.2 RG Transformation in Configuration Space (Lattice)

We shall consider a d dimensional hypercubic lattice in Euclidean space. The field φ is defined on the lattice points n, where n is a d dimensional vector of integers. On the lattice the generating functional Z is written as:

$$Z = \prod_n \int d\varphi_n \, e^{-S(\varphi)} \quad , \tag{22}$$

where n runs over all lattice sites. A RG transformation in configuration space averages out the short distance fluctuations, i.e. fluctuations over distances $O(a)$, where a is the lattice unit. We form blocks on the lattice labelled by the block index n_B (see Fig. 7). To every block we associate a block variable φ', which is

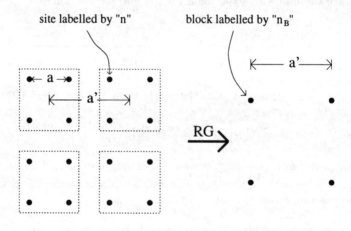

Fig. 7. *Lattice before and after block transformation.*

an average of the original fields in the block. Integrating over the fields φ keeping the block averages fixed, Z is expressed in terms of a functional integral over the block fields:

$$Z = \prod_n \int d\varphi_n \overbrace{\prod_{n_B} \int d\varphi'_{n_B} \, \delta(\varphi'_{n_B} - b \sum_{n \in n_B} \varphi_n)}^{1} e^{-S(\varphi)} =$$

$$= \prod_{n_B} \int d\varphi'_{n_B} \prod_n \int d\varphi_n \, \delta(\varphi'_{n_B} - b \sum_{n \in n_B} \varphi_n) \, e^{-S(\varphi)} =$$

$$= \prod_{n_B} \int d\varphi'_{n_B} \, e^{-S'(\varphi')} \quad ,$$

where
$$e^{-S'(\varphi')} = \prod_n \int d\varphi_n \, \delta(\varphi'_{n_B} - b \sum_{n \in n_B} \varphi_n) \, e^{-S(\varphi)} \quad . \tag{23}$$

Here n_B extends over all blocks, b is a free parameter at this stage, which corresponds to the rescaling parameter ζ in (18), and $S'(\varphi')$ is the resulting effective action after performing the block transformation. The lattice unit $a \sim (\Lambda^{\text{cut}})^{-1}$ transforms under the considered RG transformations as:

$$a \xrightarrow{RG} 2a = a' \quad .$$

The dimensionful and dimensionless correlation length transforms as in (21):

$$\xi_{df} \xrightarrow{RG} \xi_{df}$$

$$\xi \xrightarrow{RG} \frac{\xi}{2} \quad .$$

Let us close this section by discussing a basic assumption of the RG theory. The constrained functional integral in (23) seems to be technically more complicated than the original functional integral for Z. How the RG idea can help then to get quantitative predictions out of the model considered? The point is that functional integrals, or partition sums are not always very difficult. Consider for example a classical statistical system at very high temperatures. The partition sum (and other quantities) can be calculated to high precision by using the technique of high temperature expansion. The physical reason that such an expansion works is that at high temperatures the system has short range fluctuations only. The partition sum becomes really difficult to calculate at temperatures when the system has long range fluctuations, when collective behaviour occurs, when many terms in the partition sum contribute to the final result essentially. The functional integral in (23) is not expected to be difficult in this sense. The constraints expressed by the δ-functions are expected to disrupt the long-range fluctuations of the system. Actually, this is a basic assumption of the RG theory: the integrals over the high momentum (short distance) components of the system define a non-critical problem with short distance fluctuations only.

4 Fixed Point (FP), Behavior in the Vicinity of a FP

Let us see, how the action is changed by RG transformations. The transformed action $S'(\Phi')$, in general, will contain all kinds of different interactions, even if the original action $S(\Phi)$ had a simple form.

To begin with, we take an action of the general form

$$S(\Phi) = \sum_\alpha K_\alpha \theta_\alpha(\Phi) \quad ,$$

where $\theta_\alpha(\Phi)$ ($\alpha = 1, 2, \ldots$) denotes all kinds of different interaction terms like

$$\theta_\alpha(\Phi), : \int_x \Phi^2(x), \int_x \Phi^4(x), \int_x (\partial_\mu \Phi(x))^2, \int_x (\partial_\mu \Phi(x))^2 \Phi^2(x), \ldots \int_x \Phi^{18}(x), \ldots \tag{24}$$

and K_α are the corresponding dimensionless couplings (all coupling constants can be made dimensionless by multiplying with a suitable power of the cutoff Λ^{cut}). The transformed action $S'(\Phi')$ is also expanded in terms of these interaction terms

$$S'(\Phi') = \sum_\alpha K'_\alpha \theta_\alpha(\Phi') \ .$$

Thus, under repeated RG transformations the set of all couplings and the corresponding dimensionless correlation length changes step by step in the following manner:

$$\{K_\alpha\} \xrightarrow{RG} \{K'_\alpha\} \xrightarrow{RG} \{K''_\alpha\} \xrightarrow{RG} \ldots$$

$$\xi \xrightarrow{RG} \frac{\xi}{2} \xrightarrow{RG} \frac{\xi}{2^2} \xrightarrow{RG} \ldots \ .$$

The RG transformation can have *fixed points* $\{K^*_\alpha\}$ (FP) in the space of couplings which have the property

$$\{K^*_\alpha\} \xrightarrow{RG} \{K^*_\alpha\} \ .$$

As the dimensionless correlation length has to decrease at each RG step, at a FP it has to be infinity, i.e. $\xi = \infty$ (the other possibility $\xi = 0$ is not interesting for us). A set of points in the space of couplings K_α, where $\xi = \infty$, forms a hypersurface which is called *critical surface*. Since $\xi \xrightarrow{RG} \frac{1}{2}\xi$, an RG transformation drives the point $\{K_\alpha\}$ *away* from a critical surface, except when $\{K_\alpha\}$ is *on* the critical surface.

Let us consider now the behavior of the action under RG transformations in the vicinity of a FP: Assume there is a FP $\{K^*_\alpha\}$ on the critical surface. Take a point $\{K_\alpha\}$ in the neighborhood of the fixed point, i.e. $\Delta K_\alpha = K_\alpha - K^*_\alpha$ is small. Perform now an RG transformation

$$\{K_\alpha\} \xrightarrow{RG} \{K'_\alpha\} \quad , \quad \Delta K'_\alpha = K'_\alpha - K^*_\alpha \ .$$

We can now express $\Delta K'_\alpha$ in terms of ΔK_α by an expansion

$$\Delta K'_\alpha = \Delta K'_\alpha(\{K\}) = \underbrace{\Delta K'_\alpha(\{K^*\})}_{0} + \sum_\beta \frac{\partial}{\partial K_\beta} \Delta K'_\alpha(\{K^*\}) \cdot \Delta K_\beta + O\left((\Delta K)^2\right)$$

A RG transformation which takes only into account the linear term of this ΔK-expansion, is usually called *linearized RG transformation*. With the definition:

$$T_{\alpha\beta} = \left. \frac{\partial}{\partial K_\beta} \Delta K'_\alpha(\{K\}) \right|_{K=K^*}$$

the equation can be rewritten considering only the linear term in the ΔK-expansion

$$\Delta K'_\alpha = \sum_\beta T_{\alpha\beta} \Delta K_\beta \ .$$

Diagonalization of $T_{\alpha\beta}$ gives the eigenvectors h^a ($a = 1, 2, \ldots$) and the corresponding eigenvalues λ^a [2]. We distinguish three cases:

- $|\lambda^a| > 1$: The eigenvalue is called *relevant*.
- $|\lambda^a| = 1$: The eigenvalue is called *marginal*.
- $|\lambda^a| < 1$: The eigenvalue is called *irrelevant*.

The reason for these names is the following. Using the notation

$$h^a(\Phi) = \sum_\alpha h^a_\alpha \, \theta_\alpha(\Phi) \ ,$$

the action can be expanded in the eigenbasis $h^a(\Phi)$:

$$S(\Phi) = S^*(\Phi) + \sum_a c^a \, h^a(\Phi) \ ,$$

where c^a are the corresponding expansion coefficients. Repeated application of RG transformations changes the action close to the FP in the following way:

$$S(\Phi) \xrightarrow{RG} S^*(\Phi) + \sum_a \lambda^a \, c^a \, h^a(\Phi) \xrightarrow{RG} S^*(\Phi) + \sum_a (\lambda^a)^2 \, c^a \, h^a(\Phi) \xrightarrow{RG} \ldots \ .$$

What one sees is that on one hand there is a part $S^*(\Phi)$ in the action, which is unchanged by successive RG transformations. $S^*(\Phi)$ is called the *fixed point action*. On the other hand, the interaction terms associated with an irrelevant eigenvalue, the so called *irrelevant interactions* or *irrelevant operators* are suppressed by each RG transformation step, while terms associated with a relevant eigenvalue (*relevant interactions* or *relevant operators*) are enhanced. For terms belonging to marginal eigenvalues it is not immediately clear what will happen. One has to look to higher order terms in the ΔK-expansion which determine whether the corresponding interaction term will be suppressed or enhanced by RG transformations or whether this property remains in higher orders also implying the existence of a fixed line rather than a fixed point.

Let us consider for illustration the $d = 3 (= 2 + 1)$ scalar field theory (see Fig. 8). Analytical and numerical results suggest the following scenario. Two FPs (FP$^{(1)}$ and FP$^{(2)}$) can be identified which have to lie in the critical plane. FP$^{(2)}$ is positioned at the origin of the coupling space. This FP is called the *Gaussian FP*. The Gaussian FP itself (here and in models with other field content) describes massless free field theories, but its neighbourhood might reveal more interesting physical properties. As one can see in Fig. 8, at each FP there

---

[2] Actually, this step requires more care: $T_{\alpha\beta}$ is not a symmetric matrix, in general. For the associated technical complications see, for example, ref. [1].

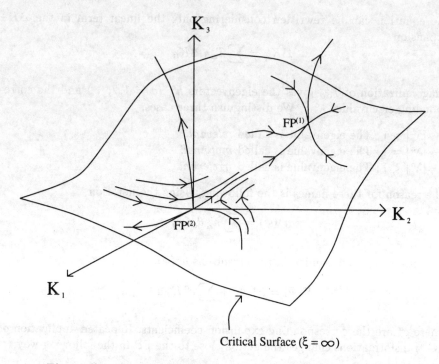

Fig. 8. *Coupling space and RG trajectories of the $d = 3$ scalar FT.*

are attractive (irrelevant) and repulsive (relevant) eigendirections with respect to RG transformations. At both fixed points there is a relevant direction running out of the critical surface. In addition, there is one trajectory leading from the Gaussian FP (FP$^{(2)}$) to FP$^{(1)}$. This shows that the dimension of the basin of attraction of FP$^{(2)}$ is one less than the one of FP$^{(1)}$. That implies that in order to reach the basin of attraction of FP$^{(2)}$ one has to tune one additional parameter.

The number of different fixed points in a d dimensional scalar field theory grows as d is decreased towards 2. On the other hand, in $d = 4$, there is only one FP identified until now.

5 Gaussian FP

5.1 The Gaussian FP in Scalar FTs

In the last section it was mentioned that in four Euclidean dimensions ($d = 4 = 3+1$) only one FP can be identified for a scalar FT. This FP is the Gaussian FP which exists in any dimension. The Gaussian FP has an enormous importance, especially in QFTs, so we shall investigate its properties in some detail. To identify the Gaussian FP in scalar FTs, we start with a simple Gaussian scalar theory, i.e. an action which is quadratic in the field

$$S = \frac{1}{2} \int_{k=0}^{\Lambda^{\text{cut}}} \tilde{\Phi}(k) \tilde{\Phi}(-k) \tilde{\rho}(k) \ .$$

Here $\tilde{\rho}(k)$ is a general function which can be expanded in increasing powers of k:

$$\tilde{\rho}(k) = \tilde{r}_0 + \underbrace{1}_{\text{convention!}} \cdot \sum_{\mu=1}^{d} k_\mu k_\mu + \tilde{\alpha} \left(\sum_{\mu=1}^{d} k_\mu k_\mu \right)^2 + \tilde{\beta} \sum_{\mu=1}^{d} k_\mu k_\mu k_\mu k_\mu + \cdots \ .$$

By rescaling the integration variable $\tilde{\Phi}(k)$ in the path integral one can arrange that the coefficient of the term $\sum_\mu k_\mu k_\mu$ is 1. We shall take this convention. As in Sect. 3 we then introduce dimensionless quantities with the help of the cutoff Λ^{cut}

$$q_\mu = \frac{k_\mu}{\Lambda^{\text{cut}}} \ , \quad |q| \leq 1$$

$$\varphi(q) = \tilde{\Phi}(k) \cdot \left(\Lambda^{\text{cut}} \right)^{\frac{d+2}{2}}$$

$$\rho(q) = \tilde{\rho}(k) \cdot \left(\Lambda^{\text{cut}} \right)^{-2}$$

$$r_0 = \tilde{r}_0 \cdot \left(\Lambda^{\text{cut}} \right)^{-2}$$

$$\alpha = \tilde{\alpha} \cdot \left(\Lambda^{\text{cut}} \right)^{2}$$

$$\beta = \tilde{\beta} \cdot \left(\Lambda^{\text{cut}} \right)^{2}$$

$$\cdots$$

yielding

$$S = \frac{1}{2} \int_{q=0}^{1} \varphi(q) \varphi(-q) \rho(q)$$

$$\rho(q) = r_0 + \sum_\mu q_\mu q_\mu + \alpha \left(\sum_\mu q_\mu q_\mu \right)^2 + \beta \sum_\mu q_\mu q_\mu q_\mu q_\mu + \cdots \ .$$

Now we perform an RG transformation as in Sect. 3.1. The generating functional is

$$Z = \int D\varphi(q) \ e^{-S(\varphi)} = \prod_{|q| \leq 1} \int d\varphi(q) \ \exp\left\{ -\frac{1}{2} \int_{q=0}^{1} \varphi(q) \varphi(-q) \rho(q) \right\} \ .$$

Following the steps in Sect. 3.1 we split the fields and the action in low and high momentum parts

$$\varphi(q) = \underbrace{\varphi_0(q)}_{0 \leq |q| \leq \frac{1}{2}} + \underbrace{\varphi_1(q)}_{\frac{1}{2} \leq |q| \leq 1}$$

$$S(\varphi_0 + \varphi_1) = S(\varphi_0) + S(\varphi_1) \ ,$$

where the last equation is valid for quadratic actions only. Integration over φ_1 gives for the generating functional (up to a multiplicative constant)

$$Z = \text{const} \cdot \prod_{|q| \leq \frac{1}{2}} \int d\varphi_0(q) \, \exp\left\{-\frac{1}{2} \int_{q=0}^{\frac{1}{2}} \varphi_0(q) \, \varphi_0(-q) \, \rho(q)\right\} =$$

$$= \text{const} \cdot \prod_{|q'| \leq 1} \int d\varphi'(q') \, \exp\left\{-\frac{1}{2} \zeta^2 \, 2^{-d} \int_{q'=0}^{1} \varphi'(q') \, \varphi'(-q') \, \rho\left(\frac{q'}{2}\right)\right\} =$$

$$= \text{const} \cdot \prod_{|q'| \leq 1} \int d\varphi'(q') \, e^{-S'(\varphi')} \ .$$

As usual we rescaled the dimensionless momenta and fields:

$$\varphi'(q') = \zeta^{-1} \varphi_0(q) \ , \quad q' = 2q \ , \quad 0 \leq |q'| \leq 1 \ .$$

So we have

$$S(\varphi) \xrightarrow{RG} S'(\varphi') = \frac{1}{2} \zeta^2 \, 2^{-d} \int_{q'=0}^{1} \varphi'(q') \, \varphi'(-q') \, \rho\left(\frac{q'}{2}\right)$$

$$\rho(q) \xrightarrow{RG} \rho\left(\frac{q}{2}\right) \zeta^2 \, 2^{-d} =$$

$$= \zeta^2 \, 2^{-d} \cdot \left\{ r_0 + \frac{1}{4} \sum_\mu q_\mu q_\mu + \frac{1}{16} \alpha \left(\sum_\mu q_\mu q_\mu\right)^2 + \frac{1}{16} \beta \sum_\mu q_\mu q_\mu q_\mu q_\mu + \ldots \right\} \ .$$

To satisfy the convention that the coefficient of the $\sum_\mu q_\mu q_\mu$-term in the action should be 1 we choose

$$\zeta = 2^{\frac{d+2}{2}}$$

yielding

$$r_0 \xrightarrow{RG} r_0' = 4 r_0 \xrightarrow{RG} r_0'' = 4^2 r_0 \xrightarrow{RG} \ldots$$

$$\alpha \xrightarrow{RG} \alpha' = \tfrac{1}{4} \alpha \xrightarrow{RG} \alpha'' = \left(\tfrac{1}{4}\right)^2 \alpha \xrightarrow{RG} \ldots$$

$$\beta \xrightarrow{RG} \beta' = \tfrac{1}{4} \beta \xrightarrow{RG} \beta'' = \left(\tfrac{1}{4}\right)^2 \beta \xrightarrow{RG} \ldots$$

$$\ldots$$

We see that the only relevant operator in the action is the one with the coefficient r_0 (eigenvalue: $\lambda = 4$), all other interactions are irrelevant, i.e.:

$$\begin{array}{ll} r_0 & \text{relevant} \ \lambda = 4 \\ \alpha & \text{irrelevant} \ \lambda = \tfrac{1}{4} \\ \beta & \text{irrelevant} \ \lambda = \tfrac{1}{4} \\ \text{all other irrelevant} \ |\lambda| < 1 \end{array}$$

On the critical surface ($r_0 = 0$) repeated RG transformations lead to an FP action which is of a *free* scalar FT of massless ($\xi = \infty$) particles:

$$S^*(\varphi) = \frac{1}{2} \int_{q=0}^{1} \varphi(q)\,\varphi(-q)\,\rho^*(q)$$

$$\rho^*(q) = q^2 \ .$$

5.2 Classification of Operators in the Vicinity of the Gaussian FP

In the previous section we considered couplings which are quadratic in the fields. Let us generalize this by considering an extended coupling constant space in the vicinity of the Gaussian FP. Add for example to the Gaussian FP action some terms like:

$$\tilde{\lambda}_4 \int_x \Phi^4(x) \ , \quad \tilde{\lambda}_6 \int_x \Phi^6(x) \ , \quad \ldots \ .$$

This leads in momentum space (using again dimensionless variables and couplings r_0, λ_4, λ_6 ($\ll 1$)) to following terms in the action

$$S = \int_q \frac{1}{2}\left(q^2 + r_0\right)\varphi(q)\,\varphi(-q) +$$

$$+ \lambda_4 \int_{q_1\ldots q_4} \varphi(q_1)\,\varphi(q_2)\,\varphi(q_3)\,\varphi(q_4)\,\delta(q_1+q_2+q_3+q_4) +$$

$$+ \lambda_6 \int_{q_1\ldots q_6} \varphi(q_1)\ldots\varphi(q_6)\,\delta(q_1+\ldots+q_6) + \ldots \ .$$

For simplifying the algebra and notations let us keep those terms only which are explicitely written out above. ¿From this expression a straightforward calculation gives the following transformation behavior of the coupling *close* to the Gaussian FP:

$$r_0 \xrightarrow{RG} (r_0 + 6\lambda_4\,c + 15\lambda_6\,c^2)\,\zeta^2\,2^{-d}$$
$$\lambda_4 \xrightarrow{RG} (\lambda_4 + 15\lambda_6\,c)\,\zeta^4\,2^{-3d}$$
$$\lambda_6 \xrightarrow{RG} \lambda_6\,\zeta^6\,2^{-5d}$$

with

$$c = \int_{k=\frac{1}{2}}^{1} \frac{1}{k^2 + r_0} \ .$$

The RG eigenvalues λ^a and the eigenvectors h^a of this linear transformation are therefore (the last column shows the engineering dimension of the highest power in the eigenvector)

$$\begin{aligned}
&\lambda^{(1)} = 2^2 &&h^{(1)} = \Phi^2 &&\dim \Phi^2(x) = d-2\\
&\lambda^{(2)} = 2^{4-d} &&h^{(2)} = \Phi^4 + a\,\Phi^2 &&\dim \Phi^4(x) = 2d-4\;.\\
&\lambda^{(3)} = 2^{6-2d} &&h^{(3)} = \Phi^6 + b_1\,\Phi^4 + b_2\,\Phi^2 &&\dim \Phi^6(x) = 3d-6
\end{aligned}$$

The general result can be stated as:

> To every simple operator $\theta(x)$ with engineering dimension d_θ there corresponds an eigenoperator, whose highest dimensional element is $\theta(x)$, and the corresponding eigenvalue is:

$$\lambda^\theta = 2^{d-d_\theta}\;.$$

Some examples are shown in Table 1 In 3 and 4 Euclidean dimensions the re-

Table 1. RG eigenvalues for a scalar field theory

$\theta(x)$	d_θ	λ^θ
$\Phi^2(x)$	$d-2$	4
$\Phi^4(x)$	$2d-4$	2^{4-d}
$\Phi^6(x)$	$3d-6$	2^{6-2d}
$(\partial^2 \Phi(x))^2$	$d+2$	$\tfrac{1}{4}$

sult above leads to the following statements concerning the relevance of special operators in linear order in the ΔK-expansion:

$$d = 3 \qquad\qquad O(\Delta K)$$

$$\Phi^2,\;\Phi^4 \qquad\qquad \text{relevant}$$

$$\Phi^6 \qquad\qquad \text{marginal}$$

all other operators irrelevant

$d = 4$	$O(\Delta K)$
Φ^2	relevant
Φ^4	marginal

all other operators irrelevant

The fate of the marginal operators Φ^6 ($d = 3$) and Φ^4 ($d = 4$) is decided by the $O\left((\Delta K)^2\right)$ terms. In both cases they become *weakly irrelevant*.

5.3 A Brief Discussion on the Yang Mills Interaction

If the form of the RG transformation respects the symmetries of the original action, then the action after the transformation will also have these symmetries.

It might happen that a symmetry of the theory is broken by the regularization. If the corresponding symmetry breaking interactions have *no* projection to the relevant, or marginal directions of the RG transformation, then the symmetry is automatically regained in the continuum limit. The term proportional to β in ρ in the previous section gives an example for this possibility: this rotation symmetry breaking perturbation dies out in the infrared (continuum) limit.

In contrast, if a gauge symmetry is broken, the corresponding non–gauge invariant interactions will be relevant and the gauge symmetry will not come back automatically in the continuum limit. In fact, at the Gaussian fixed point there exist several non–gauge invariant relevant (marginal) operators in $d = 4$ Yang–Mills theory (λ = eigenvalue of RG transformation), e.g.

	$O(\Delta K)$
$A_\mu^a A_\mu^a$	relevant ($\lambda = 4$)
$A_\mu^a A_\mu^a A_\nu^b A_\nu^b$	marginal ($\lambda = 1$)
$\partial_\mu A_\mu A_\nu A_\nu$	marginal ($\lambda = 1$)

If one restricts oneself to gauge invariant operators, it turns out that the lowest dimensional gauge invariant operator:

$$F_{\mu\nu}^a F_{\mu\nu}^a \quad \text{with} \quad d_\theta = 4$$

is *marginal* to first order and becomes *weakly relevant* to second order in the ΔK-expansion, while all other gauge invariant operators are *irrelevant*. The coupling g which is associated with this operator is the only interaction which survives the infrared limit. This coupling is growing as the momentum scale is decreased and the other way around ("asymptotic freedom"). The simple topology of the coupling space for $d = 4$ (gauge invariant) Yang–Mills theory is displayed in Fig. 9 where $g, c_1, c_2, \ldots$ denote the relevant and irrelevant couplings respectively.

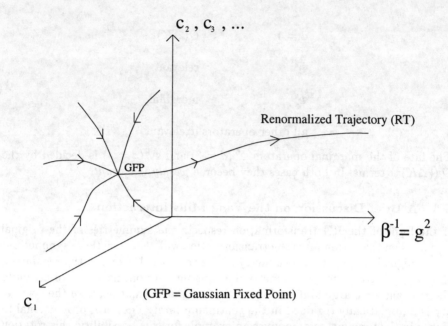

Fig. 9. *Gaussian Fixed Point in $d = 4$ Yang–Mills theory.*

6 The Perfect Lattice Regularized Free Scalar Theory: The Perfect Laplace Operator

Methods related to the RG theory are among the most powerful tools to attack difficult non-perturbative problems in statistical physics and field theory. There is now way to give even an overview on those methods and results. We shall rather discuss a specific application hoping that it is sufficiently interesting to raise your interest for digging deeper into the subject of RG. We shall discuss the possibility to construct perfect actions on the lattice. What 'perfect' means, what is it good for and what this has to do with RG – this is our subject now.

Take, for concreteness, the $d = 2$ Euclidean space continuum action

$$S = \int d^2x \frac{1}{2} \sum_{\mu=1}^{2} \partial_\mu \Phi(x)\, \partial_\mu \Phi(x), \quad . \tag{25}$$

The standard way to put this theory on the lattice is to replace the derivative by a discrete difference. Suppressing the lattice unit dependence ('$a = 1$') one obtains:

$$-\int d^2x \frac{1}{2} \sum_{\mu=1}^{2} \Phi(x) \Delta^{cont} \Phi(x) =$$

$$\stackrel{i.b.p.}{=} \int d^2x \, \frac{1}{2} \sum_{\mu=1}^{2} \partial_\mu \Phi(x) \, \partial_\mu \Phi(x) \;\to\; \sum_{n,\hat{\mu}} \frac{1}{2} (\Phi_{n+\hat{\mu}} - \Phi_n)^2 \,, \tag{26}$$

where $\hat{\mu}$ is the unit vector in 1 or 2 direction. This corresponds to the Laplace operator on the lattice

$$\left(\Delta_{standard}^{lattice} \Phi\right)_n = 4\Phi_n - \sum_{\hat{\mu}=1}^{2} (\Phi_{n+\hat{\mu}} + \Phi_{n-\hat{\mu}}) \,. \tag{27}$$

This is the familiar $4, -1, -1, -1, -1$ rule for the lattice Laplace operator. It is clear, however, that $\Delta_{standard}^{lattice}$ is only an approximation to the continuum $\Delta^{cont} = \sum_\mu \partial_\mu \partial_\mu$, and by solving $\Delta_{standard}^{lattice} \Phi = 0$ the predictions will be distorted. For small a, the distortion is $O(a^2)$ plus possibly boundary effects which can even produce $O(a)$ corrections.

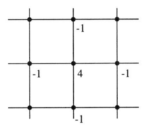

Fig. 10. *Standard Laplace operator on the lattice.*

Similarly, the corresponding field theory gives predictions which are contaminated by lattice artifacts. To see this explicitly, we determine the spectrum which in the continuum is

$$\begin{aligned}
\int dx_1 \, e^{-ipx_1} D^{cont}(x_1, x_2) &= \int dx_1 \, e^{-ipx_1} \int_{-\infty}^{\infty} \frac{d^2k}{(2\pi)^2} \frac{e^{ik_1 x_1} e^{ik_2 x_2}}{k_1^2 + k_2^2} \\
&= \int_{-\infty}^{\infty} \frac{dk_2}{2\pi} \frac{e^{ik_2 x_2}}{p^2 + k_2^2} \\
&= \frac{1}{2p} e^{-|p| x_2},
\end{aligned} \tag{28}$$

where D^{cont} is the Greens function in the continuum (obeying $\Delta^{cont} D^{cont}(x_1, x_2) = \delta(x_1) \cdot \delta(x_2)$) and $\int dx_1 \, e^{-ipx_1}$ projects into the momentum$= p$ channel. Comparing this with the expected behavior of the propagator

$$\int dx_1 \, e^{-ipx_1} D^{cont}(x_1, x_2) \stackrel{x_2 \to \infty}{\sim} e^{-x_2 E(p)}, \tag{29}$$

one observes the expected dispersion relation $E = |p|$.

Things will change, however, if we consider the propagator for the standard lattice Laplace operator (see e.g. [2]). By calculating the lattice action in Fourier space

$$\sum_{n,\hat{\mu}} \frac{1}{2} (\Phi_{n+\hat{\mu}} - \Phi_n)^2 = \int_{-\pi}^{\pi} \frac{d^2k}{(2\pi)^2} \frac{1}{2} \sum_{\mu} \left(e^{ik_\mu} - 1 \right) \left(e^{-ik_\mu} - 1 \right) \tilde{\Phi}(k)\tilde{\Phi}(-k)$$

$$= 2 \int_{-\pi}^{\pi} \frac{d^2k}{(2\pi)^2} \sum_{\mu} sin^2 \left(\frac{k_\mu}{2} \right) \tilde{\Phi}(k)\tilde{\Phi}(-k), \quad (30)$$

one can read off the propagator for $\Delta_{standard}^{lattice}$. Reintroducing a and dimensionful momenta, one gets for the propagator for small k ($k << \frac{1}{a}$):

$$\frac{1}{\frac{4}{a^2} \sum_\mu sin^2 \left(\frac{k_\mu a}{2} \right)} = \frac{1}{k^2 + O(a^2)}. \quad (31)$$

This corresponds to the dispersion relation

$$E(p) = |p| + O(p^3) \quad (32)$$

for the standard lattice action for small $|p|$. The $O((p^3))$-terms represents lattice artifacts. As one can see in Fig. 11 the standard lattice dispersion relation

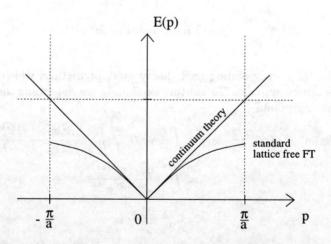

Fig. 11. *Dispersion relations for continuum theory and standard lattice theory.*

approximates the continuum relation very well for small $|p|$ but gets very bad near the edge of the Brillouin zone.

It might sound surprising, but there exists a lattice version of the free scalar theory which is free of lattice artifacts and most notably, it gives back the exact spectrum of the continuum theory. This formulation – by construction – defines a perfect Laplace operator on the lattice. Even more, this perfect lattice representation of the model is sufficiently simple to be interesting in practical calculations.

This perfect action is the FP action of the model related to some RG transformation on the lattice. To find this action, let us start with a general quadratic form in $d = 2$ dimensions:

$$S = \frac{1}{2} \sum_{n,r} \rho(r) \phi_n \phi_{n+r}, \tag{33}$$

where $\rho(r)$ are the coupling constants and $n = (n_1, n_2)$, $r = (r_1, r_2)$. The RG transformation

$$\prod_n \int d\phi_n \, e^{-\frac{1}{2} \sum_{n,r} \rho(r) \phi_n \phi_{n+r}} \prod_{n_B} \delta\left(\chi_{n_B} - \frac{1}{4} \sum_{n \in n_B} \phi_n\right) \tag{34}$$

is generalized by introducing the free parameter κ in the following way

$$\delta\left(\chi_{n_B} - \frac{1}{4} \sum_{n \in n_B} \phi_n\right) \longrightarrow \exp\left\{-2\kappa \left(\chi_{n_B} - \frac{1}{4} \sum_{n \in n_B} \phi_n\right)^2\right\}. \tag{35}$$

In the limit $\kappa \to \infty$ we recover the δ-function blocking, for finite κ it is a (legitime) generalization. We then obtain

$$\exp\left[-\frac{1}{2} \sum_{n_B, r_B} \rho'(r) \chi_{n_B} \chi_{n_B + r_B}\right] = \prod_n \int d\phi_n \exp\left[-\frac{1}{2} \sum_{n,r} \rho(r) \phi_n \phi_{n+r}\right]$$

$$\left[-2\kappa \left(\chi_{n_B} - \frac{1}{4} \sum_{n \in n_B} \phi_n\right)^2\right]. \tag{36}$$

The problem is to find a a specific set of couplings $\rho^*(r)$ such that if $\rho(r) = \rho^*(r)$, then $\rho'(r) = \rho^*(r)$ as well (fixed point). Since we have to perform Gaussian integrals only, this problem is comparatively easy to solve and we will only give the result in Fourier space

$$\frac{1}{\tilde{\rho}^*(q)} = \sum_l \frac{1}{(q + 2\pi l)^2} \prod_{i=0}^{1} \frac{\sin^2 \frac{q_i}{2}}{\left(\frac{q_i}{2} + \pi l_i\right)^2} + \frac{1}{3\kappa},$$

where $l = (l_0, l_1)$ is an integer vector and $\sum_l = \sum_{l_0 = -\infty}^{\infty} \sum_{l_1 = -\infty}^{\infty}$

and $\rho^*(r) = \int_{-\pi}^{\pi} \frac{d^2 q}{(2\pi)^2} e^{iqr} \tilde{\rho}^*(q), \quad q = (q_1, q_2). \tag{37}$

One can calculate the spectrum for this perfect Laplace operator and one obtains the astonishing result that it is exact. Not only that it is exact in the Brillouin zone, but the full spectrum is obtained. One finds a tower of poles when doing the integration in the k_2-plane. The dispersion relation reads

$$E = |p + 2\pi l_1|, \quad \text{where} \quad l_1 = 0, \pm 1, \pm 2, \ldots. \tag{38}$$

Therefore we obtain the full continuum spectrum as shown in Fig. 12.

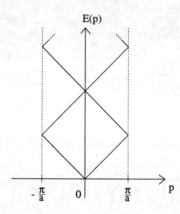

Fig. 12. *Dispersion relation for the perfect Laplace operator on the lattice.*

Table 2. *The couplings of the quadratic two–spin interaction terms at a distance $r = (r_0, r_1)$ for the optimal choice of the block transformation with $\kappa = 2$. Note that in our convention, for the standard action the only non–vanishing entry in this list would be $\rho_{ST}(1,0) = -1$.*

r	$\rho(r)$	r	$\rho(r)$
(1,0)	-0.61802	(4,0)	$-2.632 \cdot 10^{-6}$
(1,1)	-0.19033	(4,1)	$7.064 \cdot 10^{-7}$
(2,0)	$-1.998 \cdot 10^{-3}$	(4,2)	$1.327 \cdot 10^{-6}$
(2,1)	$-6.793 \cdot 10^{-4}$	(4,3)	$-7.953 \cdot 10^{-7}$
(2,2)	$1.625 \cdot 10^{-3}$	(4,4)	$6.895 \cdot 10^{-8}$
(3,0)	$-1.173 \cdot 10^{-4}$	(5,0)	$-8.831 \cdot 10^{-8}$
(3,1)	$1.942 \cdot 10^{-5}$	(5,1)	$3.457 \cdot 10^{-8}$
(3,2)	$5.232 \cdot 10^{-5}$	(5,2)	$3.491 \cdot 10^{-8}$
(3,3)	$-1.226 \cdot 10^{-5}$	(5,3)	$-3.349 \cdot 10^{-8}$
		(5,4)	$8.408 \cdot 10^{-9}$
		(5,5)	$-1.657 \cdot 10^{-10}$

The properties (most notably the interaction range) of $\rho(r)$ depend on the free parameter κ. The choice $\kappa \approx 2.$ is optimal in the sense that $\rho(r)$ decays very rapidly with growing $|r|$, like $\sim \exp(-3.44|r|)$. For this choice $\rho(r)$ is strongly dominated by the nearest neighbor and diagonal couplings while the couplings at distance > 1 are already small. As Table 2 (taken from [3]) shows, $\rho(3,3)$ is, for example, 5 orders of magnitude smaller than the nearest neighbor coupling. The choice $\kappa = \infty$ (corresponding to a block transformation with δ-function) gives a considerably larger interaction range, $\rho(r) \sim \exp(-1.45|r|)$.

7 Perfect Lattice Action in Asymptotically Free Theories

In this section we illustrate the idea of improved lattice actions. As Yang-Mills theory or QCD in 4 dimensions are technically quite complicated, we choose as a toy model the non-linear σ model in $d = 2$ dimensions for which more details can be found in [3].

7.1 Introduction to the Non-linear σ Model

The continuum action of the O(3)-symmetric non-linear σ model in Euclidean space is given by

$$S^{cont} = \frac{\beta}{2} \int d^2x \sum_{\mu=1}^{2} \partial_\mu \boldsymbol{\Phi}(\mathbf{x}) \, \partial_\mu \boldsymbol{\Phi}(\mathbf{x}),$$

$$\text{with the constraint}: \quad \boldsymbol{\Phi}^2(\mathbf{x}) = 1, \tag{39}$$

where $\boldsymbol{\Phi}$ is a 3-component vector and $\beta = 1/g$, the coupling constant g is dimensionless.

This model has some features common to Yang-Mills theories in 4 dimensions:

- The theory is asymptotically free in g.
- The dynamics of the system generates a mass gap m. We obtain a triplet of particles with mass m.
- Classical instanton solutions exist with scale invariant actions.

This model has a long history, as this interesting, but relatively simple theory is an excellent testing ground for theoretical and numerical ideas.

On the lattice one obtains the standard formulation of the action (39) by replacing the derivatives with nearest neighbor differences,

$$S^{lattice}_{standard} = \frac{\beta}{2} \sum_{n,\hat{\mu}} (\boldsymbol{\Phi}_{n+\hat{\mu}} - \boldsymbol{\Phi}_n)^2 = -\beta \sum_{n,\hat{\mu}} \boldsymbol{\Phi}_n \boldsymbol{\Phi}_{n+\hat{\mu}} + const, \tag{40}$$

where $n = (n_1, n_2)$ labels the lattice sites and $\hat{\mu}$ is the unit vector in 1 or 2-direction.

7.2 The Renormalized Coupling g_r

An interesting problem is to calculate the renormalized coupling $g_r(L)$ for this model, where L is some physical length scale. For small L ($L \ll \frac{1}{m}$) this can be done in perturbation theory, due to asymptotic freedom. The challenging task is, however, to calculate $g_r(L)$ for large L, away from the perturbative regime.

Before we will deal with this problem, we first have to define the renormalized coupling $g_r(L)$. It is easy to see that

$$g_r(L) := m(L) \cdot L \qquad (41)$$

is a reasonable definition, where $m(L)$ is the mass gap in a finite, periodic box of length L in space direction (time is kept infinite). We will provide an intuitive argument to show that the relation (41) is a possible definition of the renormalized coupling:

Consider L to be small, then only the constant modes of Φ with respect to x are important, implying

$$S = \frac{\beta}{2} \int_0^L dx \int dt \sum_{\mu=1}^{2} (\partial_\mu \Phi(x))^2 \xrightarrow{L \to 0} \frac{L}{2g} \int dt \left(\dot{\Phi}\right)^2. \qquad (42)$$

Eq. (42) defines a simple problem in quantum mechanics. It describes a quantum mechanical rotator in 3 dimensions with a moment of inertia $\Theta = 1/g$. The energy spectrum is given by

$$E = \frac{l(l+1)}{2\Theta} = \frac{l(l+1)}{2\frac{L}{g}}, \qquad l = 0, 1, 2, \ldots \qquad (43)$$

and we obtain the mass gap as the energy of the lowest excitation,

$$m(L) = \frac{g}{L} = \frac{g_r}{L} + O(g^2). \qquad (44)$$

After this detour we return to the task of calculating the renormalized coupling in the non-perturbative regime. This problem was considered in ([4]) as a preparation for the analogous highly relevant calculation in QCD. Consider, as an example, the value of $g_r(L) = 1.0505$ and calculate the renormalized coupling at the scale $2L$: $g_r(2L)$. At such big couplings the problem is non-perturbative. One has to resort to Monte Carlo simulations using lattice regularization. The discretization introduces systematical errors whose size is reduced as the lattice unit a is decreased relative to the physical scales of the problem, in our case relative to L. To this end $g_r(2L)$ is calculated on lattices with $L/a = 5, 6, \ldots, 16$ and the results are then extrapolated to $L/a = \infty$. This is a non-trivial problem even in $d = 2$. The results of this procedure are shown in Fig. 13.

The question we want to investigate is, whether we can find an improved lattice action with drastically reduced lattice artifacts.

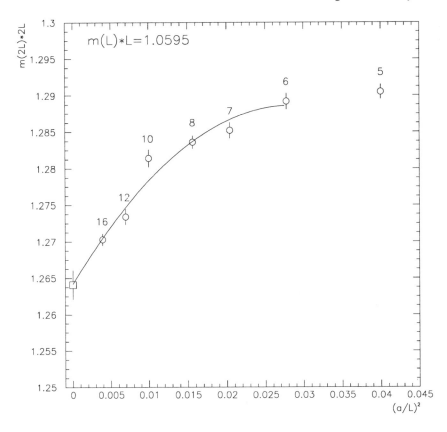

Fig. 13. *Running coupling in non-linear σ model: extrapolation to the continuum limit.*

7.3 RG Group Flow and the Perfect Action for the Non-linear σ Model

A radical solution would be to use a *perfect* lattice action which is completely free of lattice artifacts. That such perfect actions exist, follows from Wilson's renormalization group (RG) theory ([1], [6] and [7]).

Beyond the basic requirements of O(3) symmetry, locality, correct classical limit, translation and 90°–rotation symmetry, the form of the lattice action is largely arbitrary. It might contain nearest neighbor, next–to–nearest neighbor, etc., even different multi-spin interactions. Let us denote the corresponding couplings by $c_1, c_2, \ldots$. The action $\beta S(\boldsymbol{\Phi})$ is represented by a point in the infinite dimensional space of couplings $(\beta, c_1, c_2, \ldots)$. We shall consider RG transformations in configuration space, namely block transformations with a scale factor of 2. Under repeated block transformations the action moves in this coupling constant space. The expected flow diagram is sketched in Fig. 14 [7], [5]. In the $\beta = \infty$ hyperplane (correlation length $\xi = \infty$) there is a fixed point (FP) $c_1^*, c_2^*, \ldots$, whose exact position depends on the details of the block transformation. We shall use the notation $S(\boldsymbol{\Phi}; c_1^*, c_2^*, \ldots) = S^*(\boldsymbol{\Phi})$ and call $\beta S^*(\boldsymbol{\Phi})$ the FP–action. The FP has one marginal and infinitely many irrelevant directions. The marginal operator is S^* itself [7]. Actually, S^* is not exactly marginal, it is weakly relevant. The trajectory which leaves the FP along the weakly relevant direction is called the renormalized trajectory (RT). For large β the RT runs along the FP–action, but for smaller β they do not coincide anymore. It is easy to see that the points

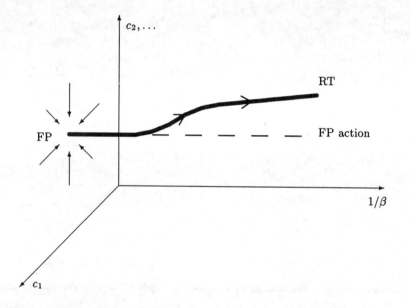

Fig. 14. *Flow of the couplings under RG transformation in the $O(N)$ non–linear σ model.*

of the RT define perfect actions. The argument goes as follows: At any given β, the point of the RT is connected to the infinitesimal neighborhood of the FP by (infinitely many steps of) exact RG transformations. Since each step increases the lattice unit by a factor of 2, *any* distance at the given β (even 1 lattice unit) corresponds to a long distance close to the FP. The infinitesimal neighborhood of the FP is in the continuum limit, there are no cut–off effects at long distances. On the other hand, for all the questions which can be formulated in terms of degrees of freedom after the transformation we get the same answer as before. So, there are no lattice artifacts at the given β on the RT at any distance.

Since an RG step is a non–critical problem, the FP–action and the actions on the RT in general are expected to be local ([1], [6] and [7]). Locality, however, allows non–negligible interactions over several lattice units in the action. For practical reasons we need more than locality. For practical applicability we have to answer positively the following questions:

- Is it possible to determine $S^*(\boldsymbol{\Phi})$ to a good precision?
- Is $S^*(\boldsymbol{\Phi})$ of sufficiently short range? Is the structure of S^* simple enough allowing a parameterization where the number of couplings remains relatively low, O(10–100)?
- Questions i. and ii. for the points of the RT.

Our pilot study shows that for the non–linear σ model, using a properly chosen RG transformation, the answer to these questions is 'yes', even if we go down to small correlation lengths. The determination of $S^*(\boldsymbol{\Phi})$ is a saddle point problem which requires minimization over classical fields. The range of interaction in $S^*(\boldsymbol{\Phi})$ depends on the RG transformation. By a proper choice of the block–transformation $S^*(\boldsymbol{\Phi})$ becomes surprisingly short ranged. $S^*(\boldsymbol{\Phi})$ contains multi-spin couplings also but its structure is relatively simple. With O(20) couplings an excellent parameterization can be obtained which works even on coarse configurations, i.e. at small correlation lengths. The problems related to $S^*(\boldsymbol{\Phi})$ can be solved partly analytically, which is a special bonus in asymptotically free theories.

7.4 Determining the Perfect Classical Action for the Non-linear σ Model

After these remarks let us look in detail at the RG transformation in configuration space. As we have seen in Sect. 3.2 (equation 23), one naturally introduces block spins as averages over the original degrees of freedom. In our case the effective action would read

$$e^{-\beta' S'(\boldsymbol{\Phi}')} = \int_{\boldsymbol{\Phi}} \delta\left(\boldsymbol{\Phi}'_{n_B} - \frac{\sum_{n\in n_B}\boldsymbol{\Phi}_n}{|\sum_{n\in n_B}\boldsymbol{\Phi}_n|}\right) e^{-\beta S(\boldsymbol{\Phi})},$$

$$\text{where}: \quad \int_{\boldsymbol{\Phi}} = \prod_n \int d\boldsymbol{\Phi}_n \delta\left(\boldsymbol{\Phi}_n^2 - 1\right). \tag{45}$$

We would like to introduce a parameter in the transformation (45), which can be tuned in order to make the fixed point action short ranged. For this purpose we write the the delta function in (45) as

$$\delta\left(\boldsymbol{\Phi}'_{n_B} - \frac{\sum_{n\in n_B}\boldsymbol{\Phi}_n}{|\sum_{n\in n_B}\boldsymbol{\Phi}_n|}\right) =$$

$$= const \cdot \lim_{\kappa \to \infty} \exp\left\{\beta\kappa\left[\boldsymbol{\Phi}'_{n_B}\sum_{n\in n_B}\boldsymbol{\Phi}_n - Y_{Nn}\left(\beta\kappa \cdot \left|\sum_{n\in n_B}\boldsymbol{\Phi}_n\right|\right)\right]\right\}. \tag{46}$$

The function Y_N (for O(N)-symmetry) obeys the relation

$$\int d\boldsymbol{\Phi}\, \delta\left(\boldsymbol{\Phi}^2 - 1\right) e^{\boldsymbol{\Phi}\mathbf{b}} = const \cdot Y_N\left(|\mathbf{b}|\right) \tag{47}$$

and ensures that the partition function is only changed up to an irrelevant constant by the corresponding RG transformation for all values of κ. Therefore κ can be taken as a free parameter of the RG transformation. In the limit $\beta \to \infty$ the RG transformation which corresponds to the smearing out of δ-function (expression (47) for finite κ) can be written as

$$e^{-\beta' S'(\boldsymbol{\Phi}')} = \int_{\boldsymbol{\Phi}} \exp\left(-\beta\left\{S(\boldsymbol{\Phi}) - \kappa \sum_{n_B} \left[\boldsymbol{\Phi}'_{n_B} \sum_{n \in n_B} \boldsymbol{\Phi}_n - |\sum_{n \in n_B} \boldsymbol{\Phi}_n|\right]\right\}\right) \tag{48}$$

where $\beta' = \beta - O(1)$, due to asymptotic freedom. Equation (48) is a saddle point problem in this limit, giving

$$S'(\boldsymbol{\Phi}') = \min_{\boldsymbol{\Phi}} \left\{ S(\boldsymbol{\Phi}) - \kappa \sum_{n_B} \left[\boldsymbol{\Phi}'_{n_B} \sum_{n \in n_B} \boldsymbol{\Phi}_n - |\sum_{n \in n_B} \boldsymbol{\Phi}_n|\right]\right\}. \tag{49}$$

The FP of the transformation is determined by the equation

$$S^*(\boldsymbol{\Phi}') = \min_{\boldsymbol{\Phi}} \left\{ S^*(\boldsymbol{\Phi}) - \kappa \sum_{n_B} \left[\boldsymbol{\Phi}'_{n_B} \sum_{n \in n_B} \boldsymbol{\Phi}_n - |\sum_{n \in n_B} \boldsymbol{\Phi}_n|\right]\right\}. \tag{50}$$

Equation (50) can be solved by stochastic methods for a given configuration $\boldsymbol{\Phi}'$. The final thing one has to do is to find a parameterization for the fixed point action.[3] We choose the parameterization

$$S^*(\boldsymbol{\Phi}) = -\frac{1}{2} \sum_{n,r} \rho(r)\left(1 - \boldsymbol{\Phi}_{n+r}\right) \tag{51}$$
$$+ \sum_{n_1,n_2,n_3,n_4} c(n_1,n_2,n_3,n_4)\left(1 - \boldsymbol{\Phi}_{n_1}\boldsymbol{\Phi}_{n_2}\right)\left(1 - \boldsymbol{\Phi}_{n_3}\boldsymbol{\Phi}_{n_4}\right) + \ldots$$

where the summations go over all the lattice points. It is a significant help in the parameterization and optimalization problem that the first two functions ρ and c in equation (51) can be calculated analytically. $\rho(r)$ are just the coefficients for the perfect Laplace operator on the lattice as discussed in Sect. 6. For the optimal choice $\kappa \approx 2$ $\rho(r)$ falls off as $\sim \exp(-3.44|r|)$. This is reasonably short ranged. The coefficients c have to be determined numerically. A good fit for the action can be obtained by using only 24 parameters. The values for an explicit parameterization of the fixed point action (51) are given in Table 3.

---

[3] Actually for Yang-Mills theories in 4 dimensions the real difficulty is to find a proper parameterization of the action with a finite number of parameters, whereas solving the fixed point equation for a given field configuration is a numerically feasible problem.

Table 3. *The couplings used for the FP-action. The diagrams indicate the form of the interaction. All couplings used for this fit can be put on a 1x1 square. The coefficients of the quadratic and quartic interactions are calculated analytically, the higher order interactions are fitted.*

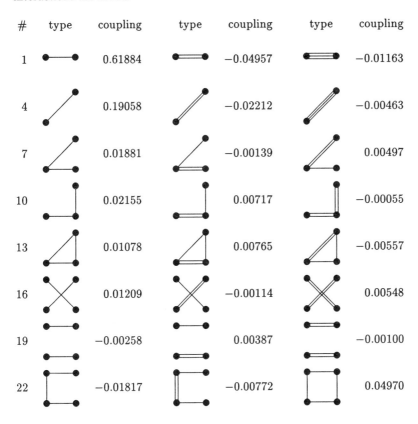

7.5 Results Obtained by Simulating the FP Action

The FP–action S^*_{FP} is defined as $\beta_{FP}S^*(\Phi)$. For large β_{FP}, the RT runs together with the FP–action (Fig. 14), therefore S^*_{FP} is the perfect action. For intermediate β_{FP} values it is not perfect anymore, but – as we shall see – it performs amazingly well. As we discussed in the previous sections, the FP–action can be determined with the help of classical calculations. For $\kappa = 2$ it is very short ranged and a relatively simple parameterization describes it well. This parameterized form (given in Table 3) can be easily simulated. Even the cluster Monte Carlo method can be generalized to such multi-coupling actions. This parametrized FP–action can be considered as a first approximation towards 'perfectness'. We consider again the problem of the running coupling in the non-linear σ model. One can simulate the FP-action on a lattice of spatial size, say, $L/a = 5$ and tune β_{FP} until $m(L)L$ become close to the prescribed value: at $\beta_{FP} = 1.0821$ one obtains $g_r(L) = 1.0578(5)$. Then measuring on a lattice with $L'/a = 2L/a = 10$

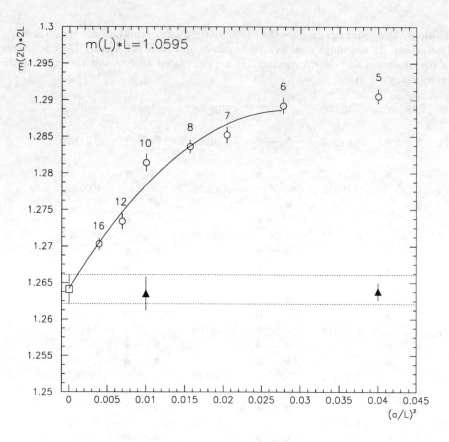

Fig. 15. *Simulating the perfect classical action.*

at the same β_{FP} value one obtained $m(2L)2L = 1.2611(9)$. In the time direction the lattice was chosen to be at least 6–times larger than the finite box correlation length $\xi(L) = 1/m(L)$ and distances larger than three times ξ were used in the fitting procedure to obtain the mass gap. At the end one shifts $g_r(2L)$ according to the slight difference between the actual (1.0578) and prescribed (1.0595) value for $g_r(L)$ leading to $g_r(2L) = 1.2638(12)$. The error from this procedure is negligible. One can repeat this calculation at $L = 10$ also, for $\beta_{FP} = 1.214$ with the results: $m(L)L = 1.0613(8)$ and $m(2L)2L = 1.2664(18)$. After shifting this gives $g_r(2L) = 1.2635(22)$. In Fig. 15 these two points are compared with the extrapolated prediction $g_r(2L) = 1.2641(20)$. No cut–off effects can be seen.[4]

---

[4] Surprisingly, even for an extremely coarse lattice, $L/a = 3$, (at $\beta_{FP} = 0.98$) one obtains $g_r(2L) = 1.2626(11)$, still with no sign of lattice artifacts.

References

[1] K.G. Wilson and J. Kogut, Phys. Rep. **C12** (1974) 75.
[2] M. Creutz: Quarks, Gluons and Lattices, Cambridge UP, Cambridge (1983).
[3] P. Hasenfratz and F. Niedermayer, Nucl. Phys. **B414** (1994) 785.
[4] M. Lüscher, P. Weisz, U. Wolff, Nucl. Phys. **B359** (1991) 221.
[5] P. Hasenfratz: Monte Carlo Renormalization Group Methods and results in QCD. In: Quarks, Leptons and their Constituents, A. Zichichi (ed.) (plenum Press, NY, 1988).
[6] K. Wilson, Rev. Mod. Phys. **47** (1975) 773, *ibid* **55** (1983) 583.
[7] K. Wilson, in Recent Developments of Gauge Theories, eds. G. 't Hooft et al. (Plenum Press, 1980).

Lattice Gauge Theory and the Structure of the Vacuum and Hadrons *

J.W. Negele[1];
Notes by H.W. Grießhammer[2] and D. Lehmann[2]

[1] Center for Theoretical Physics, Laboratory for Nuclear Science and Department of Physics, Massachusetts Institute of Technology, Cambridge, Massachusetts 02139, U.S.A.
[2] Institut für Theoretische Physik III, Universität Erlangen–Nürnberg, Staudtstr. 7, 91058 Erlangen, Germany

1 Introduction

Since Wilson's paper [1] twenty years ago, lattice gauge theory has become a powerful tool to solve, rather than to model, QCD. Its applications include the demonstration of confinement and deconfinement of quarks and gluons, the measurement of hadronic matrix elements, the exploration of hadron structure, and the study of QCD thermodynamics relevant to relativistic heavy ion collisions. The combination of improved approximations to the continuum action, improved algorithms and the rapid increase in available computer power offer the prospect of definitive, quantitative solutions in the near future.

Following the advice of Viki Weisskopf, the goal of these brief lectures is "to uncover a little rather than to cover a lot". My hope is to explain the basic concepts and stimulate interest with a few selected results, so that students can appreciate the basic ideas, understand the potential and limitations of Lattice QCD, have enough background to understand seminars and colloquia, and be prepared to read the vast literature in the field. Textbooks such as those by Creutz [2], Montvay and Münster [3] and Rothe [4] provide deeper insight, key articles are reprinted in the volume by Rebbi [5], and yearly reviews and new results may be found in the proceedings of the annual lattice conferences published as supplements to *Nuclear Physics* **B**.

2 Path Integrals

2.1 Feynman Path Integral

One of the key ideas underlying our approach is the use of path integrals. The formulation of quantum mechanics in terms of path integrals provides not only

---

* Lectures presented at the workshop "Lattice QCD and Dense Matter" organised by the Graduiertenkolleg Erlangen–Regensburg, held on October 11th–13th, 1994 in Kloster Banz, Germany.

a very elegant and physical picture of the quantum evolution as a sum over time histories but also a powerful computational framework, which eliminates the non-commutativity of the quantum operators and thus reduces the problem to a quadrature.

First we illustrate the basic idea of the Feynman path integral in the simplest case of a single quantum mechanical degree of freedom. One goes to Euclidean spacetime by rotating time to imaginary time $it \to t_{\text{imag}}$ and obtains a purely real (imaginary) time evolution operator, and later the Wiener measure for the path integral (3). The time evolution operator for imaginary time β is broken up into a large number n of time slices, and by inserting complete sets of position eigenstates one obtains for the matrix element

$$\langle x^f \mid e^{-\beta H} \mid x^i \rangle = \int dx_1 \ldots dx_{n-1} \prod_{i=0}^{n} \langle x_{i+1} \mid e^{-\varepsilon H} \mid x_i \rangle, \qquad \varepsilon = \frac{\beta}{n}, \quad (1)$$

where $x_0 = x^i$, $x_n = x^f$. We may evaluate the evolution operator between two neighboring time slices in momentum space by neglecting terms of order ε^2,

$$\langle x_{i+1} \mid e^{-\varepsilon\left(\frac{p^2}{2m}+V(x)\right)} \mid x_i \rangle = \int_{-\infty}^{\infty} \frac{dp}{2\pi} \, e^{ip(x_{i+1}-x_i) - \varepsilon \frac{p^2}{2m} - \varepsilon V(x_i)} + \mathcal{O}(\varepsilon^2) \quad (2)$$

$$= \sqrt{\frac{m}{2\pi\varepsilon}} \, e^{-\varepsilon\left[\frac{m}{2}\left(\frac{x_{i+1}-x_i}{\varepsilon}\right)^2 + V(x_i)\right]} + \mathcal{O}(\varepsilon^2),$$

so that the matrix element (1) can be expressed as the sum over all paths of the exponential of the classical action $S[x(t_{\text{imag}})]$:

$$\langle x^f \mid e^{-\beta H} \mid x^i \rangle = \int d(x_1 \ldots x_{n-1}) \, e^{-\varepsilon \sum_i \left[\frac{m}{2}\left(\frac{x_{i+1}-x_i}{\varepsilon}\right)^2 + V(x_i)\right]}$$

$$\to \int_{x(0)=x^i}^{x(t_{\text{imag}})=x^f} dx(t_{\text{imag}}) \, e^{-S[x(t_{\text{imag}})]}. \qquad (3)$$

This result generalizes straightforwardly to the case of many degrees of freedom:

$$\langle x_1^f \ldots x_N^f \mid e^{-\beta H} \mid x_1^i \ldots x_N^i \rangle = \int_{x_1^i \ldots x_N^i}^{x_1^f \ldots x_N^f} dx_k(t_{\text{imag}}) \, e^{-\varepsilon \sum_i \left[\sum_k \frac{m}{2}\left(\frac{x_k^{i+1}-x_k^i}{\varepsilon}\right)^2\right]} \times$$

$$\times e^{\varepsilon \sum_i \left[\frac{1}{2}\sum_{kl} v(x_k^i - x_l^i)\right]}. \qquad (4)$$

It should be mentioned that time-ordering naturally comes in by the time slicing procedure, hence time-ordered products are represented as follows:

$$\langle x^f | \mathcal{T} O(t_1) O(t_2) e^{-\int\limits_0^T dt_{\text{imag}} H(t_{\text{imag}})} | x^i \rangle =$$
$$= \int \mathcal{D}x(t_{\text{imag}}) \; O(t_{1,\,\text{imag}}) O(t_{2,\,\text{imag}}) \; e^{-S[x(t_{\text{imag}})]} \quad . \tag{5}$$

The Wick rotation to imaginary time in (1) interchanges the role of the Lagrangian in Minkowski space and the Hamiltonian in Euclidean space:

$$\int\limits_0^{t_0} dt \left[\frac{m}{2}\left(\frac{dx}{dt}\right)^2 - V(x) \right] \xrightarrow{it \to t_{\text{imag}}} \int\limits_0^\beta dt_{\text{imag}} \left[\frac{m}{2}\left(\frac{dx}{dt_{\text{imag}}}\right)^2 + V(x) \right] . \tag{6}$$

Although we drop the subscript of t_{imag} and will always use the term *time* for Euclidean time, remember that *the "time" is not the time* on the lattice.

2.2 Scalar Field Theory

The results of the Feynman path integral can easily be transferred to a scalar field theory. In order to properly define the path integral we need to discretize the spatial coordinates $\mathbf{r} \to a\mathbf{n}$, where the n_i are integers and a is the lattice spacing. The corresponding lattice field theory is then viewed as a quantum many-body problem with canonical coordinates $\hat\phi(\mathbf{n})$ and conjugate momentum $\hat\pi(\mathbf{n})$, $\left[\hat\phi(\mathbf{n}), \hat\pi(\mathbf{m})\right] = i\delta_{\mathbf{n},\mathbf{m}}$, and the position eigenstates $\hat x | x \rangle = x | x \rangle$ are replaced by eigenstates of the field $\hat\phi(\mathbf{n}) | \phi \rangle = \phi(\mathbf{n}) | \phi \rangle$. If we define the spatial derivative on the lattice as

$$| \nabla \phi(\mathbf{n}) |^2 := \sum_{i=1}^{3} | \phi(\mathbf{n} + \mathbf{e}_i) - \phi(\mathbf{n}) |^2 \; , \tag{7}$$

the Hamiltonian is replaced by[1]

$$\int d^3r \left\{ \frac{1}{2}\pi^2(\mathbf{r}) + \frac{1}{2} | \nabla\phi(\mathbf{r}) |^2 + V(\phi(\mathbf{r})) \right\}$$
$$\to \sum_{\mathbf{n}} \left\{ \frac{1}{2}\pi^2(\mathbf{n}) + \frac{1}{2}\sum_{i=1}^{3} | \phi(\mathbf{n}+\mathbf{e}_i) - \phi(\mathbf{n}) |^2 + V(\phi(\mathbf{n})) \right\} \quad . \tag{8}$$

---
[1] For simplicity, the lattice spacing a is set to 1 in the following. It can always be reinstated by dimensional analysis.

The first term in (8) is the kinetic energy part, the remaining terms, which will be denoted as $F[\phi]$ in the following, correspond to a sum of one- and two-body potentials. Matrix elements of the Euclidean evolution operator can now be expressed as a path integral:

$$\langle \phi_f | e^{-\sum_{\mathbf{n}}\{\frac{1}{2}\pi^2(\mathbf{n})+F[\phi(\mathbf{n})]\}} | \phi_i \rangle =$$

$$= \int_{\phi(\mathbf{n},0)=\phi_i(\mathbf{n})}^{\phi(\mathbf{n},\beta)=\phi_f(\mathbf{n})} \prod_{\mathbf{n},k} d\phi(\mathbf{n},k) e^{-\sum_{k,\mathbf{n}}\frac{1}{2}(\phi(\mathbf{n},k)-\phi(\mathbf{n},k))^2]} e^{-\sum_{k,\mathbf{n}} F[\phi(\mathbf{n},k)]} . \quad (9)$$

A general time-ordered product acquires the form

$$\langle \phi_f | \mathcal{T}O(\phi) e^{-\beta \int d^3 r \{\frac{1}{2}\pi^2 + \frac{1}{2}|\nabla\phi|^2 + V(\phi)\}} | \phi_i \rangle \to \int_{\phi_i}^{\phi_f} \mathcal{D}\phi\, O(\phi) e^{-\beta S_E[\phi]} \quad (10)$$

with the Euclidean action

$$S_E[\phi] = \sum_{\mathbf{n}} \left\{ \frac{1}{2} \sum_{i=0}^{3} (\phi(\mathbf{n}+\mathbf{e}_i) - \phi(\mathbf{n}))^2 + V(\phi(\mathbf{n})) \right\} . \quad (11)$$

The Euclidean action appears symmetric in spatial coordinates x and "time" t. Hence, it is often useful to interpret lattice physics from the viewpoint of statistical mechanics in $d+1$ dimensions rather than in terms of Hamiltonian field theory in d dimensions. One may think of any direction as "time", and one should note that the physical temperature of the $(d+1)$ dimensional statistical system is given by the length of the smallest edge on the lattice. The above results may be generalized to an arbitrary second quantized theory formulated with the help of a set of creation and annihilation operators. To this end, we need a resolution of the unity in terms of eigenstates of the annihilation operators – known as coherent states – which has to be inserted in the time slicing procedure. A detailed discussion of this approach to the path integral formulation can be found in [6] and [7].

2.3 Fermionic Path Integrals

For Fermions one must take an additional step and introduce Grassmann variables which may be viewed as anticommuting c-numbers

$$\{\psi_i, \psi_j\} = 0 , \quad \{\bar{\psi}_i, \bar{\psi}_j\} = 0 , \quad \{\psi_i, \bar{\psi}_j\} = 0 \quad (12)$$

and reflect the anticommutativity of the Fermionic field operators. By (12), the most general function of a set of $2N$ Grassmann variables is a finite linear combination of monomials

$$\bar{\psi}_1^{\alpha_1} \bar{\psi}_2^{\alpha_2} \ldots \bar{\psi}_N^{\alpha_N} \psi_1^{\alpha_{N+1}} \ldots \psi_N^{\alpha_{2N}} \quad \text{with } \sum_i \alpha_i \leq 2N , \quad (13)$$

where the α_i take on the values 0 and 1. The integral is defined by

$$\int d\psi_i = 1, \quad \int d\psi_i\, \psi_i = 0 \tag{14}$$

and coincides with the (left) derivative. Both of them may be regarded as purely algebraical definitions. Although there are a few technical details which can be found in [7], the essential point is that Grassmann path integrals have essentially the same form as for Bosons, except for a few crucial minus signs which do all the correct bookkeeping for the difference between Bosons and Fermions.

Since the Fermionic action is typically bilinear in the fields, the resulting path integrals are Gaußian and the Grassmann variables can be integrated according to the formula

$$\int \prod_i d\bar{\psi}_i\, d\psi_i\ e^{-\bar{\psi}_i M_{ij} \psi_j + \bar{\eta}_i \psi_i + \eta_i \bar{\psi}_i} = \det(M)\ e^{\bar{\eta}_i M_{ij}^{-1} \eta_j} \tag{15}$$

which is easily verified for $N = 1$ by the use of (14)

$$\int d\bar{\psi}\, d\psi\ e^{-\bar{\psi} a \psi} = \int d\bar{\psi}\, d\psi\ (1 - \bar{\psi} a \psi) = a \tag{16}$$

and should be compared with the corresponding Bosonic counterpart

$$\int \prod_i \frac{dz_i^*\, dz_i}{2\pi i}\ e^{-z_i^* M_{ij} z_j + j_i^* z_i + j_i z_i^*} = [\det(M)]^{-1}\ e^{j_i^* M_{ij}^{-1} j_j} \ . \tag{17}$$

Note that $\det(M)$ appears in (15) to the power -1 instead of 1 due to the fact that Grassmann variables transform with the inverse Jacobian under coordinate transformations. Assume the action has the form

$$S(\bar{\psi}, \psi, A) = \bar{\psi}_i\, M(A)_{ij}\, \psi_j + S_B(A) \tag{18}$$

where, for example $\bar{\psi}_i\, M(A)_{ij}\, \psi_j = \bar{\psi}\,(\not{p} - \not{A} + m)\,\psi$ and A represents the real Bosonic gauge field with an action $S_B(A) = F_{\mu\nu}^2(A)$. Then

$$\int d\bar{\psi}\, d\psi\, dA\ e^{S(\bar{\psi}, \psi, A)} = \int dA\ e^{\ln \det M(A) + S_B(A)} \ , \tag{19}$$

and we are left with an integral over the Bose field A of an effective action

$$S_{\text{eff}}(A) = \ln \det M(A) + S_B(A)\ . \tag{20}$$

In the same way, thermodynamic averages of time-ordered products of field annihilation and creation operators at space-time points $\mathbf{i} = (\mathbf{x}_i, t_i)$ can be calculated by differentiating the generating functional (15) with respect to the corresponding sources $\bar{\eta}_i$ and η_j. For example, the propagator (or contraction in the language of Wick's theorem) is:

$$\langle \mathcal{T}\, \psi_i\, \bar{\psi}_j \rangle = \int d\bar{\psi}\, d\psi\, dA\ (\psi_i\, \bar{\psi}_j)\ e^{S(\bar{\psi}, \psi, A)} = \int dA\ M^{-1}(A)_{ij}\ e^{S_{\text{eff}}(A)}\ . \tag{21}$$

3 Lattice QCD for Gluons

3.1 $U(1)$ Gauge Theory and the Wilson Action

Here we concentrate on the pure gauge field sector and start with the simplest possible gauge group, $U(1)$, corresponding to QED. The Lagrangian density for a pure $U(1)$ gauge theory with coupling to an external source $j_\mu(\mathbf{x})$ given by

$$\mathcal{L} = -\frac{1}{4} F_{\mu\nu} F_{\mu\nu} + g j_\mu A_\mu \tag{22}$$

with the field strength tensor $F_{\mu\nu}$ being the four-dimensional curl of the the vector potential A_μ, $F_{\mu\nu} = \partial_\mu A_\nu - \partial_\nu A_\mu$. Note that we will always use a Euclidean metric $g_{\mu\nu} = \delta_{\mu\nu}$, with upper and lower Lorentz indices being equivalent.

Since $F_{\mu\nu}$ is gauge invariant, $F_{\mu\nu}{}^2$ is both gauge and Lorentz invariant and the Lagrangian (22) leads to Maxwell's equations of electrodynamics.

We now consider how to approximate this continuum theory on a space-time lattice. Often in numerical analysis, one may allow discrete approximations to break fundamental underlying symmetries such as translation or rotation invariance, which only would be restored in the continuum limit. In the case of lattice gauge theory, however, since gauge invariance plays such a crucial role in defining the theory, it is desirable to enforce it exactly in the lattice action. Thus we will settle for a gauge invariant action which breaks Lorentz invariance, and simply insist on making the lattice spacing small enough that the errors are acceptably small.

Following Wilson, we define the action in terms of directed link variables assigned to each of the links between sites of the space time lattice. For $U(1)$, we define the link variable from site $\mathbf{n}$ in the μ direction to site $\mathbf{n} + \boldsymbol{\mu}$ as a discrete approximation to the integral $e^{ig \int_\mathbf{n}^{\mathbf{n}+\mu} dx_\mu A_\mu}$ (no sum over repeated indices) which we denote

$$U_\mu(\mathbf{n}) = e^{i\theta_\mu(\mathbf{n})} . \tag{23}$$

Thus $\theta_\mu(\mathbf{n})$ is a discrete approximation to $g \int_\mathbf{n}^{\mathbf{n}+\mathbf{m}} dx_\mu A_\mu$ along the direction of the link, and when the direction is reversed, $U_\mu(\mathbf{n}) \to U_\mu(\mathbf{n})^\dagger$ and $\theta_\mu(\mathbf{n}) \to -\theta_\mu(\mathbf{n})$. The link variable is then a group element of $U(1)$ and the compact variable $\theta_\mu(\mathbf{n})$ will be associated with $agA_\mu(\mathbf{x})$ in the continuum limit. With these link variables, the integral over the field variables in the path integral is replaced by the invariant group measure for $U(1)$, which is $\frac{1}{2\pi} \int_{-\pi}^{\pi} d\theta$.

The fundamental building blocks of the lattice action are products of directed link variables taken counter-clockwise around each individual plaquette of the lattice. As will be seen in the section on Wilson loops, this product is gauge invariant by construction, ensuring gauge invariance of the resulting action. A typical plaquette is sketched in Fig. 1, where $\mathbf{n}$ is an arbitrary site and μ and ν denote displacements by one site in the horizontal and vertical directions. By convention, the compact variables θ_μ and θ_ν are associated with links directed in the positive μ and ν directions so that $-\theta_\mu$ and $-\theta_\nu$ must be associated with

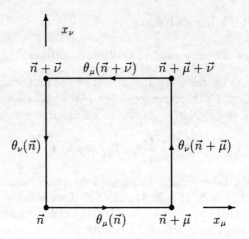

Fig. 1. *An elementary plaquette of link variables*

links in the negative μ and ν directions. The product of the four group elements around the plaquette may thus be written

$$U_\Box = e^{i\theta_\mu(\mathbf{n})} e^{i\theta_\nu(\mathbf{n}+\mu)} e^{-i\theta_\mu(\mathbf{n}+\nu)} e^{-i\theta_\nu(\mathbf{n})} \\ =: e^{i\Sigma_{\mu\nu}} \qquad (24)$$

where

$$\Sigma_{\mu\nu} = \theta_\nu(\mathbf{n}+\mu) - \theta_\nu(\mathbf{n}) - [\theta_\mu(\mathbf{n}+\nu) - \theta_\mu(\mathbf{n})] \\ =: \Delta_\mu \theta_\nu - \Delta_\nu \theta_\mu \, . \qquad (25)$$

Here, Δ_μ denotes the discrete lattice difference operator which becomes a derivative in the continuum limit, so that $\Sigma_{\mu\nu}$ is a discrete approximation to the curl on the lattice and is proportional to $\partial_\mu A_\nu - \partial_\nu A_\mu = F_{\mu\nu}$ in the continuum. Using conventional notation, the *Wilson action* is then written as

$$S = \beta \sum_\Box (1 - \operatorname{Re} U_\Box) \qquad (26)$$

$$= \beta \sum_\Box (1 - \cos \Sigma_{\mu\nu}) \, . \qquad (27)$$

Continuum QED is recovered by restoring the lattice spacing a and defining the new variables $\beta := \frac{1}{g^2}$ and $\theta_\mu(\mathbf{n}) =: agA_\mu(\mathbf{n})$, and expanding $\theta_\nu(\mathbf{n}+\mu) = \theta_\nu(\mathbf{n}) + a\partial_\mu \theta_\nu(\mathbf{n}) + \mathcal{O}(a^2)$, with the result that the leading contribution as $a \to 0$ is

$$S \approx \tfrac{1}{g^2} \sum_\Box \left(1 - \cos[a(\partial_\mu \theta_\nu - \partial_\nu \theta_\mu)]\right) \approx \tfrac{1}{g^2} \sum_\Box \left(1 - \cos(a^2 g F_{\mu\nu})\right) \\ \approx \tfrac{1}{g^2} \sum_{\mathbf{n},\{\mu\nu\}} \left(\tfrac{a^4 g^2}{2} F_{\mu\nu}^2(\mathbf{n}) + \ldots\right) \to \tfrac{1}{4} \int d^4x F_{\mu\nu}(\mathbf{x}) F_{\mu\nu}(\mathbf{x}) \, . \qquad (28)$$

In the second line, $\sum_{\{\mu\nu\}}$ denotes the sum over all pairs of μ and ν arising from the sum over plaquettes and an extra factor of $1/2$ arises in the last step due to the fact that each pair occurs twice in the double sum over repeated indices $F_{\mu\nu}F_{\mu\nu}$. The terms higher order in the lattice cutoff a vanish in the classical continuum limit and may give rise to finite renormalization of the coupling constant in quantum field theory. The lattice gauge theory defined by (26) is in a form which may be solved directly using the Metropolis or heat bath methods for updating the global action, see eg. [3].

An alternative form of lattice gauge theory which is useful in the pure gauge sector is the Hamiltonian form. Consider the generalization of the action (26) to the case of unequal lattice spacings a_s in the space direction and a_t in the time direction. By repeating the steps in (28), it is clear that in order to retain the continuum limit with unequal spacings, the action must be

$$S = \frac{a_s}{a_t}\beta \sum_{\Box_t}(1 - \cos \Sigma_{\mu\nu}) + \frac{a_t}{a_s}\beta \sum_{\Box_s}(1 - \cos \Sigma_{\mu\nu}) \qquad (29)$$

where $\Box_t$ denotes a space-time plaquette and $\Box_s$ denotes a space-space plaquette and we must have $\theta_0(\mathbf{n}) = a_t g A_0(\mathbf{n})$ and $\theta_i(\mathbf{n}) = a_s g A_i(\mathbf{n})$.

Choosing the temporal gauge in which U is set to unity on time links, so that $\theta_0 = 0$, we then obtain for the space-time plaquettes,

$$\frac{a_s\beta}{a_t}\operatorname{Re}U_\Box = \frac{a_s\beta}{a_t}\cos\left(1 + \theta_i(\mathbf{n}+\boldsymbol{\mu}_0) - 1 - \theta_i(\mathbf{n})\right) \approx \frac{a_s\beta}{2a_t}\left(\theta_i(\mathbf{n}+\boldsymbol{\mu}_0) - \theta_i(\mathbf{n})\right)^2. \qquad (30)$$

The Hamiltonian which produces the action (30) under evolution for infinitesimal time a_t is

$$H = -\frac{1}{2a_s\beta}\sum_{i,\mathbf{n}}\frac{\partial^2}{\partial\theta_i(\mathbf{n})^2} - \frac{\beta}{a_s}\sum_{\{i,j\},\mathbf{n}}(1 - \cos \Sigma_{ij}) \qquad (31)$$

where i and j run over the spatial directions and $\mathbf{n}$ runs over the spatial lattice sites. The continuum limit is verified by using $\beta = \frac{1}{g^2}$ and $\theta(\mathbf{n}) = g a_s A_i(\mathbf{n})$ and noting that the properly normalized commutation relation on the lattice

$$[A_i(\mathbf{n}), E_j(\mathbf{m})] = -i\frac{\delta_{ij}\,\delta_{\mathbf{n},\mathbf{m}}}{a_s^3} \qquad (32)$$

requires

$$E_i(\mathbf{n}) = \frac{i}{a_s^3}\frac{\partial}{\partial A_i(\mathbf{n})} \qquad (33)$$

with the result

$$H = a_s^3 \sum_{i,\mathbf{n}}\frac{1}{2}E_i^2(\mathbf{n}) - \frac{1}{g^2 a_s}\sum_{\{ij\},\mathbf{n}}\left(1 - \cos a_s^2 g\left(\partial_i A_j - \partial_j A_i\right)\right)$$

$$\approx a_s^3 \sum_{i,\mathbf{n}}\frac{1}{2}\left(E_i^2(\mathbf{n}) + B_i^2(\mathbf{n})\right) \to \int d^3x \frac{1}{2}\left(\mathbf{E}^2(\mathbf{x}) + \mathbf{B}^2(\mathbf{x})\right). \qquad (34)$$

The lattice Hamiltonian (31) may be viewed as a many-body Schrödinger equation with coordinates θ_i, and a four-body potential. The initial value Monte Carlo method [7] therefore provides a useful alternative to the usual global sampling of the Lagrangian action, and has been exploited for $U(1)$ and $SU(N)$ gauge theories [8].

In discussing the relation between the Hamiltonian and Lagrangian forms of lattice gauge theory, it is useful to examine the role of *Gauß's law* and how the presence of external charges is manifested in the theory. The basic ideas are most easily sketched in the continuum theory. Since the Hamiltonian does not constrain the charge state of the system, we must project the states appearing in the path integral onto the space satisfying $\nabla \cdot \mathbf{E} = \rho$ with a specific background charge ρ which may be accomplished by writing a functional δ function in the form $\int \mathcal{D}\chi \, e^{i \int d^3x \, dt \, \chi(\nabla \cdot \mathbf{E} - \rho)}$. Remaining in temporal gauge $A_0 = 0$ and using the form of the path integral (2) in which both the coordinate $x \leftrightarrow \mathbf{A}$ and momentum $p \leftrightarrow \mathbf{E}$ appear, the path integral for the partition function projected onto the space with external source ρ may be written

$$Z = \int \mathcal{D}\chi \, \mathcal{D}\mathbf{A} \, \mathcal{D}\mathbf{E} \, e^{\int d^3x \, dt \left[i\mathbf{E} \cdot \dot{\mathbf{A}} - \frac{1}{2}(\mathbf{E}^2 + \mathbf{B}^2) + i\chi(\nabla \cdot \mathbf{E} - \rho)\right]}$$
$$= \int \mathcal{D}\chi \, \mathcal{D}\mathbf{A} \, e^{-\int d^3x \, dt \left\{ \frac{1}{2}\left[\left(\dot{\mathbf{A}} - \nabla\chi\right)^2 + \mathbf{B}^2 - i\chi\rho\right]\right\}} \quad . \tag{35}$$

Equation (35) is an important result. Having started in temporal gauge $A_0 = 0$, we see that enforcing Gauß's law gives rise to a projection integral over an additional field χ which enters into the final action just like the original A_0 field. Indeed, renaming $\chi = A_0$ so that $\dot{A}_i - \partial_i A_0 = F_{0i}$ and writing the source as a set of point charges $\rho(x) = \sum_n q_n \delta(x - x_n)$, we obtain

$$Z = \int \mathcal{D}A_\mu \, e^{-\int d^3x \, dt \frac{1}{4} F_{\mu\nu} F_{\mu\nu}} \prod_n e^{-iq_n \int dt \, A_0(\mathbf{X}_n, t)} \quad . \tag{36}$$

Thus, the Hamiltonian path integral with projection is precisely the Lagrangian path integral with a line of $\pm A_0$ fields at the positions of the fixed external $\pm$ charges. In the case of no external charges, we may think of the Lagrangian path integral including the A_0 integral as the usual filter $e^{-\beta H}$ selecting out the ground state. In the presence of charges, the path integral augmented by lines of A_0 at the positions of the charges filters out the ground state in the presence of these sources.

3.2 $SU(N)$ Gauge Theory

The generalization to non-Abelian gauge theory is straightforward. The link variables become group elements of $SU(N)$ (Their behavior under Gauge transformations will be explained in the next section.)

$$U_\mu(\mathbf{n}) = e^{i \, ag \tilde{A}_\mu(\mathbf{n})}, \qquad \tilde{A}_\mu(\mathbf{n}) = A_\mu^c(\mathbf{n}) \frac{\lambda^c}{2}, \tag{37}$$

$$\operatorname{tr} \lambda^c = 0, \qquad \operatorname{tr} \lambda^b \lambda^c = 2\delta_{bc} \tag{38}$$

where c is a color label which runs over the $N^2 - 1$ generators λ^c (Pauli matrices for $SU(2)$, Gell-Mann matrices for $SU(3)$). The integration in the path integral is defined by the invariant group measure which we will denote by $\mathcal{D}(U)$ (for details see eg. [2][3]).

Using the same labeling conventions as in Fig. 1 with θ_μ replaced by $ag\tilde{A}_\mu$, the product of $SU(N)$ group elements around an elementary plaquette is

$$U_\Box = e^{iag\tilde{A}_\mu(\mathbf{n})} e^{iag\tilde{A}_\nu(\mathbf{n}+\mu)} e^{-iag\tilde{A}_\mu(\mathbf{n}+\nu)} e^{-iag\tilde{A}_\nu(\mathbf{n})}. \tag{39}$$

By expanding $\tilde{A}(\mathbf{n}+\mu)$ as before and making use of the Baker–Hausdorff identity $e^{a\hat{X}} e^{a\hat{Y}} = e^{a\hat{X}+a\hat{Y}+\frac{1}{2}a^2[\hat{X},\hat{Y}]+\mathcal{O}(a^3)}$ due to the non-commutativity of the generators, $U_\Box$ can be shown to reduce in the continuum limit to

$$U_\Box \approx e^{ia^2 g \tilde{F}_{\mu\nu}}$$
$$\tilde{F}_{\mu\nu} := \partial_\mu \tilde{A}_\nu - \partial_\nu \tilde{A}_\mu + ig\left[\tilde{A}_\mu, \tilde{A}_\nu\right] = F^c_{\mu\nu} \frac{\lambda^c}{2}. \tag{40}$$

The discrete lattice action for the $SU(N)$ gauge theory is then defined as

$$S(U) = \beta \sum_\Box \left(1 - \frac{1}{N} \operatorname{Re} \operatorname{Tr} U_\Box \right), \qquad \beta = \frac{2N}{g^2}. \tag{41}$$

Substitution of $U_\Box$ from (39) in the action (41) yields the desired continuum action in leading order

$$S(U) = \beta \sum_\Box \left\{ 1 - \frac{1}{N} \operatorname{Re} \operatorname{Tr} \left(1 + ia^2 g \tilde{F}_{\mu\nu} - \frac{1}{2} a^4 g^2 \tilde{F}^2_{\mu\nu} \cdots \right) \right\}$$
$$\approx \frac{1}{2} \beta a^4 g^2 \sum_{\mathbf{n},\{\mu\nu\}} \frac{1}{N} \operatorname{Tr} \left(\frac{1}{2} \lambda^c F^c_{\mu\nu}(\mathbf{n}) \frac{1}{2} \lambda^b F^b_{\mu\nu}(\mathbf{n}) \right)$$
$$\approx \beta \frac{g^2}{2N} \sum_\mathbf{n} a^4 \sum_{\{\mu\nu\}} \frac{1}{2} F^c_{\mu\nu}(\mathbf{n}) F^c_{\mu\nu}(\mathbf{n}) \tag{42}$$
$$\to \frac{1}{4} \int d^4 x\, F^c_{\mu\nu}(\mathbf{x}) F^c_{\mu\nu}(\mathbf{x}).$$

Summation over repeated indices is implied everywhere except where $\sum_{\{\mu\nu\}}$ denotes the sum over distinct pairs μ and ν.

3.3 Wilson Loops and Lines

If there were a quark field defined on a lattice, then under a local gauge transformation, the field ψ_i at each site would be multiplied by an element g_i of the gauge group,

$$\begin{aligned} \psi_i &\to g_i \psi_i, \\ \bar{\psi}_i &\to \bar{\psi}_i g_i^{-1}. \end{aligned} \tag{43}$$

The link variables $U_\mu(\mathbf{n})$ were explicitly introduced to compensate such a gauge transformation, so the link variable U_{ij} going from site i to j transforms like

$$U_{ij} \to g_i U_{ij} g_j^{-1} \ . \tag{44}$$

The only variables in the pure gauge sector are link variables, and the only gauge invariant objects which can be constructed are products of link variables around closed paths, for which the factors of g and g^{-1} combine at each site. The *Wilson loop* is therefore defined as the trace of a closed loop of link variables

$$W := \text{Tr} \prod_{ij \in c} U_{ij} = \text{Tr}\, U_{ij} U_{jk} \ldots U_{mn} U_{ni} \tag{45}$$

and specifies the rotation in color space that a quark would accumulate along the loop c from the path-ordered product $P_c\, e^{\int ig\tilde{A}}$.

Note that by (43),(44) a product of link variables transforms in the same way as a quark anti-quark pair under gauge transformations

$$\begin{aligned} U_{ik} U_{k\ell} \ldots U_{mj} &\to g_i U_{ik} U_{k\ell} \ldots U_{mj} g_j^{-1} \\ \psi_i \bar{\psi}_j &\to g_i \psi_i \bar{\psi}_j g_j^{-1} \ . \end{aligned} \tag{46}$$

Thus, as far as the gluon fields are concerned, the ends of a chain of link variables are equivalent to an external quark-antiquark source, and the presence of such a chain of link variables therefore measures the response of the gluon fields to an external quark-antiquark source.

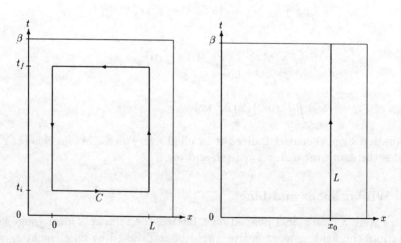

Fig. 2. *A space-time Wilson loop defined by the chain of link variables C on a finite lattice (left) and a Wilson or Polyakov line defined by the chain of link variables L on a finite lattice (right).*

Now, consider the time evolution of the system corresponding to the expectation value of the Wilson loop drawn in Fig. 2:

$$\langle W \rangle = \frac{\int \mathcal{D}(U)\, e^{-S(U)} \operatorname{Tr} \prod_c U_{ij}}{\int \mathcal{D}(U)\, e^{-S(U)}}. \tag{47}$$

Prior to the time t_i, there are no color sources present, so evolution filters out the gluon ground state in the zero charge sector, $|0\rangle = e^{-t_i H}\, |Q=0\rangle$. At time t_i, the line of link variables between 0 and L creates an external antiquark source at 0 and a quark source at L. As discussed in connection with (36), the links in the time direction between t_i and t_f maintain these sources at 0 and L. Hence, for any t between t_i and t_f, the evolution filters out the lowest gluon configuration in the presence of external quark-antiquark sources producing the state $|\psi\rangle = e^{-(t-t_i)H}\, \psi(0)\, \bar\psi(L)\, |0\rangle$. Finally, at time t_f, the external sources at 0 and L are removed by a line of links from L to 0, and the system is returned to the zero charge sector. Using Feynman's picturesque language of antiquarks corresponding to quarks propagating backwards in time, one may succinctly characterize the Wilson loop as measuring the response of the gluon fields to an external quark-like source traveling around the perimeter of the space-time loop in the direction of the arrows.

Quantitatively, if $t_f - t_i$ is large enough, the lowest gluon state in the presence of quark and antiquark sources separated by L will dominate, and $\langle W \rangle$ will be proportional to $e^{-(t_f-t_i)V(L)}$ where $V(L)$ is the static quark-antiquark potential. Physically, this potential corresponds to the potential arising in heavy quark spectroscopy, and its calculation therefore signals confinement. Furthermore, at large distances, the potential in the pure gluon sector becomes linear (since the flux tube cannot be broken by $q\bar q$ pair creation), so the Wilson loop enables direct numerical calculation of the string tension. If the Wilson loop has I links in the time direction and J links in the space direction, then

$$W(I,J) = \left\langle \operatorname{Tr} \prod_{C_{IJ}} U \right\rangle \begin{array}{c} \stackrel{I\to\infty}{\longrightarrow} e^{-aIV(aJ)} \\ \stackrel{I,J\to\infty}{\longrightarrow} e^{-a^2\sigma IJ} \end{array}. \tag{48}$$

The exponent is thus proportional to the area for large loops, and this area behavior is a signature of confinement, since it arises directly from the linearly rising potential. Indeed, the area law and hence confinement is found in the strong coupling expansion of lattice Gauge theories (see next section). Although the preceding physical argument was framed in Hamiltonian form with evolution in the time direction, it is clear that because of the symmetry of the Euclidean action, all space-time dimensions are equivalent and the area law reflects this symmetry. Although the area law behavior of large Wilson loops is clear from (48), in practical calculations on finite lattices, there are significant corrections, including a term proportional to the perimeter arising from the self-energy of the external sources and a constant arising from gluon exchanges at the corners, so that

$$-\ln W(I, J) \approx C + D(I + J) + a^2 \sigma I J \ . \tag{49}$$

To eliminate the constant and perimeter terms, the following ratio of Wilson loops having the same perimeter is calculated to cancel out the C and D terms in (49)

$$\chi(I, J) = -\ln\left(\frac{W(I, J)W(I-1, J-1)}{W(I, J-1)W(I-1, J)}\right) \approx a^2 \sigma \ , \tag{50}$$

and we will subsequently show results for this quantity.

A *Wilson* or *Polyakov line* is another form of gauge invariant closed loop which can be placed on a periodic lattice. In this case, as sketched in Fig. 2, the links are located at a fixed position in space x_0 and run in the time direction from the first time slice to the last, which by periodicity is equivalent to the first and thus renders the product gauge invariant. If the length of the lattice in the time direction is β_t (were the subscript t distinguishes it from the inverse coupling constant $\beta_g \equiv \frac{2N}{g^2}$ to which we will append a subscript g when necessary), the expectation value of the Wilson line yields the partition function for the gluon field in the presence of a single fixed quark at inverse temperature β_t and thus specifies the free energy F_{quark} of a single quark.

$$\langle \hat{L} \rangle = \frac{\int \mathcal{D}(U)\, e^{-S(U)}\, \text{Tr} \prod_L U_{ij}}{\int \mathcal{D}(U)\, e^{-S(U)}} =: e^{-\beta_t F_{\text{quark}}} \ . \tag{51}$$

This quantity will be useful as an order parameter for the deconfinement phase transition. Note that because the periodic lattice is a four-dimensional torus and L winds around the lattice once in the time direction, it is characterized by a winding number and is thus topologically distinct from a Wilson loop which has winding number 0. By the preceding argument, two lines in opposite directions, one at $x = 0$ and one at $x = L$, will produce the free energy of a quark and antiquark separated by distance L, and as $\beta_t \to \infty$ this provides an alternative means of calculating the static quark-antiquark potential.

The interaction energy can be measured by comparing sources of different temporal extent $t_f - t_i$ to cancel out end effects, and properties of the ground state may be obtained by measuring appropriate observables on intermediate time slices. For example, E^2 and B^2 in the presence of a static source can be measured using the fact that, in the continuum limit, a plaquette in the $\mu\nu$ plane may be expanded

$$\text{Tr}\, U_{\mu\nu}^\Box = \text{Tr}\, e^{ia^2 g F_{\mu\nu}} \sim 1 - \frac{1}{2} a^4 g^2 F_{\mu\nu}^2 \ . \tag{52}$$

Hence, in the presence of a source $\mathcal{O}$, where $\mathcal{O}$ denotes e.g. L or W above, the change in E^2 or B^2 relative to the vacuum is given by the space-time or space-space components of

$$\langle F_{\mu\nu}^2(\mathbf{r}) \rangle = \frac{\langle \mathcal{O} U_{\mu\nu}^\Box(\mathbf{r}) \rangle}{\langle \mathcal{O} \rangle} - \langle U_{\mu\nu}^\Box \rangle \ . \tag{53}$$

3.4 Strong Coupling Expansion

For strong coupling g, one can obtain a useful physical picture [1] of what happens when one evaluates the expectation value of a Wilson loop

$$\langle W \rangle = Z^{-1} \int \mathcal{D}(U) \, e^{-\beta_g \sum_\square \left(1 - \frac{1}{2N} \text{Tr}\left(U_\square + U_\square^\dagger\right)\right)} \, \text{Tr} \prod_C U \qquad (54)$$

by expanding the exponential in powers of the inverse coupling constant $\beta_g := \frac{2N}{g^2}$. This expansion is formally equivalent to the high temperature expansion in statistical mechanics where the inverse temperature plays the role of β_g. Note however, that the actual physical inverse temperature of our system is specified by the length of the lattice in the time direction, $\beta_t = N a_t$, and is distinct from β_g.

The structure of the expansion is revealed by considering the integrals over group elements which arise in the path integral (54). A general discussion of integration over $SU(N)$ group elements is given by Creutz [2], but for our present purposes it is sufficient to use the following two results, where Greek indices denote $SU(N)$ matrix indices, not sites

$$\int dU \, U_{\alpha\beta} = 0 \, , \qquad (55)$$

$$\int dU \, U_{\alpha\beta} U^{-1}_{\gamma\delta} = \frac{1}{N} \delta_{\alpha\delta} \delta_{\beta\gamma} \qquad (56)$$

which follow directly from the orthogonality relation for irreducible matrix representations of the group and are trivially verified for $U(1)$ for which the invariant measure is $\int dU = \frac{1}{2\pi} \int_{-\pi}^{\pi} d\theta$ and $U = e^{i\theta}$.

Now consider the diagrams which result from drawing the links in the Wilson loop $\prod_C U$ and some set of plaquettes $U_\square$ and $U_\square^\dagger$ obtained from expanding the exponential in (54). The integral (55) tells us that any diagram which has a single exposed link (that is, a single link between a pair of sites) anywhere on the lattice gives no contribution. Thus, the only non-vanishing terms in the expansion are those in which we manage to mate plaquettes from the exponential with the Wilson loop to eliminate all exposed links. The simplest way to mate two links to obtain a non-vanishing result is to place them between the same sites in opposite directions, which by (56) yields $\frac{1}{N}$. Since each plaquette brings with it a factor of β, the lowest order non-vanishing contribution to $\langle W \rangle$ is obtained by "tiling" the interior of the Wilson loop with plaquettes oriented in the opposite direction as sketched in Fig. 3 for a 3×3 loop. Note that each of the outer links of the original Wilson loop is protected, and the interior links protect each other pairwise. The leading term for an $I \times J$ Wilson loop would thus have $I \times J$ tiles, each contributing a factor $\frac{\beta}{2n}$. In addition, because of the traces in the plaquettes and Wilson loop in (54) and the δ's in (56), there is a factor of N for each of

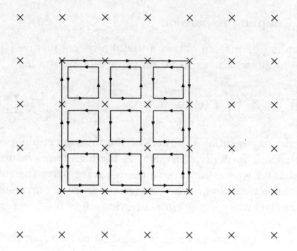

Fig. 3. A 3×3 Wilson loop tiled with plaquettes in the strong coupling expansion.

the $(I+1)(J+1)$ sites, and because of the factor $\frac{1}{N}$ in (56) there is a factor $\frac{1}{N}$ for each of the $(2IJ+I+J)$ double bonds. Hence, except for $SU(2)$, where the counting is different because the two orientations of plaquettes are equivalent, the overall contribution goes as

$$\langle W(I,J) \rangle \sim \left(\frac{\beta}{2N^2} \right)^{IJ}, \tag{57}$$

giving the lowest order contribution to the string tension

$$\sigma \sim -a^{-2} \ln \left(\frac{\beta}{2N^2} \right). \tag{58}$$

Fancier tilings are also possible if one is willing to use more tiles and thus include more powers of β. For example, one could place five tiles together to make a cubic box with an open bottom and replace one or more tiles with this box. The box could be elongated, or even grown into a tube which connects back somewhere else. Alternatively, one could replace a plaquette oriented in one direction by $(N-1)$ plaquettes oriented in the opposite direction to obtain a non-vanishing $SU(N)$ integral.

The utility of this expansion is threefold. It shows that *any* lattice Gauge theory (even QED) is confining with the linear potential in the strong coupling limes, as assumed in (48). In addition, it provides a physical picture of filling in the Wilson loop with a gluon membrane, whose vibrations and contortions represent all the quantum fluctuations of the gluon field. When observed on a particular time slice, the cross section of this surface corresponds to a color flux tube joining the quark-antiquark sources. And thirdly, in low orders, the individual terms can be calculated explicitly and provide a valuable quantitative check of numerical calculations.

3.5 Confinement Transition

The formal similarity between the Euclidean path integral for pure gauge theory and statistical mechanics in $d+1$ dimensions allows us to apply standard techniques for the symmetry analysis of phase transitions to the confinement transition. Symmetry analysis of gauge theory is treated in detail in the review by Svetitsky [9] and I will only summarize the main ideas here.

The essential idea is to follow the approach of Landau and identify an order parameter characterizing the transition, construct an effective action in terms of this order parameter, and use symmetry considerations to identify the form of the action.

The Wilson loop satisfies Landau's definition of an order parameter since, by (51), $\langle L \rangle = e^{-\beta_t F_{\text{quark}}}$. In the low-temperature confined phase, the free energy of a single quark is infinite so that $\langle L \rangle = 0$ whereas in the deconfined high-temperature phase the free energy is finite and thus $\langle L \rangle \neq 0$. Since this behavior appears superficially to be just the opposite of that of a magnet, which has magnetization $\langle M \rangle \neq 0$ at low temperature and $\langle M \rangle = 0$ at high temperature, it is important to recognize once again that the parameter which enters the lattice partition function analogously to the temperature in statistical mechanics is *not* the physical temperature. For a fixed lattice, the action has the form $e^{-\frac{2n}{g^2}S_\square}$ so that g^2 plays the role of an *effective temperature* whereas the *physical* inverse temperature is $\beta_t = N_t a$ where N_t is the (fixed) number of lattice sites in the time direction. Thus, in order for the physical temperature to increase, a must decrease, which as we shall see subsequently, means that g decreases, so that the effective temperature decreases. Hence, both a magnet and a lattice gauge system have finite order parameters when the relevant effective temperature is low and vanishing order parameters at high effective temperature.

It is useful to define an effective action $S_{\text{eff}}[L]$ such that the original partition function obtained by integrating the lattice action $S(U)$ over the group elements can be written as an integral of $S_{\text{eff}}[L]$ over the order parameter:

$$Z = \int d(U) \, e^{-S(U)} = \int dL \, e^{-S_{\text{eff}}[L]} \tag{59}$$

This effective action is constructed in the standard way by introducing a δ-function requiring that $L(x)$ be equal to the expectation value of a Wilson line at point x:

$$e^{-S_{\text{eff}}[L]} \equiv \int d(U) \, e^{-S(U)} \prod_x \delta\left(L(x) - \text{Tr} \prod_{L_x} U_{ij}\right) \tag{60}$$

I will now show that this effective action is symmetric with respect to the center of the gauge group, and that this symmetry has important implications for the confinement phase transition.

Let Z be an element of the center of the group, $Z \in \mathcal{C}$. That is, Z commutes with every element of the group. Suppose every link in the time direction originating on a particular time slice is multiplied by Z. In effect, the lattice is now no longer periodic but rather is periodic up to multiplication by the center element Z. The lattice action $S(U)$ is invariant under this multiplication by Z, since any space-time plaquette containing the affected time links is transformed as

$$U_1 U_2 U_3^\dagger U_4^\dagger \to Z U_1 U_2 U_3^\dagger Z^\dagger U_4^\dagger$$
$$= U_1 U_2 U_3^\dagger U_4^\dagger \,. \tag{61}$$

The last line shows that it is essential that Z commutes with all elements of the group. Each Wilson line, however, necessarily contains one factor of Z which may be commuted to this end of the line, so that

$$\prod_{L_x} U_{ij} \to Z \prod_{L_x} U_{ij} \,. \tag{62}$$

A matrix which commutes with every $SU(N)$ matrix must be a multiple of the unit matrix, so we may write $Z = \mathcal{Z} I$. Since the action is invariant under multiplication of an entire time slice by Z whereas the Wilson line is multiplied by Z, the effective action has the symmetry

$$S_{\text{eff}}[L] = S_{\text{eff}}[\mathcal{Z} L] \,, \quad \mathcal{Z} I \in \mathcal{C} \,. \tag{63}$$

Given this symmetry of the effective action with respect to the center, the usual symmetry arguments for analyzing phase transitions apply. The major assumption at this point is that one can integrate out the degrees of freedom at short distance scales and derive a Landau–Ginzburg action of the usual local form

$$S_{\text{eff}} \to \int dx \left\{ (\partial_i L)^2 + V(L) \right\} \tag{64}$$

where the effective potential has symmetry with respect to the center $V(L) = V(\mathcal{Z} L)$ and the reader is referred to Ref. [8] for details.

We are now ready to examine the implications of the symmetry with respect to the center of the local potential $V(L)$ for the order of the deconfinement transition. First, consider $U(1)$ gauge theory. Since the group is Abelian, the center is the whole group and $V(L)$ in symmetric under $L \to e^{i\theta} L$. This requires that V is a function of $|L|^2$ and has the generic form

$$V_{U(1)}\left(|L|^2\right) = a|L|^2 + b|L|^4 + c|L|^6 + \ldots \,. \tag{65}$$

A phase transition will occur in the region of temperature in which $a(T)$ changes sign, and we note that the value of L at which the minimum occurs will change discontinuously or continuously depending upon whether b is negative or positive. Thus, the order of the phase transition depends on the sign of b, and in particular, is not constrained by the form of (65).

Is the non-Abelian case constrained any more strongly by symmetry? It is easy to see that the center of $SU(N)$ is $Z(N)$, that is, the N roots of unity, since $\det(\mathcal{Z} I) = \mathcal{Z}^N = 1$. For $SU(2)$, this means $\mathcal{Z} = \pm 1$ and hence, $V(L) = V(-L)$.

As in the case of $U(1)$ this only requires V to be a function of $|L|^2$ leading again to the form (65) for which the order of phase transition is indeterminate. For $SU(3)$, however, we have $Z(3)$ symmetry so that $V(L) = V\left(e^{i\frac{n2\pi}{3}}L\right)$. In this case, the action may include a cubic invariant $\text{Re} L^3$ in addition to powers of $|L|^2$, so that (for real L) the action has the generic form

$$V_{SU(3)}(L) = a|L|^2 + b\,\text{Re}L^3 + c|L|^4 + \ldots \ . \tag{66}$$

In the presence of a cubic term, the position of the minimum must necessarily jump discontinuously, and we therefore conclude that the phase transition must be first order. This argument, while elegant and compelling, does depend on the formally exact effective action having the essentially local Landau–Ginzburg form (64).

There are two essential aspects of the $Z(3)$ symmetry of $SU(3)$ for our purposes. The first is the role of the Wilson line $\langle \hat{L} \rangle$ as an order parameter and measure of spontaneous symmetry breaking. In the confining phase, $\langle \hat{L} \rangle = 0$ and there is no spontaneous symmetry breaking. Thus, a plot of a Monte Carlo calculation of $\langle \hat{L} \rangle$ in the complex phase will produce a graph which is symmetric under rotation by $\frac{2\pi}{3}$, and has points equally distributed along the directions 1, $e^{\frac{2\pi}{3}i}$, and $e^{-\frac{2\pi}{3}i}$. This is analogous to a calculation of the spin in the unbroken symmetry phase of an Ising system, for which the distribution of spins is evenly divided between up and down. In the broken symmetry deconfined phase, however, $\langle \hat{L} \rangle \neq 0$, and calculated values of $\langle L \rangle$ may be clustered around any one of the three axes 1, $e^{\frac{2\pi}{3}i}$ and $e^{-\frac{2\pi}{3}i}$, just as the ordered state of the Ising system will either be concentrated around spin up or down. This behavior is clearly displayed in the Monte Carlo calculations [10] shown in Fig. 4. As the length of the lattice in the time direction β_t, corresponding to the inverse temperature, is increased from 2 to 5 lattice units, one observes a qualitative change from the broken symmetry solution in part (a) clustered around the real axis to the completely symmetric solution in part (d) which is rotationally symmetric around the origin.

The second essential result is that the $SU(3)$ confinement transition is first order. It is difficult to establish definitively the order of a phase transition by numerical calculations on finite lattices, since discontinuous transitions in the thermodynamic limit correspond to continuous transitions on finite lattices. There is, however, very strong evidence that the transition is indeed first order, including the sharpening of the transition as the lattice size is increased, hysteresis observed when a series of calculations with decreasing temperature is compared with a series calculated with increasing temperature, and two phase coexistence in which a sequence of Monte Carlo configurations in one phase persists in apparent equilibrium for a long time followed by an equally persistent sequence of configurations in the other phase.

Particularly suggestive evidence is provided by the microcanonical Monte Carlo results [11] shown in Fig. 5 for $SU(2)$ and $SU(3)$. Recall that for a van der Waals liquid-gas transition, the density is a multi-valued function of the

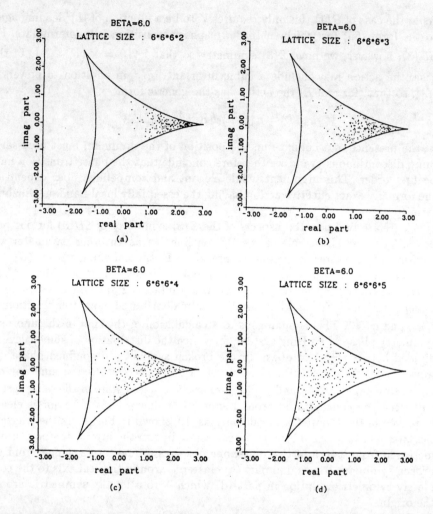

Fig. 4. Monte Carlo calculations from Ref. [5] of the order parameter $\langle L \rangle$ in the complex plane, showing symmetry restoration as the inverse temperature β_t increases from 2 to 5 lattice units.

temperature and the Maxwell construction specifies the actual discontinuity in the density occurring in the first-order transition in the thermodynamic limit. However, the S-shaped curve corresponds to metastable states which, although inaccessible in the canonical ensemble, can be explored by microcanonical calculations. For gauge theory on a lattice with a fixed number of sites, the order parameter $\langle L \rangle$ corresponds to the density order parameter, and β_g corresponds to the temperature since increasing β_g decreases g which decreases the lattice spacing a and thus increases the temperature (see next section). Figure 5 clearly demonstrates that on the same finite lattice, the Wilson line in $SU(3)$ shows the double-valued van der Waals dependence on β_g characteristic of a first-order transition whereas in $SU(2)$ there is no reentrant behavior and the transition appears second order. This result is consistent with and strongly supports the $Z(N)$ symmetry analysis.

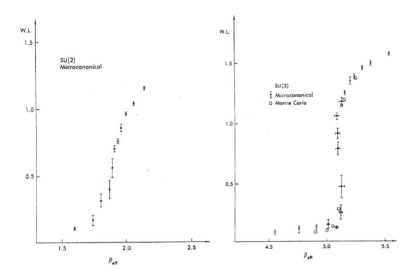

Fig. 5. *Microcanonical calculation [11] of the dependence of the Wilson line order parameter, denoted WL, on β_{eff}, which increases with increasing temperature. The reentrant behaviour for $SU(3)$ indicates a first order transition whereas the single-valued increase for $SU(2)$ indicates a second order transition.*

The deconfinement transition is much more difficult to treat in the presence of dynamical Fermions. Physically, it is clear that a Wilson line no longer serves as a rigorous order parameter since the possibility of creating quark-antiquark pairs from the vacuum can lead to screening of the external source and thus a finite rather than infinite free energy. In terms of our symmetry analysis, introduction of gauge invariant quark-gluon coupling terms of the form $\bar{\psi}_i U_{ij} \psi_j$ destroys the $Z(N)$ symmetry of the action since $\bar{\psi}_\tau U \psi_{\tau+1} \to Z \bar{\psi}_\tau U \psi_{\tau+1}$ at the time slice on which temporal link variables are multiplied by elements of the center. The hope is that nonetheless quarks are in some sense a small enough perturbation that the qualitative features of the pure gluon sector are not destroyed.

3.6 Continuum Limit and Renormalization

Pure gauge theory on a finite lattice is specified by two parameters: the dimensionless bare coupling constant g and the lattice spacing a corresponding to a momentum cutoff $p_{\max} \sim \frac{\pi}{a}$. As a is changed, the bare g must be changed to keep physical quantities fixed.

In principle, the renormalization procedure on a lattice is very simple and could be carried out as follows. First, pick an initial value of g and calculate some set of dimensionful physical observables $\langle \mathcal{O}_i \rangle$. These observables may be written in the form

$$\langle \mathcal{O}_i \rangle = a^{-d_i} \langle f_i(g) \rangle \tag{67}$$

where d_i is the dimension of the operator and f_i is the dimensionless quantity calculated on the lattice using the Wilson action with $\beta = \frac{2N}{g^2}$ and with all lengths expressed in units of the lattice spacing a. For example, we have already seen in (49) that the string tension has the form $\sigma = a^{-2}\chi$. Then, use the physical value of one operator, say $\mathcal{O}_1$, to determine the physical value of a corresponding to the selected g. Again, using the string tension example, we could define $a = \sqrt{\chi}/420\,\text{MeV}$. With this value of a determined from $\mathcal{O}_1$, all other observables $\mathcal{O}_2 \ldots \mathcal{O}_N$ are completely specified. One should then repeat this procedure for a sequence of successively smaller and smaller values of g, thereby determining the function $a(g)$ and a sequence of values for the observables $\mathcal{O}_2 \ldots \mathcal{O}_N$. If the theory is correct, then each sequence of observables $\mathcal{O}_i \; i \neq 1$ should approach a limit as $g \to 0$, and that limit should agree with nature.

In practice, it would be very difficult to carry out a series of calculations as described above to small enough g to make a convincing case. Hence, it is preferable to make use of our knowledge of the relation between the coupling constant and cutoff based on the renormalization group in the perturbative regime, and only carry out explicit lattice calculations down to the point at which the renormalization group behavior is clearly established. The foundation of the argument is the fact that the first two coefficients in the expansion of the renormalization group function $a\frac{dg}{da}$ are independent of the regularization scheme, and thus may be taken from continuum one and two loop calculations:

$$a\frac{dg}{da} = \gamma_0 g^3 + \gamma_1 g^5 + \ldots \tag{68}$$

Integration of this equation yields the desired relation

$$a(g) = \frac{1}{\Lambda_L} \left(\frac{16\pi^2}{11g^2}\right)^{51/121} e^{-\frac{8\pi^2}{11g^2}} \tag{69}$$

showing also that $g \to 0$ as $a \to 0$, where Λ_L is an integration constant and we have used the values of γ_0 and γ_1 for $SU(3)$ with no Fermions. The renormalization group behavior can be used only for asymptotically free theories, and only if a window exists in which the onset of scaling can be observed for a sequence of lattices.

The constant Λ_L governing the relation between the bare coupling constant and the lattice cutoff can be related by one-loop continuum calculations to the constants Λ_{MOM} and $\Lambda_{\overline{MS}}$, which govern the relation between the renormalized coupling constant and continuum cutoff using the momentum space subtraction procedure in Feynman gauge and the minimal subtraction procedure respectively, with the results [12]

$$\Lambda_{\text{MOM}} = 83.5\, \Lambda_L \, , \qquad \Lambda_{\overline{MS}} = 28.9\, \Lambda_L \, . \tag{70}$$

This correspondence is important for two reasons. First, the large coefficients in (70) allow us to reconcile our notion that the basic scale Λ_{QCD} is of order several hundred MeV with the fact that lattice measurements yield values of $\Lambda_L \sim 4 - 4.6$ MeV, which would otherwise appear astonishingly low. Second, it provides a quantitative consistency test, since experiments in the perturbative regime of QCD produce values of α consistent with those calculated on the lattice.

There is now substantial numerical evidence that lattice calculations in the pure gauge sector display the correct renormalization group behavior, and thus provide accurate solutions of continuum QCD. Data exist for two independent quantities, T_{tr}, the temperature of the deconfinement transition, and the string tension σ. Results for the transition temperature are shown in Fig. 6 taken from Ref. [13] based on data from Ref. [14]. The transition temperature on a lattice with N_t time slices is given by $T_{\text{tr}} = (N_t a(g_{\text{tr}}))^{-1}$ where g_{tr} is the value of the coupling for which the transition occurs. If $a(g_{\text{tr}})$ is calculated using the perturbative expression (69), then once g is small enough that the lattice theory coincides with the continuum theory, the quantity T_{tr}/Λ_L should approach a constant. As seen in Fig. 6, T_{tr}/Λ_L indeed appears constant above $\beta = \frac{6}{g^2} = 6$.

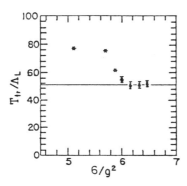

Fig. 6. *Ratio of the transition temperature T_{tr} to Λ_L as a function of inverse coupling constant.*

A second test of continuum behavior is provided by the string tension. One could display the approach to the continuum limit by plotting $\frac{\sigma}{\Lambda_L^2}$ as a function of $\beta = \frac{6}{g^2}$ and observing that, as in Fig. 6, this ratio becomes constant beyond $\beta = 6$. An alternative plot, which will also be useful for subsequent purposes is shown in Fig. 7. For the present, observe only the solid squares, which denote

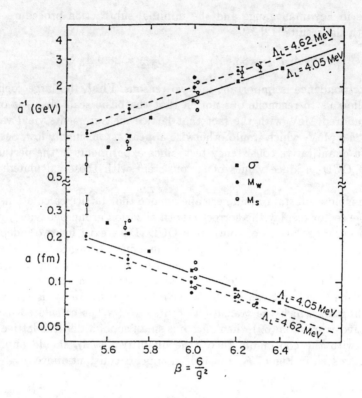

Fig. 7. Physical lattice spacing $a(g)$, determined from the string tension (squares) and hadron masses in the quenched approximation (circles). The solid and dashed lines show the dependence expected in the continuum limit from the renormalization group expression (69) for two values of Λ_L. The difference between masses calculated with Wilson Fermions (solid circles) and staggered Fermions (open circles) is discussed in the next section. Data are taken from references cited in [29].

values of a calculated from the surface tension as follows

$$a(g^2) = \left[\frac{[\sigma a^2]_{g^2}}{\sigma_{\text{expt}}}\right]^{1/2} = \frac{\sqrt{\chi}}{420 \text{ MeV}}. \tag{71}$$

For comparison, the renormalization group expression (69) is also plotted for the value $\Lambda = 4.05$ MeV which provides the best fit to the last two points. Again, one sees continuum behavior above $\beta = 6$. This graph also allows us to read off directly the values of a and a^{-1} at various values of β and compare values of a determined from the string tension with those determined from hadron masses to be discussed in a later section. One may also note that over the range of coupling constants relevant to lattice calculations, the effect of the premultiplying factor $g^{-0.84}$ in (69) is indiscernible and $\ln a$ is essentially linear in β. Although space does not permit additional figures pertinent to the continuum limit, it is also instructive to look directly at the renormalization group function $a\frac{dg}{da}$, and a useful graph combining string tension and transition temperature calculations is shown in [15].

More on renormalization on the lattice and the continuum limit can be found eg. in the lectures by P. Hasenfratz at this workshop [16].

In summary, I believe it is reasonable to regard lattice gauge theory in the pure gauge sector to be quite satisfactory. There are no glaring conceptual or computational problems, and all the numerical evidence to date suggests that one obtains an excellent approximation to the continuum theory for β_g above 6. In contrast, we will now see that full QCD including Fermions is more problematic at both the conceptual and computational levels.

4 Lattice QCD with Quarks

4.1 Naive Lattice Fermions and Doubling Problem

The simplest Hermitean finite difference expression for a Fermionic Hamiltonian that has the desired symmetries would be:

$$S_F^{\text{naive}} = a^4 \sum_{\mathbf{n}} \frac{1}{2a} \sum_{\mu} \left[\bar{\psi}(\mathbf{n}) \gamma_\mu U_\mu(\mathbf{n}) \psi(\mathbf{n} + \mathbf{a}_\mu) - \bar{\psi}(\mathbf{n} + \mathbf{a}_\mu) \gamma_\mu U_\mu^\dagger(\mathbf{n}) \psi(\mathbf{n}) \right] +$$

$$+ a^4 \sum_{\mathbf{n}} \bar{\psi}(\mathbf{n}) m \psi(\mathbf{n}) . \tag{72}$$

It has the proper continuum limit as can be seen by expanding the link variables and fields

$$S_F^{\text{naive}} \approx \int d^4x \left[\frac{1}{2a} \sum_\mu \left\{ \bar{\psi}\gamma_\mu (1 + igA_\mu)(1 + a\partial_\mu)\psi - \bar{\psi}(1 + a\overleftarrow{\partial}_\mu)\gamma_\mu(1 - igaA_\mu)\psi \right\} \right]$$

$$+ \int d^4x \, \bar{\psi} m \psi \quad \longrightarrow \quad \int d^4x \, \bar{\psi} [\gamma_\mu (\partial_\mu + igA_\mu) + m] \psi , \tag{73}$$

where Euclidean γ-matrices satisfying $\{\gamma_\mu, \gamma_\nu\} = 2\delta_{\mu\nu}$ have been used.

However, the above naive lattice action (72) has an unexpected problem that shows already up in the one dimensional case. Consider the Hamiltonian of free ($A_\mu = 0$) massless quarks in one space dimension,

$$H^{\text{naive}} = a \sum_n \psi^\dagger(n) \, \alpha \, \frac{1}{i} \left(\frac{\psi(n+1) - \psi(n-1)}{2a} \right) , \qquad \alpha = \gamma_0 \gamma_1 = \sigma_3 . \tag{74}$$

In order to obtain a Hermitean Hamiltonian it is essential to use a symmetric next to nearest neighbour difference approximation to the first derivative. We now transform the field operators to momentum space by writing the Fourier sum

$$\psi(n) = \frac{1}{\sqrt{Na}} \sum_{p=-\frac{N}{2}}^{\frac{N}{2}} \psi_{k_p} e^{ik_p na}, \qquad k_p = \frac{p\pi}{Na} \tag{75}$$

where it is understood that for a lattice with N sites and periodic boundary conditions, the sum over momenta in the first Brillouin zone extends over the N momenta. The Hamiltonian is diagonal

$$H^{\text{naive}} = \sum_{p=-\frac{N}{2}}^{\frac{N}{2}} \psi_{k_p}^\dagger \alpha \frac{\sin(k_p a)}{a} \psi_{k_p} \tag{76}$$

and thus has the eigenvalue spectrum

$$E_k = \pm \frac{\sin ka}{a} \stackrel{k \to 0}{\sim} \pm k\left(1 - \frac{(ka)^2}{3} + \ldots\right) \tag{77}$$

with eigenfunctions

$$\Psi_k^\pm(n) = e^{ikna} \chi^\pm, \qquad \chi^+ := \begin{pmatrix} 1 \\ 0 \end{pmatrix}, \qquad \chi^- := \begin{pmatrix} 0 \\ 1 \end{pmatrix}. \tag{78}$$

The comparison of the continuum spectrum for a massless Dirac particle $E_k = \pm k$ and the lattice spectrum in the top of Fig. 8 displays the species doubling problem. In the region of small k values, the lattice spectrum (77) yields a good approximation to the linear physical spectrum, and the range of linearity increases as $a \to 0$. However, at the edge of the Brillouin zone, there is a second region in which the spectrum also goes to zero linearly. In fact, for every physical mode Ψ_k, there is a precisely degenerate unphysical mode $\Psi_{\frac{\pi}{a}-k}$. Since the partition function blindly counts and weights all modes according to their energies, it is clear that all Fermion loops will be overcounted by a factor of 2 in all physical observables. Note also that since the velocity is $v = \frac{dE}{dk}$, the lattice spectrum necessarily mixes right-moving and left-moving modes.

The origin and structure of the doubled states is simple. The degenerate partner to the state $\Psi_k(n) = e^{ik na}\chi$ is the state $\psi_{\frac{\pi}{a}-k}(n) = e^{in\pi}e^{-ik na}\chi$, that is, a sawtooth mode in which every other lattice site has an extra factor of -1. The real part of a low k mode Ψ_k and its sawtooth partner $\Psi_{\frac{\pi}{a}-k}$ are sketched for a half wavelength in the middle section of Fig. 8. Note that although there are sufficient points in the half wavelength of Ψ_k to yield an accurate integral with any smooth function, there is no way that the rapidly oscillating wave function $\Psi_{\frac{\pi}{a}-k}$ can represent a mode with momentum near $\frac{\pi}{a}$. Thus, we need some way to eliminate these modes so that they play no role in the continuum limit. The origin of the degeneracy of the physical mode with its sawtooth partner is the symmetric difference approximation to the derivative $\psi' \sim \frac{\psi(n+1)-\psi(n-1)}{2a}$ in the naive Hamiltonian (74), which clearly is impervious to the minus signs $e^{-in\pi}$ and thus yields the same magnitude for the derivatives of ψ_k and $\psi_{\frac{\pi}{a}-k}$. The origin

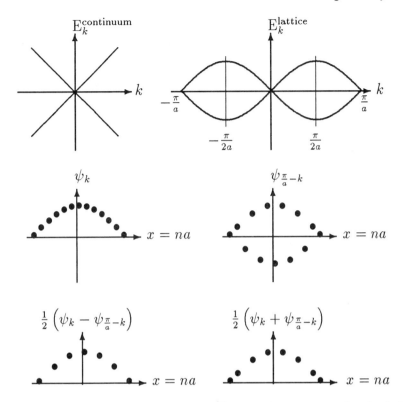

Fig. 8. Fermion doubling in one dimension. The top plots compare the physical continuum spectrum with the spectrum of the lattice Hamiltonian. The middle plots show a half wavelength of the real part of the non-vanishing component of a physical wave function (left) and its degenerate unphysical sawtooth partner (right). The bottom plots show the linear combinations corresponding to staggered Fermions.

of this symmetric difference, in turn, is Hermiticity, since expressions involving only nearest neighbor differences like $\psi^\dagger n(\psi(n+1) - \psi(n))$ are non-Hermitean and yield complex eigenvalues.

The doubling we have discussed for simplicity in one dimension arises analogously in each of the four Euclidean dimensions of the naive Fermion action, (72), so that we obtain $2^4 = 16$ lattice modes for each physical mode. Again, specializing to the massless case, the momentum space action corresponding to (72) is

$$S_F^{\text{naive}} := \bar{\psi} M \psi = \sum_k \bar{\psi}_k \sum_\mu \gamma_\mu \frac{\sin(k^\mu a)}{a} \psi_k \ , \qquad (79)$$

so that the inverse propagator is

$$\langle T \bar{\psi} \psi \rangle^{-1} = M(k) = \sum_\mu \gamma_\mu \frac{\sin(k^\mu a)}{a} \ . \qquad (80)$$

This propagator replicates the physical behavior in the region of $k^\mu \sim 0$ fifteen times around points on the edge of the four-dimensional Brillouin zone at which one or more of the components $k^\mu \approx \frac{\pi}{a}$.

We are now prepared to understand both the features giving rise to the doubling problem and the generality of the problem [17]. Whereas the specific function $\sin(k^\mu a)$ in (76) and (79) is the result of using the lowest-order Hermitean difference formula for the derivatives in the continuum action, the most general form of the chiral symmetric, Hermitean action derived from discrete derivatives on a periodic lattice with the correct continuum limit is

$$S_F = \sum_k \bar{\psi}_k \sum_\mu \gamma_\mu P^\mu(k) \psi_k \qquad (81)$$

where $P^\mu(k)$ is real for Hermiticity, $P^\mu(k) \stackrel{k \to 0}{\to} 0$ for the correct continuum limit, and $P^\mu(k)$ is periodic under $k^\mu \to k^\mu + \frac{2\pi}{a}$ and continuous for local discrete difference formulae on a lattice. Note that chiral symmetry requires the form $\bar{\psi}\gamma_\mu P^\mu \psi$, so that under a chiral transformation $\psi \to e^{i\alpha\gamma_5}\psi$, the two sign changes from γ_0 in $\bar{\psi}$ and γ_μ leave the action invariant. Since $P^\mu(k)$ is real, continuous and periodic in k^μ with period $\frac{2\pi}{a}$, it must cross the axis at some intermediate point, so that this general discrete action has the doubling observed in (79) in each of the four Euclidean directions yielding 15 spurious low-mass excitations for each physical excitation. A rigorous version of these arguments is known as the *Nielsen–Ninomiya no-go theorem* [18], which proves using homotopy theory that one cannot avoid Fermion doubling in a lattice theory which is simultaneously Hermitean, local and chirally symmetric.

An additional aspect of Fermion doubling is the absence of the axial anomaly. The unphysical doubler Fermions couple to an external axial current with the opposite chiral charge and effectively cancel the axial anomaly arising from the physical Fermions.

4.2 Wilson Fermions

One of the ways out of the no-go theorem is to give up chiral symmetry and, following Wilson, add a second derivative term to the Hamiltonian. In one dimension, combining the naive Hamiltonian H^{naive} (74) with a perturbation of the form

$$\begin{aligned} H' &= -\frac{a}{2i} a \sum_n \psi^\dagger(n)\gamma_0 \left(\frac{\psi(n+1) - 2\psi(n) + \psi(n-1)}{a^2} \right) \\ &\stackrel{a \to 0}{\longrightarrow} -\frac{a}{2i}\int dx\, \bar{\psi}(x)\psi''(x) \stackrel{a \to 0}{\to} 0 \end{aligned} \qquad (82)$$

yields the *Wilson Hamiltonian*

$$H_W = H^{\text{naive}} + r H' = \sum_k \psi_k^\dagger \left[\alpha \frac{\sin(ka)}{a} - r\frac{\gamma_0}{i}\frac{(\cos(ka)-1)}{a}\right]\psi_k. \qquad (83)$$

Using a representation with $\alpha = \sigma_3$ and $-i\gamma_0 = \sigma_1$, we obtain the energy spectrum

$$E^2 = \left(\frac{\sin(ka)}{a}\right)^2 + \left[\frac{r}{a}(\cos(ka) - 1)\right]^2 \tag{84}$$

with the limits

$$E \longrightarrow \begin{cases} \pm k\left(1 - \frac{1}{6}k^2a^2 + \frac{r^2}{8}k^2a^2\right) & (k \to 0) \\ \pm\frac{2r}{a} & (k \to \frac{\pi}{a}) \end{cases}. \tag{85}$$

Thus, for fixed r, the mode for $k \sim 0$ has the correct continuum limit whereas the sawtooth mode for $k \sim \frac{\pi}{a}$ becomes infinitely massive and decouples from the theory.

In four Euclidean dimensions, the corresponding Wilson action is

$$S_W = -a^4 \sum_{\mathbf{n}} \frac{1}{2a} \sum_{\mu} \left[\bar{\psi}(\mathbf{n})(r - \gamma_\mu) U_\mu(\mathbf{n}) \psi(\mathbf{n} + \mathbf{a}_\mu) + \bar{\psi}(\mathbf{n} + \mathbf{a}_\mu)(r + \gamma_\mu)\right.$$

$$\left. U_\mu^\dagger(\mathbf{n}) \psi(\mathbf{n})\right] + a^4 \sum_{\mathbf{n}} \left(m + \frac{4r}{a}\right) \bar{\psi}(\mathbf{n}) \psi(\mathbf{n}) \tag{86}$$

$$\tag{87}$$

and the propagators for the spurious modes acquire masses which diverge as $\frac{r}{a}$ as in the one-dimensional case.

The Wilson action manifestly breaks chiral symmetry for $m = 0$, since under the transformation $\psi \to e^{i\alpha\gamma_5}\psi$, $\bar{\psi}\psi \to \bar{\psi}e^{i2\alpha\gamma_5}\psi$. As long as the contribution of the symmetry breaking term can be made arbitrarily small, its presence does not interfere with the physics of spontaneous symmetry breaking.

Wilson Fermions provide a framework for completely solving the doubling problem which yields the correct physics in the limit of sufficiently small lattice spacing a. The primary disadvantages are associated with the lack of explicit chiral symmetry for finite a.

4.3 Staggered Fermions

An alternative way of treating Fermions on the lattice is the use of *staggered* or *Kogut–Susskind Fermions* [19]. This method does not avoid the no-go theorem, but rather reduces the number of doubled (unphysical) Fermion species by thinning out the Fermion degrees of freedom and has the advantage of maintaining explicit chiral symmetry. The basic idea is to transform the Fermion field operators to a new representation in which the naive Fermion action is diagonal in the Dirac indices, so that the naive Fermions represent N_D copies of the new Fermions, where N_D is the number of Dirac components. By keeping only one of these N_D copies, one effectively thins the degrees of freedom by $1/N_D$.

A convenient choice for the transformation of the Fermion field in four space-time dimensions is

$$\psi(\mathbf{n}) = \gamma_0^{n_0}\gamma_1^{n_1}\gamma_2^{n_2}\gamma_3^{n_3}\,\chi(\mathbf{n}) \;,$$
$$\bar{\psi}(\mathbf{n}) = \bar{\chi}(\mathbf{n})\,\gamma_3^{n_3}\gamma_2^{n_2}\gamma_1^{n_1}\gamma_0^{n_0} \;.$$
(88)

To see how this transformation renders the naive action (72) diagonal, consider a typical term:

$$\bar{\psi}(\mathbf{n})\,\gamma_2\,\psi(\mathbf{n}+\mathbf{a}_2) = \bar{\chi}(\mathbf{n})\,\gamma_3^{n_3}\gamma_2^{n_2}\gamma_1^{n_1}\gamma_0^{n_0}\gamma_2\gamma_0^{n_0}\gamma_1^{n_1}\gamma_2^{(n_2+1)}\gamma_3^{n_3}\,\chi(\mathbf{n}+\mathbf{a}_2)$$
$$= (-1)^{n_0+n_1}\bar{\chi}(\mathbf{n})\,\chi(\mathbf{n}+\mathbf{a}_2)$$
(89)

where the factor $(-1)^{n_0+n_1}$ arises from anticommuting γ_2 through the product $\gamma_0^{n_0}\gamma_1^{n_1}$ and the remaining γ matrices combine pairwise to unity. By the same argument,

$$\bar{\psi}(\mathbf{n})\,\gamma_\mu\,\psi(\mathbf{n}+\mathbf{a}_\mu) = \eta_\mu(\mathbf{n})\,\bar{\chi}(\mathbf{n})\,\chi(\mathbf{n}+\mathbf{a}_\mu)$$
(90)

where

$$\eta_\mu(\mathbf{n}) \equiv (-1)^{\sum_{\nu=0}^{\mu-1} n_\nu} \;.$$
(91)

The naive action may thus be written

$$S_F = a^4 \sum_{\mathbf{n}} \left\{ \frac{1}{2a}\sum_\mu \eta_\mu(\mathbf{n})\,\bar{\chi}(\mathbf{n}) \left[U_\mu(\mathbf{n})\,\chi(\mathbf{n}+\mathbf{a}_\mu) - U_\mu(\mathbf{n}-\mathbf{a}_\mu)\,\chi(\mathbf{n}-\mathbf{a}_\mu) \right] \right.$$
$$\left. +\bar{\chi}(\mathbf{n})\,m\,\chi(\mathbf{n}) \right\}$$
(92)

where $\chi(\mathbf{n})$ now represents any one of the N_D Dirac components and may thus be regarded as a scalar. By (88), it is clear that the components of χ are specific linear combinations of the doubled Fermion fields.

It is particularly simple to see how staggered Fermions work in the case of free, massless particles in one dimension described by the Hamiltonian (76). The Dirac equation has two components, and in this case the staggered solutions χ for a given k are just the sum or difference of the physical mode Ψ_k and the sawtooth mode $\Psi_{\frac{\pi}{a}-k}$ and are sketched in the bottom portion of Fig. 8. Note that for each of the staggered solutions, half of the points, either the even or odd points, correspond to the physical Fermion mode and the other half of the points are identically zero. Thus, the thinning of the degrees of freedom corresponds to having essentially doubled the lattice spacing. The maximum momentum which can be sustained on the lattice with spacing $2a$ is $\frac{\pi}{2a}$, so that for both the even site mode and the odd site mode only the portion of the spectrum in the upper right portion of Fig. 8 between $-\frac{\pi}{2a}$ and $\frac{\pi}{2a}$ contributes and there are no spurious low-mass modes. In the special case of one space dimension where there are two Dirac components and two naive Fermion modes, staggered Fermions completely resolve the doubling problem.

If one were to solve Hamiltonian field theory in three space dimensions, reduction of the 2^3 naive Fermions by a factor of $1/4$ for the four Dirac components would leave two species of staggered Fermions which one could regard as two degenerate flavors corresponding to up and down quarks. In Lagrangian field theory in $3+1$ dimensions, the 16 naive Fermions are only reduced to four flavors of staggered Fermions.

The principle advantage of staggered Fermions is the residual chiral symmetry of the lattice action. The mass is thereby protected from renormalization and it is possible to define a lattice axial current. So as to avoid spurious chiral symmetry breaking due to the presence of a chiral symmetry breaking term in the lattice action as in the case of Wilson Fermions, it is useful to use staggered Fermions. The chiral order parameter is calculated by evaluating

$$\begin{aligned} \langle \bar{\psi}\psi \rangle &= Z^{-1} \int \mathcal{D}\bar{\psi}\mathcal{D}\psi\mathcal{D}(U) \, e^{-\bar{\psi}M(U)\psi - S(U)} \\ &= Z^{-1} \int \mathcal{D}(U) \, e^{\ln \mathrm{Det}\, M(U) - S(U)} M^{-1}(U) \; . \end{aligned} \quad (93)$$

It serves as an order parameter for the case of zero quark mass and is also used as an indicator of a phase transition at finite quark mass. The price one pays for the staggered Fermions is the necessity of having an integer multiple of four flavors. Thus one cannot directly study the case of three flavors, where the presence of a cubic invariant implies a first-order transition (cf. the section on the confinement transition), or the physically relevant case of two flavors of light quarks.

The primary disadvantage of staggered Fermions is therefore the existence of four flavors in four dimensions. In addition, the lattice resolution is cut in half relative to Wilson Fermions, and physical operators are complicated, non-local combinations of χ fields. Hence, for most purposes in studying hadronic physics, it will be desirable to use Wilson Fermions.

4.4 Hopping Parameter Expansion

Consider Wilson Fermions with the action (86) and define the hopping parameter κ,

$$\kappa := \frac{1}{2ma + 8r} = \frac{1}{2Ma} \quad (94)$$

and rescaled Fermion fields

$$\Psi := \left(Ma^4\right)^{1/2} \psi \; , \quad (95)$$

where $M = m + \frac{4r}{a}$ enters as a mass term in (86). The action then has the form

$$S_W = \sum_{\mathbf{n}} \left\{ \bar{\Psi}(\mathbf{n})\Psi(\mathbf{n}) - \kappa \sum_{\mu} \Big[\bar{\Psi}(\mathbf{n})(r - \gamma_\mu) U_\mu(\mathbf{n}) \Psi(\mathbf{n} + \mathbf{a}_\mu) + \right. \\ \left. \bar{\Psi}(\mathbf{n} + \mathbf{a}_\mu)(r + \gamma_\mu) U_\mu^\dagger(\mathbf{n}) \Psi(\mathbf{n}) \Big] \right\} \; . \quad (96)$$

The fields have been scaled such that the diagonal term is now unity, and the hopping parameter κ specifies the strength of the nearest-neighbour coupling via link variables. The partition function for an $SU(N)$ gauge theory in the presence of Fermions with the action (96) reads in a slightly schematic form

$$Z = \int d\bar{\Psi}\, d\Psi\, d(U)\, e^{-\bar{\Psi}(1+\kappa U)\Psi - \beta \sum_\square (1-\frac{1}{N}\operatorname{Re}\operatorname{Tr} U_\square)} . \qquad (97)$$

Just as the strong coupling expansion provided insight into solutions of lattice QCD in the pure gauge sector, the hopping parameter expansion provides analogous insight into solutions in the presence of Fermions. In the strong coupling regime, $\beta = \frac{2N}{g^2}$ is small and the U's distributed according to $\exp\{-\beta S(U)\}$ are nearly random. Thus, the average of κU is in some sense small, and we may expand the exponential in (97) in the hopping parameter κ. The Fermionic content of the partition function may then be written schematically as

$$Z(U) = \int d\bar{\Psi} d\Psi\, e^{-\sum_\mathbf{n} \bar{\Psi}_\mathbf{n} \Psi_\mathbf{n}} \sum (\bar{\Psi}\kappa U\Psi)(\bar{\Psi}\kappa U\Psi)\ldots(\bar{\Psi}\kappa U\Psi) . \qquad (98)$$

Using Wick's theorem, this integral is equal to the sum of all contractions, where because the matrix in the exponent is the unit matrix, the contraction $\langle \Psi_\mathbf{m} \bar{\Psi}_\mathbf{n} \rangle$ is just $\delta_{\mathbf{m},\mathbf{n}}$. Thus, each factor κU which connects one site to an adjacent site must be connected to another hopping term κU emanating from the new site, and the net result after integrating out the Fermions is the sum of all possible closed chains of κU in which the U's are oriented head to tail

$$Z(U) = \sum \kappa^k U_{\mathbf{n}_1 \mathbf{n}_2} U_{\mathbf{n}_2 \mathbf{n}_3} \ldots U_{\mathbf{n}_k \mathbf{n}_1} . \qquad (99)$$

The full partition function is then the integral over gauge fields of all such loops weighted by the gluon action.

Consider now the hopping parameter expansion for a meson propagator (see (105) in the subsequent section)

$$\left\langle \bar{\Psi}(\mathbf{x},t)\,\Gamma\,\Psi(\mathbf{x},t)\, \bar{\Psi}(0,0)\,\Gamma\,\Psi(0,0) \right\rangle =$$

$$= Z^{-1} \int d\bar{\Psi} d\Psi d(U)\, e^{-\sum \bar{\Psi}\Psi}\, e^{-\sum \bar{\Psi}\kappa U \Psi} \times \qquad (100)$$

$$\times e^{-\beta \sum_\square (1-\frac{1}{N}\operatorname{Re}\operatorname{Tr} U_\square)} \left(\bar{\Psi}\Gamma\Psi\right)_{(\mathbf{x},t)} \left(\bar{\Psi}\Gamma\Psi\right)_{(0,0)}$$

where $\bar{\Psi}\Gamma\Psi$ represents a combination of Fermion fields of the appropriate flavors and γ matrices to create or annihilate the desired meson state. As before, we expand $\exp\{-\sum \bar{\Psi}\kappa U\Psi\}$ and apply Wick's theorem to obtain all contractions of $\bar{\Psi}$ and Ψ. The lowest order (in κ) non-vanishing contribution is obtained by creating two straight chains of U's, one from $(0,0)$ to $(\mathbf{x},t)$ and the other from $(\mathbf{x},t)$ to $(0,0)$. Higher-order contributions are obtained by elongating these two chains to form any closed path including the points $(0,0)$ and $(\mathbf{x},t)$ and by adding any additional number of separate closed loops of U's. The complete propagator is the sum over all such loops of U's

$$\left\langle \bar{\Psi}\,\Gamma\,\Psi_{(\mathbf{x},t)}\,\bar{\Psi}\,\Gamma\,\Psi_{(\mathbf{0},0)} \right\rangle = Z^{-1} \int d(U)\, e^{-\beta \sum_{\square} \left(1 - \frac{1}{2N}\operatorname{tr}\left(U_{\square} + U_{\square}^{\dagger}\right)\right)} \tag{101}$$

$$\times \sum_{\text{loops}} \kappa^{N_{\text{links}}} (U \ldots U)_{\mathbf{0}0,\mathbf{x}t} (U \ldots U) \ldots (U \ldots U)\,.$$

Note that the remnants of the Fermions at this stage of the calculation are just Wilson loops, again underscoring our previous interpretation of Wilson loops as the world lines of quarks. The simple quark model of the meson is described by the sum over all time histories of the loop $(U\,U \ldots U)_{\mathbf{0}0,\mathbf{x}t}$ representing a quark and antiquark propagating from $(\mathbf{0},0)$ to $(\mathbf{x},t)$ and the excitation of quark-antiquark pairs out of the vacuum is described by the additional quark loops $(U\,U \ldots U)$. The integral over the gauge fields in (101) now proceeds precisely as in the case of the Wilson loop discussed in the pure gauge sector. Each of the closed loops in (101) must be tiled with plaquettes, with the lowest-order contribution corresponding to the minimum tiling required to eliminate all exposed links and higher-order contributions corresponding to more elaborate surfaces. The general structure which emerges is thus a sum over Fermion loops covered or connected with gluon membranes, with the partition function dictating the optimal compromise between short Fermion paths and minimal membranes favored by small β and small κU and the higher entropy of longer Fermion paths and complicated surfaces. The physical picture of a meson state which emerges when one observes the configuration on a single time slice between $(\mathbf{0},0)$ and $(\mathbf{x},t)$ is a quark-antiquark pair connected by a flux tube. When one works out the explicit factors for $SU(N)$, one can also see the $1/N$ expansion emerge naturally, with the dominant contributions arising from planar diagrams with no additional Fermion loops. Propagators for baryons are similar to those for mesons, with three chains of U's starting at the point $(\mathbf{0},0)$, corresponding to three quarks of the appropriate flavors, and terminating at $(\mathbf{x},t)$. The surface between these chains must again be tiled with plaquettes, and the leading membrane contribution in this case, when cut on a single time slice, corresponds to a Y configuration of flux connecting the three quarks.

Instead of expanding the Fermionic action $\bar{\Psi} M(U) \Psi = \bar{\Psi}(1 + \kappa U)\Psi$ in powers of κ, one can alternatively write the result of integrating out the Fermions directly in terms of $M(U)$

$$\langle \Psi\Psi \ldots \Psi\bar{\Psi}\bar{\Psi}\ldots\bar{\Psi} \rangle =$$
$$= Z^{-1} \int d(U) d\bar{\Psi} d\Psi\, e^{-\bar{\Psi} M(U)\Psi - S(U)} \left(\Psi\Psi\ldots\Psi\bar{\Psi}\bar{\Psi}\ldots\bar{\Psi}\right)$$
$$= Z^{-1} \int d(U)\, e^{\ln \operatorname{Det} M(U) - S(U)} \sum_{\text{contractions}} M^{-1}(U)\, M^{-1}(U) \ldots M^{-1}(U)\,.$$
$$\tag{102}$$

If one now expands $M^{-1} = (1 + \kappa U)^{-1}$ and $\ln \text{Det} \, M(U) = \text{tr} \ln(1 + \kappa U)$ in κ, one observes that one obtains the previous hopping parameter expansion with all the quark propagators joining the Ψ's and $\bar{\Psi}$'s in $\langle \Psi \Psi \ldots \Psi \bar{\Psi} \bar{\Psi} \ldots \bar{\Psi} \rangle$ arising from the M^{-1}'s and all the additional closed quark loops arising from expansion of $\ln \text{Det} \, M$. As before, the quark lines thus obtained are tiled with plaquettes of the proper orientation from $S(U)$.

This knowledge of the role of the various terms in (102) allows us to understand the physics of the so-called *quenched approximation*, which might more properly be called the valence quark approximation. The quenched approximation corresponds to omitting the term $\ln \text{Det} \, M(U)$ in the exponent of (102) when performing the integral over gauge fields $\int \mathcal{D}(U)$. From the preceding argument, this approximation omits all time histories in which dynamical quark loops are excited out of the Fermi sea. Hence, only valence quarks connecting the field operators in $\langle \Psi \Psi \ldots \Psi \bar{\Psi} \bar{\Psi} \ldots \bar{\Psi} \rangle$ are included, and the integral over gluon fields incorporates the QCD interactions of these valence quarks to all orders. Technically, the motivation for making the quenched approximation is the fact that the stochastic evaluation of the path integral is immensely more difficult when the non-local term $\ln \text{Det} \, M(U)$ is included in the action than when one must only treat the local term $S(U)$.

Note that the quarks can still travel back and forth in time on the spacetime lattice so, for example, chiral logs are still taken into account and hadrons are still dressed at large distances by meson clouds [20]. The renormalization group calculations for the continuum limit have to be performed with the number of flavors $n_f = 0$ yielding stronger asymptotic freedom. So for example, the $\frac{\alpha}{r}$ potential at short distance is slightly too weak and hence meson wave functions at short distance are too small and the resulting decay constants are too weak.

4.5 Hadronic Observables

To discuss hadronic observables, it is convenient to consider idealized calculations in which one filters in the pure gauge sector for a long enough time to filter out the QCD vacuum state $|0\rangle$. (In practice, calculations usually do not completely filter out the ground state in the pure gauge sector, so the expressions here must have $|0\rangle$ replaced by sums over a sequence of states having the quantum numbers of the vacuum. The basic idea, however, is correct and the resulting formulae are more transparent.) One then creates a state $J^\dagger |0\rangle$ of the desired quantum numbers by acting on the vacuum with an appropriate source J, filters the lowest eigenstate with these quantum numbers by evolution in Euclidean time, and projects onto specified momentum states as required. A local field operator we may use for a π^+ at space-time point $\mathbf{x} = (\mathbf{x}, t)$, for example, is given by

$$J_\pi(\mathbf{x}) = \bar{d}(\mathbf{x}) \gamma_5 u(\mathbf{x}) \tag{103}$$

where u and d denote up and down quark fields, respectively, and the γ_5 makes the operator pseudoscalar. As an alternative to this local point source, it may be preferable in some applications susceptible to large stochastic errors to use non-local extended sources, especially if the physical particle is considerably larger than the lattice spacing. One will save computer time and improve accuracy of the calculation, if the overlap of the state created by the source and the physical hadron state is as large as possible.

For the proton, a local field with the correct transformation properties is

$$J_\alpha^P(\mathbf{x}) = \epsilon^{ijk} u_\alpha^i(\mathbf{x}) \left[u_\beta^j(\mathbf{x}) \, (C\gamma_5)^{\beta\gamma} \, d_\gamma^k(\mathbf{x}) \right] \tag{104}$$

where $C = \gamma_2\gamma_4$ is the charge conjugation matrix and i, j, k denote color labels, and α, β, γ denote Dirac indices. Several comments concerning this operator may be helpful [21]. Since quark and antiquark have opposite parity, the combination $uC\gamma_5 d$ transforms as a Lorentz scalar so that J_α^P transforms like u_α and thus as a spin 1/2 Dirac spinor. The color variables are explicitly antisymmetrized and the antisymmetry of the Grassmann variables combined with the fact that the local operator carries no orbital angular momentum assures the symmetry of the spin-flavor wave function. The non-relativistic limit also agrees with the non-relativistic quark model, since for the upper components $uC\gamma_5 d = u(-i\sigma^2)d = -u_\uparrow d_\downarrow + u_\downarrow d_\uparrow$ and the symmetrized state $S\{u_\uparrow (u_\uparrow d_\downarrow - u_\downarrow d_\uparrow)\}$ is the $SU(6)$ proton wave function.

Expectation values of operators are calculated by using two widely separated sources $J, J^\dagger$ to create a state of the desired quantum numbers, placing the operator to be measured in between where the lowest state has been filtered out, and projecting the momentum.

An important quantity derived from these hadronic sources is the vacuum vacuum correlation function of hadronic currents (Fig. 9(a))

$$\langle 0 | \mathcal{T} J_h(\mathbf{x}, t) J_h^\dagger(0, 0) | 0 \rangle \tag{105}$$

which is calculated for the π^+ by evaluating

$$\langle 0 | \mathcal{T} J_{\pi^+}(\mathbf{x}, t) J_{\pi^+}^\dagger(0, 0) | 0 \rangle =$$

$$= \int \mathcal{D}(\bar{\psi}\psi)\mathcal{D}(U) e^{-\bar{\psi} M(U)\psi - S(U)} \bar{d}(\mathbf{x}, t)\gamma_5 u(\mathbf{x}, t)\bar{u}(0, 0)\gamma_5 d(0, 0) = \tag{106}$$

$$= \int \mathcal{D}(U) e^{\ln \text{Det }(U) - S(U)} M^{-1}(U)_{\mathbf{x}t, 00}\gamma_5 M^{-1}_{00, \mathbf{x}t}\gamma_5$$

on a mesh of temporal extent from $-T$ to $T + t$ with hard wall boundary conditions at the time boundaries.

Its spatial integral is the two point function $C_{J_h J_h}(t)$, whose physical content becomes clear by transforming $J_h(\mathbf{x}, t)$ with translation operators in $\mathbf{x}$ and t and inserting a complete set of hadronic states $|np\rangle$, where n denotes the nth intrinsic excited state of the hadron and p denotes its momentum. Thus,

$$C_{J_h J_h}(t) = \int d^3x \langle 0 | e^{t\hat{H} - i\mathbf{x}\cdot\hat{\mathbf{P}}} J_h(0,0) e^{-t\hat{H} + i\mathbf{x}\cdot\hat{\mathbf{P}}} \sum_n \int d^3p \frac{|np\rangle\langle np|}{2E_{np}} J_h^\dagger(0,0) | 0 \rangle \;.$$

$$\tag{107}$$

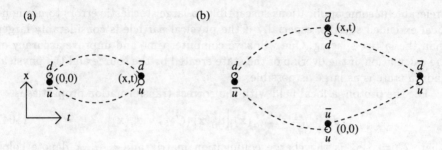

Fig. 9. Sketches of two hadronic observables. (a): Vacuum correlation function for the π^+ (105). (b): Quark density density correlation function for the π^+ (110). The dashed lines denote the propagators connecting quark creation and annihilation operators.

The integral over **x** projects onto zero momentum and for large times t only the lowest hadron survives, $\sum_n e^{-tE_{n,0}}|n0\rangle\langle n0| \to e^{-tM_h}|h\rangle\langle h|$. Hence,

$$C_{J_h J_h}(t) \to \frac{e^{-M_h t}}{2M_h}|\langle h|J_h|0\rangle|^2 \quad \forall t \to \infty \qquad (108)$$

which enables us to read off the hadron mass from the large t decay and provides the factor $|\langle h|J_h|0\rangle|$ required to normalize other quantities. In practice, since the two-point correlation function coincides with the Bethe–Salpeter wave function at zero separation discussed subsequently, it is not necessary to calculate it as a separate entity. Operationally, as seen in (106), the calculation is straightforward. Having generated a set of gauge field configurations sampling $e^{-S(U)}$ in the quenched case or $e^{\ln \text{Det } M(U) - S(U)}$ in the case of full dynamical quarks, the combination of propagators $M^{-1}(U)_{\mathbf{x}t,00}\gamma_5 M^{-1}(U)_{00,\mathbf{x}t}\gamma_5$ is simply averaged over these gauge fields.

Results of hadron masses (Table 1) in the quenched approximation agree very well with experiment in all channels except for the η', where closed loops which are not contained in the quenched approximation are known to give the main contribution to the mass.

Table 1. Calculated hadron mass ratios on lattices up to $32^2 \times 30 \times 40$ in the quenched approximation using Wilson Fermions with inverse coupling up to $\frac{6}{g^2} = 6.17$. Results are extrapolated in quark mass, lattice spacing, and volume and are compared with the experimental values. [22]

ratio	m_N/m_ρ	m_Δ/m_ρ	m_{K^*}/m_ρ	m_Φ/m_ρ	m_{Ξ^*}/m_ρ	m_Ω/m_ρ
lattice	1.216(104)	1.565(122)	1.166(16)	1.333(32)	2.055(65)	2.296(89)
expt.	1.222	1.604	1.164	1.327	1.996	2.177

Vacuum correlation functions have a number of other appealing features [23]. In many channels, they have been determined phenomenologically by using dispersion analysis of e^+e^- hadron production [24] and τ-decay experimental data. When they are defined at equal time, they may be calculated on the lattice and in the interacting instanton approximation [25][26] as well as by using sum rules [27] (cf. the next chapter). They complement bound state hadron properties in the same way scattering phase shifts provide information about the nucleon-nucleon force complementary to that provided by the properties of the deuteron. Just as nucleon-nucleon scattering allows one to explore the spin-spin, spin-orbit and tensor components of the nuclear force at different spatial separation in much more detail than deuteron observables which reflect the composite effect of all channels and ranges, so also the interaction or "scattering" of virtual quarks and antiquarks from meson sources at different spatial separations allows one to obtain much more detailed information about quark interactions for different channels and spatial separations than the composite effects reflected in hadron bound states.

Much of the richness of the study of these correlation functions derives from the different physics involved at different spatial separations. By asymptotic freedom, at extremely short distances the interactions between quarks must become negligible, and for dimensional reasons must fall as x^{-6}. For slightly larger distances, where interactions are small but non-negligible, one should be able to use the leading terms in the Wilson operator product expansion to describe the deviation. At still larger distances, the full complexity of non-perturbative QCD comes into play, and one may use this region to test and refine QCD motivated models such as the interacting instanton approximation (cf. below). Finally, at very large separation, the decay of the correlation functions is governed by the lightest hadron mass in the relevant channel as seen above.

As a result of this diverse range of physics at different spatial separations, it is clear that definitive lattice calculations of correlation functions provide an exceedingly useful supplement to accessible experimental data in allowing one to quantitatively explore and improve approximations based on the operator product expansion, sum rules, and interacting instantons.

The lattice calculations discussed subsequently [23] were performed on a $16^3 \times 24$ lattice in the quenched approximation with Wilson Fermions at an inverse coupling $6/g^2 = 5.7$, corresponding to a physical lattice spacing defined by the proton mass of approximately $a = 0.17$ fm. The spectral density function is parameterized phenomenologically as:

$$f(s) = \lambda^2 \delta(s - M) + f_c(s)\theta(s - s_0) \tag{109}$$

where λ is the strength of the coupling of the current to the lowest bound state or resonance with mass M, f_c is the lowest order perturbative result for the spectral function, and s_0 is the continuum threshold. This spectral function been fitted to the lattice correlation functions and yields good agreement with phenomenology and provides another way to calculate hadron masses. The values for parameters of the spectral function are shown in Table 3 and vacuum correlation functions for the π, ρ and nucleon are shown in Fig. 10.

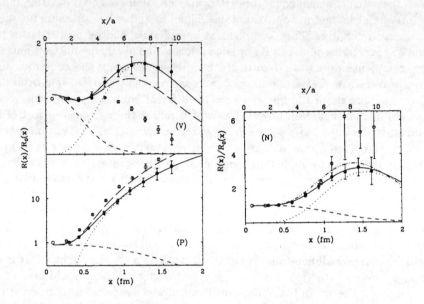

Fig. 10. Vacuum correlation functions for the vector channel (denoted by V), the pseudoscalar channel (denoted by P), and the nucleon channel (denoted by N). Extrapolated lattice data are denoted by the solid points with error bars. Fits to the lattice data using the phenomenological form (109) are given by the solid curves, with the continuum and resonance components denoted by short dashed and dotted curves respectively. The empirical results determined by dispersion analysis of experimental data in ref. [24] are shown by the long dashed curves. The open circles denote the results of the random instanton vacuum model of ref. [26]. The results from the QCD sum rule calculation of ref. [27] in (N) are indicated by the dot-dot-dashed lines. [23]

The lattice result for the ρ correlation function (Fig. 10(V)) is reasonably close to the phenomenological result obtained by Shuryak [24] from a dispersion analysis of $e^+e^- \to$ even number of π's. The fact that the phenomenological result lies below the lattice result follows from the fact that resonance peak scales as λ^2/M_ρ^4 and the lattice mass lies below experiment while the coupling constant agrees with the phenomenological value. The result of the instanton model [26] is qualitatively similar, although lower than phenomenology. The most salient physics result in the ρ channel is the fact that although the free correlator falls by four orders of magnitude, the ratio of the interacting to non-interacting correlators remains close to one. Although the ratio must approach unity very close to $x = 0$ by asymptotic freedom and there is no leading order 't Hooft instanton induced interaction in this channel, the ratio remains close to unity for much larger distances than any simple arguments suggest. This feature, which has been called superduality, indicates that the net QCD interaction is extremely weak in this channel, presumably because of a high degree of cancellation.

The pseudoscalar (π) channel (Fig. 10(P)) exhibits the strongest attraction and the most dramatic dependence on the quark mass, reflecting the special rôle of the pion as a Goldstone boson. Note that because of the light pion mass, the peak of the resonance occurs far outside of the range in which the data is fit. Nevertheless, the extracted mass and coupling constant agree well with the empirical results.

For the nucleon channel, one observes that the lattice results are quite consistent with the sum rule result of ref. [27], shown by the dot-dot-dash curve in Fig. 10(N). In addition, although there are substantial statistical errors at large distance, the random instanton vacuum model is also close to the lattice results [26].

A more detailed analysis, including the discussion of lattice errors and artifacts, can be found in [23]. In addition, other observables that have been calculated on the lattice include form factors (recent results in [22]) and magnetic moments [28] [29], and all show reasonable agreement with experiment.

To characterize the quark distribution inside hadrons, one may calculate the quark density density correlation function (Fig. 9(b)) [30][31][32]

$$\langle h|\rho^{q_1}(\mathbf{x},t)\rho^{q_2}(0,t)|h\rangle, \tag{110}$$

where $\rho^q = \bar{q}\gamma^0 q$, $\bar{q}\gamma^5 q$. Physically, this correlator specifies the probability to find two quarks q_1, q_2 at a spatial separation $\mathbf{x}$ inside a hadron h. As emphasized in [33], the correlation function in a hadron measures contributions from all the multiquark-antiquark components of the Fock space, and thus, in principle, provides valuable complementary information to that of the wave function considered in the next section. There, plots of correlations function in hadrons on the lattice will also be discussed (Fig. 17).

5 Insight into Hadron Structure

Thus far, we have considered the calculation of physical observables and their agreement with experiment to show that lattice QCD provides a valid nonperturbative solution of continuum QCD. However, if lattice QCD only serves as a black box to provide numbers which agree with experiment, we will still be far from understanding QCD. Hence, one of the major thrusts of my recent research has been to use lattice QCD as a tool to gain insight into hadron structure.

5.1 Wave Functions

Wave functions play a central rôle in our understanding of many-body systems in non-relativistic Quantum Mechanics. Examples include band theory, Hartree

Fock, BCS, and Laughlin wave functions for condensed matter systems, and Jastrow, independent pair, Faddeev, and shell model wave functions for nuclei. It is thus natural to seek analogous understanding of hadronic structure in terms of quark and gluon wave functions.

Specifying the wave functions of bound states is complicated. One cannot just focus attention on some Fock space component of the wave function and ask, for example, what the probability is of finding a quark-antiquark pair in a meson separated by some distance **x**. Rather one must also specify the gluonic component of the wave function either implicitly or explicitly.

Consider, for example, the $q\bar{q}$ wave function $\Psi(\mathbf{x})$, that is, the probability amplitude to find a $q\bar{q}$ pair separated by a distance $x = |\mathbf{x}|$ inside a meson. To measure it on the lattice [34], one creates a state with the correct mesonic quantum numbers and evolves it in Euclidean time as described above in order to filter out the mesonic ground state $|h\rangle$. One then may calculate (Fig. 11(a))

$$\Psi_n(\mathbf{x}) := \langle 0|\bar{q}(\mathbf{x},0)\Gamma q(0,0)|h\rangle \tag{111}$$

where Γ depends on the Dirac structure of the ground state hadron $|h\rangle$ and $|0\rangle$ is the QCD vacuum in the absence of Fermions.

As it stands, this expression is gauge dependent: it is not physically well defined and numerical evaluation on the lattice would yield zero because of the integration over all group elements without gauge fixing. If one chooses a specific gauge, this quantity becomes well defined, but it is essential to appreciate the extent to which the gauge choice specifies the gluon configuration and thus affects the physical result.

For example, if we select the Coulomb gauge, we actually specify that each quark is surrounded by the gluons corresponding to the static Coulomb field between the antiquark at **x** and the quark at the origin (Fig. 11(d)). Although it is hard to write the gluon configuration in nonabelian QCD, it is instructive to note that in the Abelian case, one can write down the following explicit expression for the photon field when we specify $A_0 = 0$ and Coulomb gauge:

$$\Psi_c(\mathbf{x}) := \langle 0|\bar{q}(\mathbf{x})\Gamma e^{\int d^3y\, \mathbf{E}_{\text{static}}(\mathbf{y})\cdot \mathbf{A}(\mathbf{y})} q(0)|h\rangle , \tag{112}$$

where $\mathbf{E}_{\text{static}}$ is the static Coulomb field of a pair of opposite charges separated by a distance x. The implied photonic component of $\Psi_c(\mathbf{x})$ is thus

$$\langle 0|\exp\int d^3y\, \mathbf{E}_{\text{static}}(\mathbf{y})\cdot \mathbf{A}(\mathbf{y}) ,$$

which is sketched in Fig. 11(d). The lattice wave function would then specify the probability that an electron-positron pair were separated by a given distance and that the photons were in the configuration given by the static Coulomb field. Clearly this wave function could be small because the leptons are unlikely to be found at that separation, because the photons are not in the configuration specified by the Coulomb field, or some combination of the two.

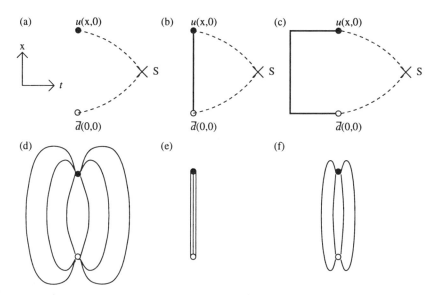

Fig. 11. *Alternative definitions of mesonic wave functions (top row) and the corresponding gluonic configuratons (bottom row). The meson source is denoted by S, products of link variables U are indicated by solid lines, and quark propagators are denoted by dashed lines. (a), (d) Coulomb gauge wave function (111/112); (b), (e) axial gauge wave function (113); (c), (f) adiabatic wave function (114).*

Another alternative is to choose an axial gauge with $A_x = 0$ along the line joining the quark and antiquark. This choice corresponds to the gauge invariant wave function:

$$\Psi_s(\mathbf{x}) := \langle 0|\bar{q}(\mathbf{x})\Gamma\,\mathrm{P}e^{i\int_0^x dx'\,A_x(x')}\,q(0)|h\rangle. \tag{113}$$

As shown in (Fig. 11(b)), the quark and antiquark are connected by a product of link variables U along the path connecting them. Physically, the gluonic component $\langle 0|\mathrm{P}\exp i\int_0^x dx'\,A_x(x')\rangle$ is that of an infinitely thin string of glue (Fig. 11(e)). The wave function is now to be interpreted as the probability amplitude to find a quark and antiquark separated by distance x times its overlap with a string of gluons in a hadron, and this latter will cause a considerably stronger fall-off of the wave function Ψ_s with increasing distance than that of Ψ_c (Fig. 12).

Physically, there is no reason to expect the Coulomb gauge wave function to be an accurate approximation, and clearly a string wave function is too localized to give a large overlap with the true wave function. The most realistic definition of the gluon wave function that we have been able to calculate is the adiabatic wave function, corresponding to the ground state configuration of gluons in the presence of a static $q\bar{q}$ pair at separation x (Fig. 11(c),(f)):

$$\Psi_a \propto \langle \Omega(\mathbf{x})|\bar{q}(\mathbf{x})\Gamma q(0)|h\rangle \tag{114}$$

$$|\Omega(\mathbf{x})\rangle := \lim_{n\to\infty} U^0_{\mathbf{0}t,\mathbf{0}t+n} U^y_{\mathbf{0}t+n,\mathbf{y}t+n} U^{0\,\dagger}_{\mathbf{y}t,\mathbf{y}t+n}|0\rangle \ .$$

When the temporal links are extended sufficiently far, the ground state flux tube is projected out, and in practice, it is sufficient to use 2 to 4 time steps [34].

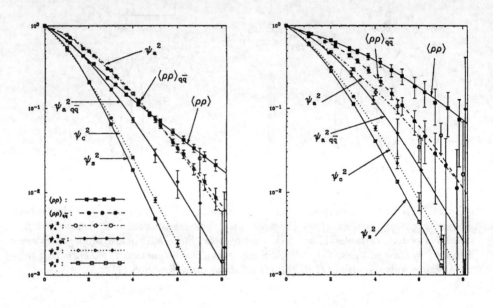

Fig. 12. *Comparison of density-density correlation functions and $|\Psi|^2$ for π (left) and ρ (right) mesons as a function of $q\bar{q}$ separation in units of 0.2 fm in the quenched approximation with the Wilson action on a 16^4 lattice at $6/g^2 = 5.7$. [34]*

A comparison of the squares of each of the wave functions discussed above with each other and with density-density correlation functions is shown in Fig. 12. The first point to note is the fact that the overlap between the true wave function at any spatial separation and the gluon wave function specified by the gauge choice or adiabatic condition increases slightly as one goes from Ψ_s to Ψ_c and substantially when one goes to Ψ_a, showing that the adiabatic wavefunction is much more physical than either of the fixed gauge choices. Ideally, one would like to compare the overlap of the adiabatic wave function with the square of the full quark-antiquark component of the exact wave function. A first step in this direction is comparison of $|\Psi_a|^2$ with the density-density correlation function $\langle \rho\rho \rangle$. However, the density-density correlation function contains the contributions of all Fock-space components, so a more relevant comparison is with the density-density correlation function $\langle \rho\rho \rangle_{q\bar{q}}$, which has been projected onto the quark-antiquark subspace. On the lattice, this projection is accomplished by calculating propagators from the left and right sources with hard-wall boundary conditions on the time slice containing the density operators, so that the quark and antiquark world lines can only cross the central time slice once. Comparison of $\langle\rho\rho\rangle$ with $\langle\rho\rho\rangle_{q\bar{q}}$ (normalized to one at the origin) in Fig. 12 shows that $q\bar{q}$ pairs indeed play a large role in dressing the hadron at large distances. Furthermore, comparison of the adiabatic wave function $|\Psi_a|^2$ with the projected density-density correlation function $\langle\rho\rho\rangle_{q\bar{q}}$ indicates that the adiabatic approximation is reasonably accurate.

To summarize, these results show a large and significant effect of the intrinsic gluon component in the definition of a hadron wave function and that the overlap with the gluons in the hadron increases substantially when going from Ψ_s or Ψ_c to Ψ_a. Furthermore, a $q\bar{q}$ pair in the adiabatic gluon ground state is quite close to the projected density-density correlation function, $\langle\rho\rho\rangle_{q\bar{q}}$.

5.2 Instantons on the Lattice

Another way in which we seek to use the lattice as a tool to understand hadron structure is to use it to explore the role of instantons. In principle, since we are numerically evaluating a path integral, it should be possible to use the numerical results to identify those configurations which are most important in the sense that they dominate the path integral. In cases that can be solved analytically, we know that the dominant configurations are those corresponding to fluctuations around the paths which make the action stationary. In QCD, although one can treat instantons analytically in the dilute gas approximation, there is no known way to study their role analytically in the true QCD ground state. Hence, we will use the lattice to find the stationary configurations closest to the configurations determined numerically which sample the gluon ground state, and thereby reveal the instanton content of these vacuum configurations.

Having extracted the instanton content of the gluon configurations, it is then possible to study the role of instantons in hadron structure by comparing observables calculated with all gluonic excitations and those calculated with the instantons alone. There is a large body of theoretical and phenomenological evidence from the work of Shuryak et al. [25][26] and Diakonov and Petrov [35] suggesting that for light quarks, propagation of quarks between the zero-modes associated with each instanton in the vacuum dominates the physics. The lattice provides an ideal laboratory to quantitatively test this picture and compare the phenomenological parameters of instanton models with those of QCD.

In this section, all calculations are again performed in the quenched approximation, in which the gluon configurations are not influenced by quarks and may therefore be fully described independent of the specific hadronic observable being investigated.

Before studying instantons on the lattice and comparing observables in an instanton background with the exact lattice results [36], it is useful to review briefly the basic features of instantons and the instanton gas model. More thorough introductions to instantons in continuum QCD may be found in recent lecture notes [37] and [38] and references therein.

Consider, as a pedagogical example, paths a particle can take in a periodic potential (Fig. 13(a)) with minima at points $x_i = iL$, $i \in Z$ equivalent to the problem of a particle in a continuous potential on a circle with circumference L.

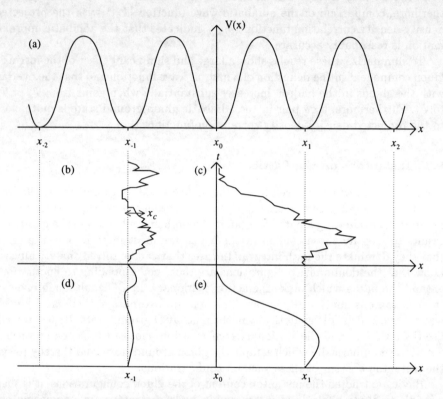

Fig. 13. Particle in a periodic potential (a) and representatives of two classes of contributions to the path integral: (b): Path about one well, indicating also the local minimization procedure (cooling) for a point x_c on the path; (c): Path connecting two wells. (The large fluctuation into the well about x_2 may be interpreted as a close kink-anti-kink pair.); (d),(e): The result of several cooling steps on the paths (b) and (c).

Classically, a particle with energy $E < V_{\max}$ cannot leave its well about the "classical ground state" x_i, but already the semiclassical WKB (stationary phase) approximation gives a nonzero probability for the particle to tunnel from the well about x_i to the one about x_{i+1} (x_{i-1})(Fig. 13(c)). In the picture of a particle on a circle, this corresponds to a path winding once around the torus in the positive (negative) orientation, and so the path between x_i and x_{i+1} (x_{i-1}) is said to have winding number +1 (−1). Such a path is topologically stable because of its invariance under any continuous perturbation of the paths. The path of winding number +1 (−1) which has the highest probability weight in the path integral, i.e. the least action, is the *kink* (anti-kink). The analog of the kink in QCD in four dimensions is the instanton.

One may then characterize two classes of configurations which contribute to the path integral. Paths within a well, i.e. which are contractible on the circle, describe the perturbative regime. Paths which connect different wells, winding at least once about the circle, and which can be decomposed into a number of kink and anti-kink paths plus fluctuations describe intrinsically non-perturbative effects (Fig. 13(c)). This characterization becomes increasingly accurate as the energy difference between $V_{\max}$ and the particle energy increases.

The Yang Mills equations of motion allow for infinitely many topologically distinct classical ground states besides $A_\mu = 0$, i.e. pure gauge configurations which carry different topological winding number $n \in Z$. QCD-instantons (anti-instantons) are the classical self-dual (anti-self-dual) solutions to the Euclidean Yang Mills equations, localized both in space and time and having radial extent ρ. They have winding number $+1$ (-1) and interpolate between two topologically distinct but neighboring vacua. They have the minimum Euclidean action $S_0 = \frac{8\pi^2}{g^2}$ and topological winding number

$$Q_0 = -\frac{1}{32\pi^2} \int d^3x\, dt\, \epsilon^{\mu\nu\rho\sigma} F_{\mu\nu} F_{\rho\sigma} = \pm 1 \ . \tag{115}$$

A configuration of N instantons and $\bar{N}$ anti-instantons has winding number Q and action S

$$Q = N - \bar{N} \ , \quad S \approx (N + \bar{N})S_0 \ . \tag{116}$$

In the instanton gas model of Shuryak et al. [25][26] and Diakonov and Petrov [35], the QCD vacuum is characterized by a dense, stable distribution of instantons. In the simplest version [26], the instantons and anti-instantons are distributed randomly in space and have uniform size ρ and density n. The instanton size and density are determined phenomenologically.

In order to isolate the contribution of instantons to a given gluonic configuration, one must remove the short-range perturbative fluctuations and extract the underlying semiclassical stationary solution. This may be achieved by locally relaxing or *cooling* the configuration, a method which is most easily described in the context of the periodic potential.

Each point on the given path is varied such that the action along the path is minimized, keeping all other points fixed. Proceeding this way from point to point along the path, one minimizes the action locally (Fig. 13(b)). When this procedure is performed several times, the short wavelength, local fluctuations associated with perturbative gluons in QCD are removed most rapidly, while topologically stabilized instanton excitations are removed much more slowly. Thus, as the number of cooling steps increases, the configuration becomes more and more dominated by its instanton content (Fig. 13(d)/(e)).

To monitor the instanton content in QCD (and decide how much cooling is necessary), one measures the topological charge density. The simplest discrete approximation to (115), which is adequate for a sufficiently smooth configuration, at lattice point **n** given by [36]

$$Q(\mathbf{n}) = -\frac{1}{32\pi^2}\epsilon^{\mu\nu\rho\sigma}\operatorname{Re}\operatorname{tr}[U_{\mu\nu}(\mathbf{n})U_{\rho\sigma}(\mathbf{n})] \ . \tag{117}$$

Its integral will be the number of instantons minus anti-instantons. In the dilute limit the topological susceptibility

$$\chi = \left(\sum_{\mathbf{n}} Q(\mathbf{n})\right)^2 \to N + \bar{N} \tag{118}$$

gives the number of instantons plus anti-instantons in a given configuration and may be compared with the minimum the action of an instanton configuration can reach (116), $S_{\min} \approx (N + \bar{N})S_0$, in order to determine when the cooling procedure has reached the dilute limit.

To provide a picture of how cooling extracts the instanton content of a thermalized gluonic configuration, Fig. 14 shows the action density $S(1,1,z,t)$ and topological charge density $Q(1,1,z,t)$ for a typical slice of a gluon configuration before cooling and after 25 and 50 cooling steps. As one can see, there is no recognizable structure before cooling. Large, short wavelength fluctuations of the order of the lattice spacing dominate both the action and topological charge density. After 25 cooling steps, three instantons and two anti-instantons can be identified clearly. The action density peaks are completely correlated in position and shape with the topological charge density peaks for instantons and with the topological charge density valleys for anti-instantons. Note that both the action and topological charge densities are reduced by more than two orders of magnitude, so that the fluctuations removed by cooling are several orders of magnitude larger than the topological excitations that are retained. From Fig. 14(e,f) we see that further cooling to 50 steps results in the annihilation of the nearby instanton - anti-instanton pair but retains the well separated instantons and anti-instanton.

One should note that as useful as cooling is, it does have limitations. Whereas instantons on a manifold are stable, they are not necessarily stable on a discrete lattice. For the Wilson action, the action of an isolated instanton decreases as a function of ρ (instead of being independent of ρ as in the continuum) with the result that eventually, an isolated instanton will "fall through the lattice" after extended cooling. Also, prolonged cooling results in the annihilation of overlapping instanton - anti-instanton pairs, leaving only well separated instantons and anti-instantons. Thus, both large instantons, which are likely to overlap other anti-instantons, and small instantons, which quickly fall through the lattice, are preferentially removed from the ensemble. This effect is seen in Fig. 15, which shows the distribution of instanton sizes on a 16^4 lattice. Fortunately, for a substantial range of ρ a window exists in which instantons are reasonably impervious to cooling.

With this orientation, consider the ensemble averages of observables as a function of the number of cooling steps shown in Fig. 16. As expected, the action is

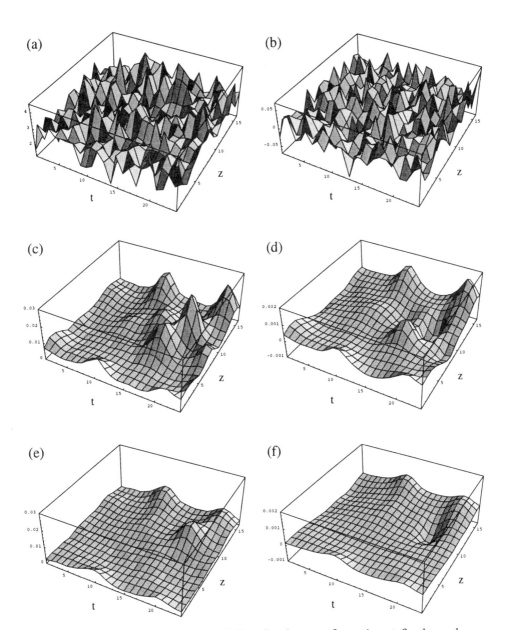

Fig. 14. Cooling history for a typical slice of a gluon configuration at fixed x and y as a function of z and t. The left column shows the action density $g^2 S(1,1,z,t)/6$ before cooling (a), after cooling for 25 steps (c) and after 50 steps (e). The right column shows the topological charge density $Q(1,1,z,t)$ before cooling (b), after cooling for 25 steps (d) and after 50 steps (f). [36]

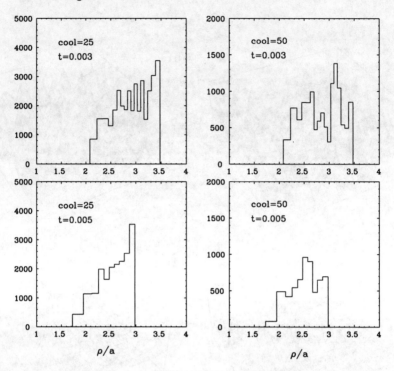

Fig. 15. *Distribution of instantons as a function of size ρ, using 19 configurations. The number of cooling steps is shown at each figure together with the value of the threshold t determining whether two adjacent lattice points belong to the same instanton cluster.* [36]

dominated by the short range fluctuations which are very strongly damped by cooling. Thus, the action decreases by several orders of magnitude in the first few steps. In contrast, the topological charge and susceptibility are much less affected by cooling. At cooling step 25, the averaged total action in units of a single instanton action is ~ 65 whereas $\langle Q^2 \rangle$ is $\sim 25 \pm 10$ throughout the cooling process. This difference indicates that there are sufficient nearby instanton - anti-instanton pairs in each configuration that the dilute regime where $\langle Q^2 \rangle \approx N + \bar{N}$ has not yet been reached. Since the nearby pairs continue to annihilate under further cooling, we only expect a clear plateau for the topological charge but not for the action in this region of cooling. It is only when the configurations are composed of well isolated instantons that plateaus for both action and topological charge would start to emerge. Here, this is expected to happen beyond 50 cooling steps, where $\langle S \rangle / S_0$ and $\langle Q^2 \rangle$ are nearly equal.

The combined information from Figs. 14, 15 and 16 suggests the following qualitative description of the cooled configurations. The configurations cooled with 25 steps are comprised of smooth, clearly recognizable instantons and anti-instantons and still retain many nearby pairs. The configurations cooled with

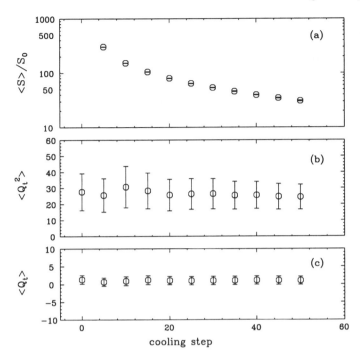

Fig. 16. *Mean values of three observables as a function of number of cooling steps for 19 configurations. (a) Total action in units of a single instanton action $S_0 = 8\pi^2/g^2$. The uncooled value $\langle S \rangle/S_0 = 20211$ is far off scale and is not plotted. (b) Topological charge squared (118). (c) Topological charge (117). [36]*

50 steps consist of more dilute instantons with their total action starting to be dominated by the well isolated peaks. The small instanton contribution is also suppressed, giving rise to potential systematic error. One may regard the configurations cooled with 25 steps as providing a more complete description of the instanton content of the original configurations, and I will therefore emphasize them in our subsequent calculation of hadronic properties.

By cooling, the instanton content of 19 gluon configurations was extracted from a $16^3 \times 24$ lattice at $\frac{6}{g^2} = 5.7$ [36]. The string tension σa^2 monitoring confinement was measured on a 4×7 Wilson loop. The size of the Wilson loop is relevant, since the local minimization of the action corresponds to replacing each link by the sum of staples made up of the other three links of each plaquette to which the original link contributes. Thus, each cooling step replaces a Wilson loop by a bundle of loops smeared by at most one lattice site. So, as long as the number of cooling steps is much smaller than the size of the loop, one must still see confinement. However, once the number of steps is larger than the loop size, there is nothing to prohibit the string tension from going to zero and it appears to do so.

A significant conceptual issue in comparing observables calculated using cooled configurations with uncooled results is how to change the renormalization of the bare mass and coupling constant as the gluon configurations are cooled. Clearly, as the fluctuations corresponding to gluon exchange are filtered out, the gluonic contribution to the physical mass and coupling constant change significantly. One may use the physical pion and nucleon masses to determine the hopping parameter κ and lattice size a for the cooled configurations. As will be seen below, a changes by $\sim 16\%$ after 25 cooling steps when the nucleon mass is used to set the scale, and within errors, the rho mass remains unchanged after cooling with this value of a. The other extreme would be to keep a fixed at the uncooled value and thus display what remains in the original path integral when only instantons are retained. This constant a would also be consistent with the constant topological susceptibility. It is a remarkable result that these two extremes differ by only 16%, so that even if one took the most conservative possible view of not changing the scale, the qualitative results would still not be changed significantly.

Table 2. *Summary of properties of cooled configurations. The symbols $S, \sigma, a, \rho, n,$ and χ denote the action, string tension, lattice spacing, instanton size, instanton density, and topological susceptibility.*

Cooling steps	$\langle S \rangle / S_0$	σa^2	a (fm)	ρ (fm)	n (fm^{-4})	χ (MeV^4)
0	20 211	0.18	0.168			
25	64	0.05	0.142	0.36	1.64	$(177)^4$
% of uncooled value	0.3	27	84			
50	31	0.03	0.124	0.35	1.33	$(200)^4$
Instanton Model [26]				0.33	1.0	$(180)^4$

Table 2 shows the result of cooling and a comparison with the phenomenological values used in the instanton gas model [26]. It shows the dramatic decrease in the action and the string tension, indicating a very strong reduction in the perturbative and confinement effects. Although instantons do not contribute significantly to confinement, quarks will still be bound through the attractive interactions arising from the 't Hooft interaction, or equivalently, through the zero modes for massless quarks associated with each instanton and anti-instanton. The close agreement between the parameters of the instanton gas model [26] and our results clearly suggests that we should obtain results similar to this model when we calculate hadron properties in the instanton configurations determined on the lattice.

One extremely important result is the fact that vacuum correlation functions of hadron currents calculated with all gluon excitations and only instantons agree very closely. This is demonstrated by the close agreement of the hadron spectral function parameters determined from two point vacuum correlation functions for uncooled and cooled configurations shown Table 3. Note also the good agreement between cooled and uncooled lattice results with phenomenology and the instanton gas model.

Table 3. *Hadron Parameters determined from vacuum correlation functions for uncooled and cooled configurations. λ denotes the coupling of the current to the ground state, cf. [23] and $\sqrt{s_0}$ denotes the continuum threshold (109). [36]*

Channel	Source	M (GeV)	λ	$\sqrt{s_0}$ (GeV)
Vector	lattice (cool=00)	0.72 ± 0.06	$(0.41 \pm 0.02\, GeV)^2$	1.62 ± 0.23
(ρ)	lattice (cool=25)	0.65 ± 0.03	$(0.385 \pm 0.004\, GeV)^2$	1.38 ± 0.05
	lattice (cool=50)	0.70 ± 0.05	$(0.410 \pm 0.005\, GeV)^2$	1.42 ± 0.04
	instanton[a]	0.95 ± 0.10	$(0.39 \pm 0.02\, GeV)^2$	1.50 ± 0.10
	phenomenology[b]	0.78	$(0.409 \pm 0.005 GeV)^2$	1.59 ± 0.02
Pseudoscalar	lattice (cool=00)	0.156 ± 0.01	$(0.44 \pm 0.01\, GeV)^2$	< 1.0
(π)	lattice (cool=25)	0.140^d	$(0.341 \pm 0.010\, GeV)^2$	1.05 ± 0.15
	lattice (cool=50)	0.140^d	$(0.475 \pm 0.015\, GeV)^2$	1.80 ± 0.18
	instanton[a]	0.142 ± 0.014	$(0.51 \pm 0.02\, GeV)^2$	1.36 ± 0.10
	phenomenology[b]	0.138	$(0.480 GeV)^2$	1.30 ± 0.10
Nucleon	lattice (cool=00)	0.95 ± 0.05	$(0.293 \pm 0.015\, GeV)^3$	< 1.4
	lattice (cool=25)	0.938^e	$(0.281 \pm 0.004\, GeV)^3$	1.47 ± 0.13
	lattice (cool=50)	0.938^e	$(0.297 \pm 0.004\, GeV)^3$	1.54 ± 0.11
	instanton[a]	0.960 ± 0.030	$(0.317 \pm 0.004\, GeV)^3$	1.92 ± 0.05
	Sum Rule[c]	1.02 ± 0.12	$(0.324 \pm 0.016\, GeV)^3$	1.5
	phenomenology[b]	0.939	—	1.44 ± 0.04
Delta	lattice (cool=00)	1.43 ± 0.08	$(0.326 \pm 0.020\, GeV)^3$	3.21 ± 0.34
	lattice (cool=25)	1.06 ± 0.04	$(0.285 \pm 0.002\, GeV)^3$	1.91 ± 0.08
	lattice (cool=50)	1.05 ± 0.09	$(0.298 \pm 0.003\, GeV)^3$	2.22 ± 0.06
	instanton[a]	1.440 ± 0.070	$(0.321 \pm 0.016\, GeV)^3$	1.96 ± 0.10
	Sum Rule[c]	1.37 ± 0.12	$(0.337 \pm 0.014\, GeV)^3$	2.1
	phenomenology[b]	1.232	—	1.96 ± 0.10

[a] Instanton Liquid Model by Shuryak et al. [b] Phenomenology estimated by Shuryak and from the particle data book. [c] QCD sum rule by Belyaev and Ioffe [27]. [d] Used to fix the quark mass. [e] Used to fix the lattice constant.

A second extremely important result is the close agreement of cooled and uncooled hadron density-density correlation functions (110) in the ground state of the pion, ρ, and nucleon as shown in Fig. 17. The striking result for both the ρ and the nucleon is the fact that the spatial distribution of quarks is essentially unaffected by cooling – instantons alone govern the gross structure of these hadrons, as indeed they also governed vacuum correlation functions of hadron currents in these same channels.

The only case in which a noticeable change is brought about by cooling is in the short distance behavior of the ground state of the pion. This difference

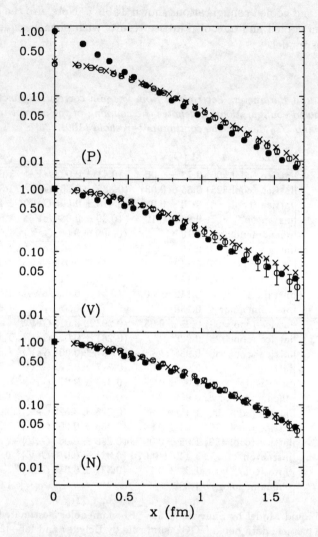

Fig. 17. *Comparison of uncooled and cooled hadron density-density correlation functions for the pion (P), ρ (V), and nucleon (N). The solid circles denote the correlation functions calculated with uncooled QCD [23], the open circles with error bars show the results for 25 cooling steps, and the crosses denote the results for 50 cooling steps. The rho and pion results are compared for $M_\pi^2 = 0.16$ GeV^2, and the nucleon results are compared for $M_\pi^2 = 0.36$ GeV^2. The separation is shown in physical units using values of a from Table 2. All correlation functions are normalized to 1 at the origin, except for the cooled pion correlation functions, which are normalized to have the same volume integral as the uncooled pion result. Errors for the uncooled results and for 50 steps, which have been suppressed for clarity, are comparable to those shown for 25 steps.* [36]

is understandable since in the physical pion, in addition to instanton-induced interactions, there is also a strong attractive hyperfine interaction arising from perturbative QCD which, combined with the attractive $1/r$ interaction, gives rise to the Coulomb cusp in the uncooled density. Despite this difference at the origin, which receives small phase space weighting, when the correlation functions are normalized to the same volume integral as in Fig. 17, one observes that the overall size and long distance behavior do not change appreciably with cooling. In contrast, in the ρ the combined effect of the hyperfine interaction and the $1/r$ interaction is much weaker, both because the hyperfine interaction is repulsive and because it is three times weaker. It is noteworthy that the cooled density-density correlation functions shown in Fig. 17 for the π, ρ, and nucleon are comparable within error bars. This uniformity strongly suggests that instantons set the overall spatial scale for these hadrons.

The conclusion from these results is that instantons do indeed play a dominant role in light quark propagation in the vacuum and in the low energy structure of hadrons. The picture which emerges, consistent with the instanton gas approximation, is that a light quark propagating in the QCD vacuum doesn't really respond to the details of the huge, short-wavelength fluctuations seen in the top of Fig. 14, but rather hops between the localized quark states corresponding to the zero modes associated with the instantons which become visible in the lower panels of Fig. 14.

Although I believe these results provide substantial evidence for the role of instantons, there are several open questions which are the subject of current investigation. As discussed, the cooling we used is an imprecise filter, and we are currently refining it to avoid the problem of small instantons falling through the mesh and to decrease the amount of instanton anti-instanton annihilation. In addition, the quenched approximation is being eliminated by repeating the calculation with dynamical quarks. If the same qualitative behavior remains with these two improvements, the dominant role of instantons will be clearly established by lattice QCD.

6 Summary

As indicated at the outset, these lectures could only provide an elementary introduction to lattice QCD and an extremely limited survey of results. With this introduction you are now prepared to undertake the much more detailed treatments in the books by Creutz [2], Rothe [4], and Montvay and Münster [3]. I hope these lectures will enable all of you to appreciate the usefulness of lattice calculations, follow the major research developments in this area and perhaps even motivate some of you to contribute to them.

References

[1] K.G. Wilson: *Phys. Rev.* **D10**(1974), 2445.
[2] M. Creutz: *Quarks, Gluons and Lattices*, Cambridge University Press, Cambridge, UK, 1983.
[3] I. Montvay, G. Münster: *Quantum Fields on a Lattice*, Cambridge University Press, Cambridge Monographs on Mathematical Physics, Cambridge, UK, 1994.
[4] H.J. Rothe: *Lattice Gauge Theories: An Introduction*, World Scientific, Singapore, 1992.
[5] C. Rebbi: *Lattice Gauge Theories and Monte Carlo Simulations*, World Scientific, Singapore, 1983.
[6] J.W. Negele: *QCD and Hadrons on a Lattice*, in: *Hadrons and Hadronic Matter, Proceedings of the NATO Advanced Study Institute at Carghese*, eds. D. Vautherin, F. Lenz, J.W. Negele, Plenum, New York, 1990.
[7] J.W. Negele, H. Orland: *Quantum Many-Particle Systems*, Addison Wesley, Reading, MA, 1988.
[8] S. Chin, J.W. Negele, S.E. Koonin: *Annals of Physics* (NY) **157** (1984) 190; S.A. Chin, O.S. van Roosmalen, E.A. Umland, S.E. Koonin: *Phys. Rev.* **D31** (1985) 3201.
[9] B. Svetitsky: *Physics Reports* **132** (1986) 1.
[10] R. Gupta, A. Patel: *Nucl. Phys.* **B226** (1983) 152.
[11] J. Kogut, J. Polonyi, H.W. Wyld, J. Shigemitsu, D.K. Sinclair: *Nucl. Phys.* **B251** (1985) 311.
[12] R. Dashen and D.J. Gross: *Phys. Rev.* **D23** (1981) 2340.
[13] B. Svetitsky: *Nucl. Phys.* **A461** (1987) 71c.
[14] S.A. Gottlieb, J. Kuti, D. Toussaint, A.D. Kennedy, S. Meyer, B.J. Pendleton, R.L. Sugar: *Phys. Rev. Lett.* **55** (1985) 1958.
[15] A. Hasenfratz, P. Hasenfratz, U. Heller, F. Karsch: *Phys.Lett.* **143B** (1984) 193.
[16] P. Hasenfratz: Quantum Field Theory, Renormalization Group and Lattices, in this volume.
[17] L.H. Karsten, J. Smit: *Nucl. Phys.* **B183** (1981) 103.
[18] H.B. Nielsen, M. Ninomiya: *Nucl. Phys.* **B185** (1981) 20.
[19] J. Kogut, L. Susskind: *Phys. Rev.* **D11** (1975) 395.
[20] M. Burkardt, J.M. Grandy, J.W. Negele: *Annals of Physics* **238**(1995), 441.
[21] M. Lissia: Ph.D. Dissertation, Massachusetts Institute of Technology (1989).
[22] D. Weingarten: in *Lattice 93, Proceedings of the International Symposium on Lattice Field Theory*, *Nucl. Phys.* **B**(Proc. Suppl.) **34** (1994), 29.
[23] M.-C. Chu, J.M. Grandy, S. Huang, J.W. Negele: *Phys. Rev.* **D48**(1993), 3340.
[24] E. Shuryak, *Rev. Mod. Phys.* **65**(1993), 1.
[25] E.V. Shuryak: *Nucl. Phys.* **B203**(1982), 93; **B203**(1982), 116; **B203**(1982), 140; **B302**(1988), 559; **B302**(1988), 599; **B319**(1989), 521; **B319**(1989), 541; **B328**(1989), 85; **B328**(1989), 102; in: *Lattice 93, Proceedings of the International Symposium on Lattice Field Theory*, *Nucl. Phys.* **B**(Proc. Suppl.) **34** (1994), 107.
[26] E.V. Shuryak, J.J.M. Verbaarschot: *Nucl. Phys.* **B410**(1993), 55; T. Schäfer, E.V. Shuryak, J.J.M. Verbaarschot:*Nucl. Phys.* **B412**(1994), 143.
[27] B.L. Ioffe: *Nucl. Phys.* **B188**(1981), 317; V.M. Belyaev, B.L. Ioffe: Zh. Ekap. Teor. Fiz. **83**(1982), 976 [Sov. Phys. JEPT **56**(1982), 547].
[28] D.W. Leinweber, R.M. Woloshyn: *Phys. Rev.* **D43**(1991), 1629.
[29] M. Fukugita, N. Ishizuka, H. Mino, M. Okawa, A. Ukawa: *Phys. Rev.* **D47**(1993), 4739.

[30] K. Barad, M. Ogilvie, C. Rebbi: *Phys. Lett.* **B143**(1984), 222; *Ann. Phys.* (NY) **168**(1986), 284.
[31] M.-C. Chu, M. Lissia, and J.W. Negele: *Nucl. Phys.* **B360**(1991), 31.
[32] M. Lissia, M.-C. Chu, J.W. Negele, J.M. Grandy: *Nucl. Phys.* **A555**(1993), 272.
[33] S. Huang, J.W. Negele, J. Polonyi: *Nucl. Phys.* **B307** (1988) 669.
[34] K.B. Teo, J.W. Negele: in :*Lattice 93, Proceedings of the International Symposium on Lattice Field Theory, Nucl. Phys.* B(Proc. Suppl.) **34** (1994), 390.
[35] D.I. Dyakanov, V.Y. Petrov: *Nucl. Phys.* **B245**(1984), 259; **B272**(1986), 457.
[36] M.-C. Chu, J.M. Grandy, S. Huang, J.W. Negele: *Phys. Rev.* **D 49**(1994), 6039
[37] R. Jackiw: *Topological Effects on the Physics of the Standard Model*, in this volume.
[38] D.I. Diakonov: *The QCD Instanton Vacuum*, in: *Strong Fields. Selected Topics* , GK-Notes 1-92, Universitäten Erlangen-Nürnberg und Regensburg, 1992.

Topological Effects on the Physics of the Standard Model[*]

R. Jackiw[1];
Notes by H.W. Grießhammer[2], O. Schnetz[2], G. Fischer[3], and S. Simbürger[3]

[1] Center for Theoretical Physics, Laboratory for Nuclear Science and Department of Physics, Massachusetts Institute of Technology, Cambridge, MA 02139, USA
[2] Institut für Theoretische Physik III, Universität Erlangen–Nürnberg, Staudtstr.7, 91058 Erlangen, Germany (Section 1)
[3] Institute for Theoretical Physics, University of Regensburg, Universitätsstraße 31, 93040 Regensburg, Germany (Section 2)

1 Four Dimensional Gauge Theories and Instantons

1.1 Notation

Nonabelian gauge theories deal with matrix-valued vector potentials which can be decomposed with respect to a basis of the Lie algebra of a gauge group G:[1]

$$A_\mu(x) = A_\mu^a(x) T^a \ . \tag{1}$$

The group generators are in my notation [1] the (for $SU(N)$ N^2-1) antihermitean, traceless matrices T^a obeying the normalisation condition and algebra

$$T^{a\dagger} = -T^a \ , \quad \operatorname{tr} T^a T^b = -\tfrac{1}{2}\delta^{ab} \ , \tag{2}$$

$$[T^a, T^b] = f^{abc} T^c \tag{3}$$

with f^{abc} the totally antisymmetric, real structure constants.

The covariant derivative, field strength tensor and Lagrangean of the Yang–Mills field are given by

$$\nabla_\mu = \partial_\mu + A_\mu \text{ acting on a representation of } G, \tag{4}$$

$$D_\mu = \partial_\mu + [A_\mu, \cdot\,] = T^a \left(\delta^{ac} \partial_\mu + f^{abc} A_\mu^b \right) \text{ acting on a rep. of the Lie alg.}, \tag{5}$$

$$F_{\mu\nu} \equiv F_{\mu\nu}^a T^a := [\nabla_\mu, \nabla_\nu] = \partial_\mu A_\nu - \partial_\nu A_\mu + [A_\mu, A_\nu] \ , \tag{6}$$

$$F_{\mu\nu}^a = \partial_\mu A_\nu^a - \partial_\nu A_\mu^a + f^{abc} A_\mu^b A_\nu^c \ , \tag{7}$$

$$\mathcal{L}_{YM} = -\tfrac{1}{4} F^{\mu\nu\, a} F_{\mu\nu}^a = \tfrac{1}{2} \operatorname{tr} F^{\mu\nu} F_{\mu\nu} \ . \tag{8}$$

---

[*] Lectures presented at the workshop "TOPICS in Field Theory" organised by the Graduiertenkolleg Erlangen–Regensburg, held on October 12th–14th, 1993 in Kloster Banz, Germany
[1] Summation over repeated indices is understood, as is the use of the natural system of units $\hbar = c = 1$.

Note that we have scaled the potentials so that the coupling constant is absorbed into A_μ. To make contact with the conventions used in perturbation theory (eg. [2]), one should substitute $-i\frac{\lambda^a}{2}$ for T^a, $-igA_\mu^a\frac{\lambda^a}{2}$ for A_μ, where λ^a are the (hermitean) Gell–Mann matrices, and in addition replace $F^{\mu\nu\,a}$ by $-igF^{\mu\nu\,a}$. The Lagrangean density (8) remains unchanged.

Under a gauge transformation $g(x) \in G$ at a point x in spacetime, the fields transform as

$$A_\mu \to {}^g\!A_\mu := g^{-1}\left(A_\mu + \partial_\mu\right)g \tag{9}$$
$$F_{\mu\nu} \to {}^g\!F_{\mu\nu} := g^{-1}F_{\mu\nu}g \ , \tag{10}$$

which shows that the Lagrangean (8) remains unchanged.

The equations of motion (transforming covariantly under gauge transformations)

$$D_\mu F^{\mu\nu} = 0 = \partial_\mu F^{\mu\nu\,a} + f^{abc}A_\mu^b F^{\mu\nu\,c} \tag{11}$$

show that, due to the self-coupling in the second term, the theory is not free even in the absence of matter. Indeed, in most what follows we will not bother with matter fields.

From the definition (6) of the field strength tensor one finally obtains the Bianchi identity:

$$\varepsilon^{\mu\nu\rho\sigma}D_\nu F_{\rho\sigma} = 0 \ . \tag{12}$$

1.2 Canonical Quantisation

As in Maxwell theory, a straightforward quantisation of nonabelian gauge theories is impossible due to the absence of a momentum conjugate to A_0^a:

$$\frac{\partial \mathcal{L}_{YM}}{\partial \dot{A}_0^a} = 0 \tag{13}$$

There is a variety of ways to handle this problem. In QED, one introduces a "transversal" Dirac function in order to obtain canonical commutation relations which are consistent with Gauß' law $\partial \cdot \mathbf{E} = 0$ [3], but this procedure obscures the physics in Yang–Mills theory since from Gauß' law the transversality of the gauge bosons does not follow (see Section 1.3). Technically even more involved is a constraint quantisation following Dirac [4].

If we do not want to use the path integral formalism, the simplest way of quantisation is to perform a classical gauge transformation yielding the Weyl gauge $A_0 = 0$ before quantising [1]. One finds for the momentum conjugate to $\mathbf{A}(\mathbf{x})$ the chromoelectric field

$$\Pi_i^a = \frac{\partial \mathcal{L}_{YM}}{\partial \dot{A}^{i\,a}} = -F_{0i}^a = -\dot{A}_i^a = -E_i^a \ , \tag{14}$$

and therefore postulates the canonical equal time commutation relations

$$[A_i^a(\mathbf{x}), \Pi_j^b(\mathbf{y})] = i\delta_{ij}\delta^{ab}\delta^{(3)}(\mathbf{x}-\mathbf{y}) = [E_j^b(\mathbf{y}), A_i^a(\mathbf{x})] \ , \tag{15}$$
$$[A_i^a(\mathbf{x}), A_j^b(\mathbf{y})] = 0 = [\Pi_i^a(\mathbf{x}), \Pi_j^b(\mathbf{y})] \ . \tag{16}$$

The Hamiltonian equations of motion obtained from the Hamilton operator

$$H = \frac{1}{2}\int d^3x \left[\mathbf{E}^a(\mathbf{x})\mathbf{E}^a(\mathbf{x}) + \frac{1}{2}F^a_{ij}(\mathbf{x})F^a_{ij}(\mathbf{x})\right] \qquad (17)$$

reproduce the generalised Ampère's law as the spatial components of (11)

$$i[H, A^a_i(\mathbf{x})] = \dot{A}^a_i(\mathbf{x}) = E^a_i(\mathbf{x}) \;,\;\; i[H, E^a_i(\mathbf{x})] = \dot{E}^a_i(\mathbf{x}) = \left(D_j F^{ji}\right)^a(\mathbf{x}) \;\;(18)$$
$$\Rightarrow D_\mu F^{\mu i}(\mathbf{x}) = 0 \;,$$

but the time component of (11), the generalised Gauß' law $G(\mathbf{x}) := \mathbf{D}\cdot\mathbf{E}(\mathbf{x}) = 0$, is absent, as it is an equation at fixed time.

Note that the resulting theory (without Gauß' law) has its own right, but it is not clear whether it is renormalisable, and Lorentz invariance is surely lost. Rather than imposing it, one regains Gauß' law by the following considerations:

Going to the Weyl gauge before quantisation does not fix the gauge completely. One can still perform residual, time independent gauge transformations, in particular infinitesimal ones,

$$\delta \mathbf{A}(\mathbf{x}) = \mathbf{D}\beta(\mathbf{x}) + \mathcal{O}(\beta^2) \;, \qquad (19)$$

which are symmetries of H. Since

$$i\left[\int d^3y\, \beta^a(\mathbf{y})G^a(\mathbf{y}), A_i(\mathbf{x})\right] = \delta A_i(\mathbf{x}) \;, \qquad (20)$$

$$i\left[H, \int d^3x\, \beta^a(\mathbf{x})G^a(\mathbf{x})\right] = 0 \;, \qquad (21)$$

Gauß' law is the generator of the infinitesimal gauge transformations and commutes with the Hamilton operator. It also obeys the commutation relations of group generators,

$$i\left[G^a(\mathbf{x}), G^b(\mathbf{y})\right] = f^{abc}G^c(\mathbf{x})\delta^{(3)}(\mathbf{x}-\mathbf{y}) \;, \qquad (22)$$

which means that there exist in general only as many independent constants of motion associated with the G^a's as there are linearly independent matrices T^a which can be diagonalised simultaneously, namely $N-1$ in SU(N).

One can think of the G^a's as generators of a symmetry of H we just discovered, without any reference to the Lagrangean (8) we started with. Imposing as a constraint on physical states

$$G^a(\mathbf{x}) \mid \text{phys}\rangle = 0 \;, \qquad (23)$$

one regains Gauß' law and therefore the complete quantum theory of the Lagrangean (8). Note that since $[H, G^a(\mathbf{x})] = 0$, the sector of physical states is invariant under time development.

All topological effects of the quantum theory can be uncovered by looking at the Gauß' law operator in a theory which is carefully quantised in this way, as can be seen from experience [1].

An analogy of the above situation is known from rotation invariant Hamilton operators in quantum mechanics. In the s-wave sector, the angular momentum operators **J** as generators of this symmetry have to annihilate the states one allowes for:

$$\mathbf{J} \mid \text{s-wave}\rangle = 0 \ . \tag{24}$$

Setting $\mathbf{J} = 0$ is inconsistent since its components do not commute with each other. In contradistinction to this example, the Gauß' law operators have a continuous spectrum and hence in looking at their zero eigenvalues one obtains non-normalisable states.

A note on the procedure: We first quantised the theory and then imposed the constraint on physical states. In general, reversing this order will yield a different result to order $\hbar$, none of the two ways being a priori right or wrong.

Furthermore it is not trivial that choosing the Weyl gauge and quantising commute with each other. Again, one example for that is the rotation invariant Hamilton operator in quantum mechanics [4]: Quantising first yields a centrifugal barrier proportional to $j(j+1)/r^2$, while first going to polar coordinates one misses the barrier. It is only reintroduced if one observes that the momentum conjugate to r, $-i\partial/\partial r$ is not hermitean, and the true canonical momentum is $-i\left(\frac{\partial}{\partial r} + \frac{1}{r}\right)$. If both procedures do not commute, the transformation eliminating A_0 would induce a curvature, and the momentum $-E_i^a$ would not be self-adjoint as is the case for the central force potential. Instead, one would have to hermitise it, $\Pi_i^a = -E_i^a + f_i^a(\mathbf{A})$, so that the components of the chromoelectric field do not commute with each other, thus revealing the curvature in the "Christoffel symbols" $f_i^a(\mathbf{A})$.

In both cases, one prefers to take the procedure for which one regains the classical theory for $\hbar \to 0$. The problem is that one doesn't know whether – due to confinement – a classical limit to the quantum Yang–Mills theory exists at all. However, in QED the classical limit exists and – what is more – one can show that all quantisation methods yield the same result. One therefore can expect this to hold in Yang–Mills theory, too. At least the induction of a curvature by the Weyl gauge can be ruled out, since ghosts decouple in the path integral version when choosing an axial gauge.

1.3 The Schrödinger Representation

In the Schrödinger representation,

$$-E_i^a(\mathbf{x}) = \Pi_i^a(\mathbf{x}) = -i\frac{\delta}{\delta A_i^a(\mathbf{x})} \ , \tag{25}$$

one obtains as fixed time Schrödinger equation for energy eigenstates

$$\int d^3x \left[-\frac{1}{2}\frac{\delta^2}{\delta A_i^a(\mathbf{x})\delta A_i^a(\mathbf{x})} + \frac{1}{4}F_{ij}^a(\mathbf{x})F_{ij}^a(\mathbf{x})\right]\Psi_E[\mathbf{A}] = E\Psi_E[\mathbf{A}] \ , \tag{26}$$

and Gauß' law constraint (23) on physical states reads

$$\left[\partial_i \frac{\delta}{\delta A_i^a(\mathbf{x})} + f^{abc} A_i^b(\mathbf{x}) \frac{\delta}{\delta A_i^c(\mathbf{x})}\right] \Psi_{\text{phys}}[\mathbf{A}] = 0 \ . \qquad (27)$$

In the abelian theory ($f^{abc} = 0$) one considers $\Psi[\mathbf{A}]$ to be a functional of the Fourier transform of $\mathbf{A}(\mathbf{x}) = \mathbf{A}_T(\mathbf{x}) + \boldsymbol{\partial} A_L(\mathbf{x})$ decomposed into its transverse ($\boldsymbol{\partial} \cdot \mathbf{A}_T(\mathbf{x}) = 0$) and longitudinal part[2]. Gauß' law reads after applying the chain rule

$$k_i k_i A_L(\mathbf{k}) = 0 \ , \qquad (28)$$

and hence physical states can be an arbitrary functional of the transverse components of $\mathbf{A}$ only, independent of its longitudinal degrees of freedom. This can also be seen from the fact that an abelian gauge transformation $A_i(\mathbf{x}) \to A_i(\mathbf{x}) + \partial_i \beta(\mathbf{x})$ leaves the transverse components untouched and changes only the longitudinal ones. Therefore the choice of the Coulomb gauge for free QED is unavoidable in the Hamiltonian formulation. In Yang–Mills theories, the Coulomb gauge is no natural choice since from Gauß' law (27) one cannot conclude that the wave functional depends on $\mathbf{A}_T$ only.

Free QED can even be solved this way [1]: Looking at the Schrödinger equation

$$\frac{1}{2} \int d^3 x \left[-\frac{\delta^2}{\delta A_i(\mathbf{x}) \delta A_i(\mathbf{x})} + A_i(\mathbf{x}) h_{ij} A_i(\mathbf{x}) \right] \Psi_E[\mathbf{A}] = E \Psi_E[\mathbf{A}] \ , \qquad (29)$$

$$h_{ij} := -\boldsymbol{\partial}^2 \delta_{ij} + \partial_i \partial_j \ , \qquad (30)$$

one constructs the gauge invariant ground state in analogy to the harmonic oscillator as

$$\Psi_0[\mathbf{A}] \propto \exp -\frac{1}{2} \int d^3 x d^3 y \ A_i(\mathbf{x}) \omega_{ij}(\mathbf{x}, \mathbf{y}) A_j(\mathbf{y}) \qquad (31)$$

$$\propto \exp -\frac{1}{4} \int d^3 x d^3 y \ F^{ij}(\mathbf{x}) \frac{1}{\sqrt{-\boldsymbol{\partial}^2}} F^{ij}(\mathbf{y}) \ , \qquad (32)$$

$$\omega_{ij}(\mathbf{x}, \mathbf{y}) := h_{ij} \int d^3 k \ e^{-i\mathbf{k} \cdot (\mathbf{x}-\mathbf{y})} \frac{1}{|\mathbf{k}|} = -\frac{2}{\pi^2 |\mathbf{x}-\mathbf{y}|^4} \left(\delta_{ij} - 2 \frac{x_i}{|\mathbf{x}|} \frac{y_j}{|\mathbf{y}|} \right) \qquad (33)$$

with the infinite vacuum energy $E_0 = \frac{1}{2} \text{tr} \, \omega$.

Since $\Psi_0[\mathbf{A}]$ depends on transverse fields only, Gauß' law is automatically satisfied, and the vacuum state of the free theory is unique. One can now construct excited states like the one photon state

$$\Psi_1[\mathbf{A}] := A_i^T(\mathbf{p}) \Psi_0[\mathbf{A}] \ , \quad A_i^T(\mathbf{p}) = \left(\delta_{ij} - \frac{p_i p_j}{\mathbf{p}^2} \right) \int d^3 x \ e^{i \mathbf{p} \cdot \mathbf{x}} A_j(\mathbf{x}) \ . \qquad (34)$$

---

[2] We neglect the zero mode of $\mathbf{A}$.

The Schrödinger representation offers an alternative way to derive Gauß' law. As remarked above, states should be invariant against infinitesimal spatial gauge transformations: $\Psi[\mathbf{A} + \mathbf{D}\beta] \stackrel{!}{=} \Psi[\mathbf{A}]$, so that expanding around $\Psi[\mathbf{A}]$ yields

$$\int d^3x \, (D_i\beta)^a \, \frac{\delta}{\delta A_i^a} \Psi[\mathbf{A}] = 0 \;, \tag{35}$$

and one recovers (27) after partial integration.

1.4 Large Gauge Transformations and the θ Angle

Gauß' law (23) as generator of infinitesimal gauge transformations annihilates physical states, and therefore physical states are invariant under infinitesimal gauge transformations and all gauge transformations that can be built up by iterating infinitesimal ones, called *small gauge transformations*. The question arises whether all gauge transformations are small or whether there exist *large gauge transformations*, i.e. if there are solutions to eqs.(26,27) which obey Gauß' law but are not gauge invariant:

$$\Psi[{}^g\mathbf{A}] \neq \Psi[\mathbf{A}] \;. \tag{36}$$

Let's turn to the question of boundary conditions for the fields. Assuming the absence of monopoles, all position dependent observables should vanish faster than $\frac{1}{|\mathbf{x}|^2}$ for $|\mathbf{x}| \to \infty$. This means that going to spatial infinity one finds a unique physical vacuum. Strictly speaking, the vector potentials have only to approach a pure gauge configuration at spatial infinity, but one can show that there exists always a regular gauge transformation after which

$$\lim_{|\mathbf{x}| \to \infty} |\mathbf{x}| \mathbf{A}(\mathbf{x}) = 0 \;, \tag{37}$$

simultaneously reducing the set of possible gauge transformations to those that do not violate this condition:

$$\lim_{|\mathbf{x}| \to \infty} g(\mathbf{x}) = \text{const.} \tag{38}$$

These boundary conditions have been used to derive eqs.(20,21).

The last requirement identifies all points at spatial infinity so that g is uniquely defined there, and one compactifies the Euclidean space R^3 to the sphere S^3 when considering g.

One may investigate whether the maps $g(\mathbf{x}) : S^3 \to G$ can be decomposed into different classes. All maps in a given class can be deformed into each other and differ only by small gauge transformations. The classes are separated by topologically nontrivial, large gauge transformations. The set of all classes clearly forms a group, called the third homotopy group of G, $\Pi_3(G)$ [5]. If $\Pi_3(G) = 1$, as is the case in QED, only small gauge transformations exist, and all of these can be continuously deformed to the map $S^3 \to 1$. For any semisimple Lie group G, particularly for SU(N), it has been shown that $\Pi_3(G) = Z$, the additive group of integers, and hence large gauge transformations do exist. One can indeed show the existence of large gauge transformations without bothering with such topological considerations [6], [7], as we will explain now.

There exists a functional of **A** which satisfies Gauß' law but is not gauge invariant, known as the integral over the Chern–Simons three form:

$$W[\mathbf{A}] = -\frac{1}{16\pi^2}\int d^3x\, \varepsilon^{ijk}\mathrm{tr}\left[A_i\left(F_{jk} - \frac{2}{3}A_j A_k\right)\right] = \qquad (39)$$

$$= -\frac{1}{8\pi^2}\int d^3x\, \varepsilon^{ijk}\mathrm{tr}\left[A_i\left(\partial_j A_k + \frac{2}{3}A_j A_k\right)\right]\ .$$

Since

$$\frac{\delta W[\mathbf{A}]}{\delta A_i^a(\mathbf{x})} = \frac{1}{16\pi^2}\varepsilon^{ijk}F_{jk}^a(\mathbf{x}) + \frac{1}{16\pi^2}\int d^3y\, \varepsilon^{ijk}\partial_j\left[\delta^{(3)}(\mathbf{x}-\mathbf{y})A_k^a(\mathbf{y})\right] \qquad (40)$$

and the surface term vanishes due to (38), $W[\mathbf{A}]$ fulfills Gauß' law (27) because of the Bianchi identity (12):

$$D_i\frac{\delta W[\mathbf{A}]}{\delta A_i^a(\mathbf{x})} = 0\ . \qquad (41)$$

On the other hand,

$$W[{}^g\mathbf{A}] - W[\mathbf{A}] = n(g) - \frac{1}{8\pi^2}\int d^3x\, \varepsilon^{ijk}\partial_i\mathrm{tr}\left[(\partial_j g)g^{-1}A_k\right]\ , \qquad (42)$$

$$n(g) := \frac{1}{24\pi^2}\int d^3x\, \varepsilon^{ijk}\mathrm{tr}\left[(g^{-1}\partial_i g)(g^{-1}\partial_j g)(g^{-1}\partial_k g)\right]\ ,$$

where with the boundary conditions eqs.(37,38) the surface term vanishes again. $n(g)$ is in general a nonzero integer and corresponds to the winding number of the map $g : S^3 \to G$, as can be seen most easily for $G =$ SU(2)$\cong S^3$. As one can imagine, there are infinitely many ways to map spheres on spheres which are not continuously deformable into each other and can be labeled by the number of times one sphere is wrapped around the other. This winding number is additive:

$$n(g_1 g_2) = n(g_1) + n(g_2) + \text{ a vanishing surface term }\ . \qquad (43)$$

As an example, one representative of each class can be obtained by considering the following gauge transformations obeying the boundary conditions eqs.(37,38), where σ^i are the Pauli matrices which for SU(N) only have to be embedded into the higher groups:

$$g(\mathbf{x}) = \exp i\sigma \cdot \frac{\mathbf{x}}{|\mathbf{x}|}f(|\mathbf{x}|)\ :\ f(0) = 0\ ,\ \lim_{|\mathbf{x}|\to\infty}f(|\mathbf{x}|) = n\pi\ . \qquad (44)$$

Assuming physical states to be eigenstates of all unitary operators $\Omega_n[\beta]$ implementing gauge transformations $g_n(\mathbf{x}) = e^{i\beta(\mathbf{x})}$ of winding number n, we see that

$$\Omega_n[\beta]\Psi[\mathbf{A}] = \Psi[{}^{g_n}\mathbf{A}] = e^{-i\theta n(g_n)}\Psi[\mathbf{A}]\ , \qquad (45)$$

because $\Omega_0[\beta]$ describes a gauge transformation generated by (a succession of) infinitesimal ones, and hence $\Psi[\mathbf{A}]$ is invariant under it by virtue of Gauß' law (23). θ is the Yang–Mills vacuum angle [6], [7], a new, hidden parameter in the quantum theory, which has been derived without any approximations here. Its effects will be examined in greater detail later.

It is tantalising to observe that $\exp(\pm 8\pi^2 W[\mathbf{A}])$ solves the non-abelian functional Schrödinger equation (26) with zero eigenvalue (even in QED). Unfortunately, this solution is divergent for large $\mathbf{A}$ and hence not normalisable[3]. On top of that, it lacks any physical meaning; yet one can use it to show that the gauge invariant state

$$\Phi[\mathbf{A}] := e^{i\theta W[\mathbf{A}]}\Psi[\mathbf{A}] \quad : \quad \Omega_n[\beta]\Phi[\mathbf{A}] = \Phi[\mathbf{A}] \tag{46}$$

is an eigenstate to the same energy eigenvalue as the original state and obeys a Schrödinger equation which reads:

$$\int d^3x \left[\left(-i\frac{\delta}{\delta A_i^a(\mathbf{x})} + \frac{\theta}{16\pi^2}\varepsilon^{ijk}F_{jk}^a(\mathbf{x}) \right)^2 + \frac{1}{2}F_{ij}^a(\mathbf{x})F_{ij}^a(\mathbf{x}) \right] \Phi[\mathbf{A}] = E\Phi[\mathbf{A}] \ . \tag{47}$$

By that, one moved the θ angle from the state to a Hamilton operator which can be obtained from the Lagrangean

$$\int d^3x\, \mathcal{L}_\theta = \int d^3x\, \mathcal{L}_{YM} - \frac{\theta}{16\pi^2}\int d^3x\, \varepsilon_{\mu\nu\rho\sigma}\mathrm{tr}\,[F^{\mu\nu}F^{\rho\sigma}] = \tag{48}$$

$$= \int d^3x\, \mathcal{L}_{YM} + \theta\frac{d}{dt}W[\mathbf{A}] \ ,$$

where in order to derive the last line one used that the Chern–Simons term is related to the Chern–Pontryagin density [5] via

$$\frac{1}{16\pi^2}\varepsilon_{\mu\nu\rho\sigma}\mathrm{tr}\,[F^{\mu\nu}F^{\rho\sigma}] = \frac{1}{8\pi^2}\partial_\mu\varepsilon^{\mu\nu\rho\sigma}\mathrm{tr}\,\left[A_\nu\left(\partial_\rho A_\sigma + \frac{2}{3}A_\rho A_\sigma\right)\right] \tag{49}$$

and that the surface terms at spatial infinity do not contribute due to eqs.(37,38). Therefore one can make three observations:

(i) The θ angle can be removed from the gauge variant states $\Psi[\mathbf{A}]$ making them gauge invariant (46), but only on the expense of breaking the invariance of the Lagrangean under large gauge transformations, changing $\mathcal{L}_{YM}$ to $\mathcal{L}_\theta$ by adding a Lorentz invariant, but P and T violating term.

The additional term in (48) is independent of the choice $A_0 = 0$, and therefore the occurence of the angle θ does not depend on choosing the Weyl gauge before quantisation. It is a new, unremovable hidden parameter in the theory, and no principle is known which requires it to be zero. The unique classical Yang–Mills theory gives rise to a θ-family of quantum theories.

---

[3] Compare to all $E \neq (n + \frac{1}{2})\omega$ – solutions of the quantum mechanical harmonic oscillator: They also diverge for large x.

(ii) There is no remnance of the Yang–Mills angle in the equations of motion, nor in the Hamilton operator obtained from $\mathcal{L}_\theta$ via the procedure described above, as long as one writes it in terms of the vector potential and chromoelectric field. Yet since under a large, time dependent gauge transformation $\int d^3x\, \mathcal{L}_\theta$ changes by a total time derivative $\theta \frac{d}{dt} n(g)$, gauge invariant quantum states acquire a phase in the temporal developement between two states that are connected by g_n, as is familiar from quantum mechanics.

(iii) The previous point is connected with the fact that the momentum conjugate to $A_i^a(\mathbf{x})$ in $\mathcal{L}_\theta$ is no longer $-E_i^a(\mathbf{x})$ (14), but (cf. (47))

$$\Pi_i^a(\mathbf{x}) = -E_i^a(\mathbf{x}) - \frac{\theta}{16\pi^2} \varepsilon^{ijk} F_{jk}^a(\mathbf{x}) \; . \tag{50}$$

Therefore the components of the electric field do not commute with each other, and a connection is introduced in the physical Hilbert space thus revealing its nonzero curvature.

1.5 QED in Two-Dimensional Spacetime

There is an intriguing example of the occurence of a new hidden parameter [8], [9], [1], [10] in two dimensions. The Hamilton operator and Gauß' law of QED are in the Schrödinger representation given by (cf. eqs.(26,27)):

$$H = \frac{1}{2} \int dx\, E^2(x) = -\frac{1}{2} \int dx\, \frac{\delta^2}{\delta A(x) \delta A(x)} \; , \tag{51}$$

$$\frac{d}{dx} \frac{\delta}{\delta A(x)} \Psi[A] = 0 \; . \tag{52}$$

Therefore, $\Psi[A]$ is a function of the zero mode of A only:

$$\Psi[A] = f\left(\int_{-\infty}^{\infty} dx\, A(x) \right) \; . \tag{53}$$

The wave functional solving both the Schrödiner equation and Gauß' law is

$$\Psi[A] = \exp -iE_0 \int dx\, A(x) \; , \tag{54}$$

where applying $E(x) = i\frac{\delta}{\delta A(x)}$ (25) shows that E_0, due to Gauß' law the only observable, is the zero mode of the electric field. The energy density is finite and given by $\frac{1}{2} E_0^2$.

In analogy to the discussion above, compactifying the space R^1 to S^1 by requiring all field fluctuations to vanish at spatial infinity[4] amounts to the following boundary condition on the gauge transformations allowed:

$$e^{-i\Lambda(\infty)} = e^{-i\Lambda(-\infty)} \; . \tag{55}$$

---

[4] Note that one may not demand physical observables to vanish at infinity since then $E_0 = 0$ and the wave functional (54) is 1.

Again, $e^{-i\Lambda(x)}$ has a well defined value at spatial infinity.

We again ask whether there exist large gauge transformations, i.e. transformations which are not generated by Gauß' law. The mappings $g(x) : S^1 \to U(1) \cong S^1$ decompose obviously into different classes, labeled by the number of times one circle winds around the other. Hence, under a gauge transformation in QED

$$A(x) \to A(x) - \frac{d}{dx}\Lambda(x) ,\qquad (56)$$

the zero mode

$$\int dx\, A(x) \to \int dx\, A(x) - \Delta\Lambda , \quad \Delta\Lambda := \Lambda(\infty) - \Lambda(-\infty) = 2\pi n , \quad n \in Z \quad (57)$$

changes by 2π times the winding number n (55). If $n \neq 0$, the unitary operator implementing the gauge transformation is not

$$\exp i \int dx\, (\frac{d}{dx}E(x))\Lambda(x) ,\text{ but } \Omega[\Lambda] = \exp -i \int dx E(x)\frac{d}{dx}\Lambda(x) , \qquad (58)$$

because the surface term in which the two expressions differ cannot be dropped. The effect of such gauge transformations on $\Psi[A]$ can easily be calculated:

$$\Omega_n[\Lambda]\Psi[A] = e^{-in\theta}\Psi[A] , \quad \theta := 2\pi E_0 . \qquad (59)$$

So the θ angle emerges as a constant electric background field which cannot be changed within the theory since $[H, E_0] = 0$, and whose different values therefore separate different worlds.

The operator which is invariant under small gauge transformations, but changes under large ones is the zero mode of the vector potential (57), cf. (39):

$$W[A] = \frac{1}{2\pi}\int dx\, A(x) \quad : \quad \Omega_n[\Lambda]W[A]\Omega_n^\dagger[\Lambda] = W[{}^\Lambda A] = W[A] + n . \qquad (60)$$

In order to construct the Schrödinger equation for gauge invariant states, cf. (46),

$$\Phi[A] := e^{\frac{i\theta}{2\pi}\int dx\, A(x)}\Psi[A] , \qquad (61)$$

one has to move the θ angle to the Hamiltonian and Lagrangean (cf. (48)):

$$\mathcal{L}_\theta = \frac{1}{2}E^2(x) - \frac{\theta}{2\pi}E(x) . \qquad (62)$$

The momentum conjugate to $A(x)$ is given by

$$\Pi(x) := \frac{\partial \mathcal{L}_\theta}{\partial \dot{A}(x)} = \dot{A}(x) + \frac{\theta}{2\pi} . \qquad (63)$$

Since $\mathcal{L}$ changes by a total time derivative under these operations, there is again no remnance of θ in the equations of motion, yet physical states acquire a phase under time developement.

If one incorporates fermions into the theory,

$$\mathcal{L} = \mathcal{L}_\theta + \bar{\psi}\left(i\gamma^\mu \nabla_\mu - m\right)\psi \ , \tag{64}$$

one notes that in the Schwinger model ($m=0$) $\mathcal{L}$ changes under a chiral redefinition of the fermionic fields due to the axial anomaly (see Section 1.8) ($\gamma^5 = -\gamma^{5\dagger}$) [11], [10], [1]:

$$\psi \to e^{\alpha \gamma^5}\psi \ : \ \mathcal{L} \to \mathcal{L} + \frac{\alpha}{\pi} E(x) \ . \tag{65}$$

Since it can be eliminated by re-defining the fermionic fields $2\alpha = \theta$, the Yang–Mills vacuum angle is physically irrelevant in that case.

Yet as soon as $m \neq 0$, this chiral redefinition is impossible and the θ angle is physical [8], [9], giving the value of the background electric field, on which e.g. the number of stable particles and the spacing between successive isosingulet states crucially depend.

If one would embed two dimensional QED into a larger theory, the background field might be determined by the new theory, dynamically fixing θ; but no such mechanism has been found so far.

1.6 A Physical Picture of θ Vacua and Instantons

Before deriving the axial anomaly in four dimensions and showing that the value of θ is unobservable in QCD in the presence of massless fermions by the same mechanism as in twodimensional QED, we compare the situation in QCD with a well-known quantum mechanical example.

A physical picture of the vacuum θ angle [12], [13], [6], [7] emerges when one looks at a particle in a periodic potential (Figure 1):

$$L = \tfrac{1}{2}\dot{x}^2 - V(x) \ , \ \ H = \tfrac{1}{2}p^2 + V(x) \tag{66}$$
$$V(x+a) = V(x) \ , \ \ p = \dot{x} \ . \tag{67}$$

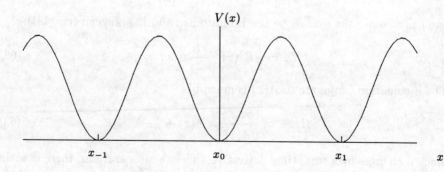

Fig. 1. *Particle in a Periodic Potential*

The discrete displacement as implemented by the translation operator

$$\Omega_n : \Omega_n x \Omega_n^\dagger = x + na \ , \ n \in Z \tag{68}$$

is a symmetry of the system. Ω_n should be compared to the operator $\Omega_n[\beta]$ implementing large gauge transformations in the physical Hilbert space of QCD. The infinite degeneracy of the classically stable "ground state" solutions at $x_n : V(x_n) = 0$ corresponds to an infinite number of classical gauge field configurations $\mathbf{A}(\mathbf{x}) = g_n^{-1} \partial g_n$ which are "pure gauge" and therefore have zero kinetic and potential energy but are topologically distinct from the trivial vacuum $\mathbf{A} = 0$ because of their nonzero winding numbers n.

In the interpretation of Floquet's (Bloch's) Theorem via the tight binding approximation of solid state physics, this degeneracy is removed in Quantum Mechanics by a nonzero tunneling probability from one x_n-"vacuum" to another. If the wave function $\Psi_n(x)$ is an approximate solution of least energy to one well of the potential, localised around the n-th minimum x_n, the superposition

$$\Psi_\theta(x) = \sum_n e^{-in\theta} \Psi_n(x) \tag{69}$$

is an eigenfunction to Ω_n (cf. 45)

$$\Omega_n \Psi_\theta(x) = e^{-in\theta} \Psi_\theta(x) \ , \tag{70}$$

and the ground state energy now depends on the Bloch momentum θ.

How can one describe the tunneling process just sketched in classical mechanics? Of course, there exists no classical zero energy solution which interpolates between different classical minima. Yet going to imaginary time $t \to -i\tau$, one interchanges the rôle of Hamiltonian and Lagrangean

$$L \to L_I = \frac{1}{2}\left(\frac{\partial x}{\partial \tau}\right)^2 + V(x) \tag{71}$$

$$H \to H_I = \frac{1}{2}\left(\frac{\partial x}{\partial \tau}\right)^2 - V(x) \tag{72}$$

and thus obtains a classical solution in imaginary time

$$\frac{\partial x}{\partial \tau} = \pm\sqrt{2V(x)} \tag{73}$$

that maintains zero energy throughout the interpolation between two different classical vacua x_n, x_m. Such a solution is called "instanton". The instanton action is given by

$$S_I = \int d\tau L_I = \int_{x_n}^{x_m} dx \sqrt{2V(x)} \tag{74}$$

which is closely connected to the tunneling amplitude through the potential barrier in real time as given by the WKB approximation

$$P_{n \to m}^{\text{WKB}} \propto \exp - \int_{x_n}^{x_m} dx \sqrt{2V(x)} . \qquad (75)$$

In Yang Mills theory, instantons are classical solutions of least energy interpolating between two classical vacua of different winding number, localised both in space and time, as explicit construction shows. They can be constructed [12] in the same way as above by going to imaginary time and solving

$$H_I = \tfrac{1}{2} \int d^3x \left[\mathbf{E}^a(\mathbf{x})\mathbf{E}^a(\mathbf{x}) - \tfrac{1}{2} F_{ij}^a(\mathbf{x}) F_{ij}^a(\mathbf{x}) \right] \qquad (76)$$

$$\Rightarrow \quad F_{\mu\nu}^a = \pm \tfrac{1}{2} \varepsilon_{\mu\nu\rho\sigma} F^{\rho\sigma\,a} . \qquad (77)$$

So instantons are classical (anti)selfdual solutions to the Euclidean Yang Mills equations with zero energy.

The tunneling amplitude between two vacua which can only be connected by a large gauge transformation of winding number n is (eqs.(74,75))

$$\exp - \int d\tau L_I(\tau) = \exp - \int d^4x \frac{1}{4} \varepsilon_{\mu\nu\rho\sigma} \text{tr}\, F^{\mu\nu} F^{\rho\sigma} = \exp - \frac{8\pi^2}{g^2}|n| , \qquad (78)$$

where we reintroduced the coupling constant g as described in Section 1.1. Note the interplay between the instanton action, the Chern–Pontryagin density (49) and the winding number of the gauge transformation g_n.

A word of caution is in order here: The analogy between the tunneling process in solid state physics and the connection of different classical QCD vacua by instantons should not be pushed too far. After all, the occurence of a physically measurable Bloch momentum is connected to Ω_n being a "physical" transformation, namely implementing spatial displacement. The gauge transformation $\Omega_n[\beta]$ is unobservable. The Bloch momentum can also be changed, while there is – as indicated – no way to change the vacuum θ angle, which moreover becomes physically irrelevant in certain situations, e.g. the chiral limit, as has been hinted on in the previous section and we shall see now.

1.7 The Axial Anomaly

In the two dimensional example we gave in Section 1.5 it was shown that there exists a connection between the chiral symmetry of massless fermions and the θ-angle.

In this section we continue to discuss topological aspects of the standard model with a more detailed analysis of the chiral symmetry [1]. Therefore we consider the quark sector of a four-dimensional gauge theory. The Lagrangean density is

$$\mathcal{L}_{\text{quark}} = \bar{\psi} i \left(\slashed{\partial} + \slashed{A} \right) \psi , \qquad (79)$$

where A_μ^a describes a nonabelian background gauge field.

On the classical level this Lagrangean has the global chiral symmetry

$$\psi \to e^{\alpha\gamma_5}\psi, \qquad \bar\psi \to \bar\psi e^{\alpha\gamma_5} \qquad \left(\gamma_5^\dagger = -\gamma_5\right). \tag{80}$$

Since $\{\gamma_5, \gamma_\mu\} = 0$ we get for the classical theory $\mathcal{L}_{\text{quark}} \to \mathcal{L}_{\text{quark}}$ under this transformation.

The Noether current connected to the chiral symmetry is

$$j_5^\mu = i\bar\psi\gamma^\mu\gamma_5\psi \tag{81}$$

which is classically conserved

$$\partial_\mu j_5^\mu = 0. \tag{82}$$

This can easily be verified to be a consequence of the equation of motion

$$i\left(\slashed\partial + \slashed A\right)\psi = 0. \tag{83}$$

For a quantum theory the situation is different. Expressions like $\mathcal{L}_{\text{quark}}$ in (79) or j_5^μ in (81) are not well defined. The product of two field operators at the same space-time point is singular and requires regularisation. This is most easily seen from the quantisation relation

$$\{\psi(x), \psi^\dagger(y)\}_{x^0 = y^0} = \delta^{(3)}(\mathbf{x} - \mathbf{y}). \tag{84}$$

The regularisation may be carried out using point splitting. However it has to be done carefully since the introduction of a further parameter may spoil the symmetries of the theory. Nevertheless it is possible to regularise the theory in a way that the local gauge invariance is maintained. This is necessary since the gauge symmetry is a fundamental intrinsic property of the theory and it should not be spoiled.

The requirement of keeping the gauge symmetry restricts the freedom how to regularise. Therefore one has to take into account that other, less important symmetries may be violated within the regularisation procedure. For such a symmetry the corresponding currents are not conserved. The symmetry is said to be broken by an anomaly. One example is the axial symmetry which is spoiled by quantisation according to the axial anomaly.

Since j_5^μ has no gauge group label it is a gauge singlet current. In this sense we call the axial anomaly also "abelian anomaly".

We will proceed with a discussion of this anomaly. We take the expectation value of j_5^μ with respect to the perturbative fermionic vacuum.

$$\langle 0_F | j_5^\mu(x) | 0_F \rangle \equiv \langle j_5^\mu(x) \rangle = \langle \bar\psi(x) i\gamma^\mu\gamma_5\psi(x) \rangle \equiv \langle \bar\psi(x) \Gamma_5^\mu \psi(x) \rangle \tag{85}$$

This can be done without loosing information since we expect the result for $\partial_\mu j_5^\mu$ to have no fermion operator component[5]. The result may be regarded as

---

[5] This is confirmed by the path integral approach, which gives the same result as our calculations (see eg. [14], p. 100).

the amplitude for a quark to interact at the space-time point x with Γ_5^μ, to propagate in the gauge background field and to return to x^6. The background field coupling can be treated as two point interaction. So the propagation in the background field may be calculated perturbatively to get a power series in the backgroud field A_μ shown in figure 2.

$$\langle j_5^\mu(x)\rangle = \langle \bar\psi(x)\Gamma_5^\mu \psi(x)\rangle \tag{86}$$

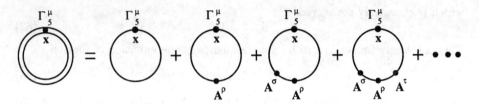

Fig. 2. *Power series expansion for $\langle j_5^\mu(x)\rangle$*

When calculating $\partial_\mu \langle j_5^\mu \rangle$ we recognise that the first term on the right hand side does not contribute since it is x-independent[7]. The second term does not contribute since it is linear in A and we expect the background field to be invariant under charge conjugation. If the background field stems from the Feynman integral of a physical theory this property is guaranteed[8]. With the same reasoning the third order term in A vanishes, as does every odd order in A.

With each interaction of the background field one gets an extra fermion propagator $S_F = 1/(\not{p} + i\varepsilon)$ and the amplitude becomes more convergent. So, by power counting, terms of fourth and higher order are finite. Their amplitudes can not contribute to $\partial_\mu \langle j_5^\mu \rangle$ since j_5^μ is classically conserved order by order, and for finite amplitudes we can apply the classical result.

Therefore the only diagram that can give rise to a non-vanishing $\partial_\mu \langle j_5^\mu \rangle$ is the second order contribution in A_μ. We will focus on it in the following. Its contribution to $\langle j_5^\mu \rangle$ is given by (the trace goes over color as well as spinor indices)

$$\langle j_5^\mu(x)\rangle_{A^2} = i\int d^4z_1 d^4z_2 \; \text{tr}\,[\Gamma_5^\mu S_F(x-z_1)$$
$$\not{A}(z_1)S_F(z_1-z_2)\not{A}(z_2)S_F(z_2-x)] \tag{87}$$

---

[6] Or close to it, when we are applying point splitting to regularise the theory.

[7] To be precise each of the lower order contributions of $\langle j_5^\mu \rangle$ is singular. Therefore it gets a more complicated x- and A_μ-dependence as a consequence of a gauge invariant regularisation prescription. For details see [16], [15].

[8] From experiments we know that, in contrary to the two dimensional theory (s. below), the charge conjugation symmetry is not dynamically broken.

$$= \int d^4z_1 d^4z_2 A^a_\rho(z_1) A^b_\sigma(z_2) \int d^4p d^4q e^{i(p+q)x} e^{ipz_1} e^{iqz_2} T^{\mu\rho\sigma}_{ab}(p,q) ,$$

where $T^{\mu\rho\sigma}_{ab}(p,q)$ is given by the triangle graph shown in figure 3

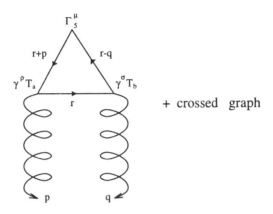

+ crossed graph

Fig. 3. The triangle graph

$$T^{\mu\rho\sigma}_{ab} = -ie^2 \int \frac{d^4r}{(2\pi)^4} \, \text{tr} \, \gamma^\mu \gamma_5 \frac{1}{\slashed{r}+\slashed{p}+i\varepsilon} \gamma^\rho T_a \frac{1}{\slashed{r}+i\varepsilon} \gamma^\sigma T_b \frac{1}{\slashed{r}-\slashed{q}+i\varepsilon}$$
$$+ \quad (\rho \leftrightarrow \sigma , \quad p \leftrightarrow q)$$
$$= \frac{1}{2}\delta_{ab} ie^2 \int \frac{d^4r}{(2\pi)^4} \, \text{tr} \, \gamma^\mu \gamma_5 \frac{1}{\slashed{r}+\slashed{p}+i\varepsilon} \gamma^\rho \frac{1}{\slashed{r}+i\varepsilon} \gamma^\sigma \frac{1}{\slashed{r}-\slashed{q}+i\varepsilon}$$
$$+ \quad (\rho \leftrightarrow \sigma , \quad p \leftrightarrow q) \equiv -\frac{1}{2}\delta_{ab} T^{\mu\rho\sigma}(p,q) . \qquad (88)$$

The integral is linearly divergent which reflects the fact that $j^\mu_5(x)$ was not properly regularised. This has the consequence that a shift in the integration variable $r \to r + a$ changes the value of the (finite part of the) integral by a surface term[9]. This can easily be seen in a one dimensional analogon: Consider the integral

$$\Delta(a) = \int_{-\infty}^{\infty} (f(x+a) - f(x)) \, dx \qquad (89)$$

where f is an analytic function.

We expand $f(x+a)$ in a Taylor series at the point x and perform the integral with the result

$$\Delta(a) = a(f(\infty) - f(-\infty)) + \frac{a^2}{2}(f'(\infty) - f'(-\infty)) + \ldots \qquad (90)$$

---

[9] The Feynman rules do not describe how to introduce the loop integration variable. Each of the choices $r + a$ are a priory possible.

If the integral would be convergent or at most logarithmically divergent then, of course, $0 = f(\pm\infty) = f'(\pm\infty) = \ldots$ and the integral vanishes.

However if the integral is linearly divergent we only have $0 = f'(\pm\infty) = f''(\pm\infty) = \ldots$ and we get for $\Delta(a)$ the surface contribution

$$\Delta(a) = a\left(f(\infty) - f(-\infty)\right) \tag{91}$$

which is in general non zero.

The same applies to the four-dimensional integral $T^{\mu\rho\sigma}$. The surface term can be calculated quite easily. We start with the first part of $T^{\mu\rho\sigma}$ and get

$$\Delta_1(a) = -ie^2 \int \frac{d^4r}{(2\pi)^4}\, \mathrm{tr}\, [\gamma^\mu \gamma_5 \left(\exp\left(a_\alpha \frac{\partial}{\partial r_\alpha}\right) - 1\right) \frac{1}{\not{r} + \not{p} + i\varepsilon}\gamma^\rho$$
$$\cdot \frac{1}{\not{r} + i\varepsilon}\gamma^\sigma \frac{1}{\not{r} - \not{q} + i\varepsilon}]$$
$$= -ie^2 a_\alpha \int \frac{d^4r}{(2\pi)^4} \frac{\partial}{\partial r_\alpha} (1 + \mathcal{O}(r^{-1}))\, \mathrm{tr}\, [\gamma^\mu \gamma_5 \frac{1}{\not{r} + \not{p} + i\varepsilon}\gamma^\rho$$
$$\cdot \frac{1}{\not{r} + i\varepsilon}\gamma^\sigma \frac{1}{\not{r} - \not{q} + i\varepsilon}]\ .$$

Now we Wick-rotate to Euclidean space-time ($t \to ix_4$) and use "one quarter" of the four-dimensional Gauß theorem

$$\int_M d^4r \frac{\partial}{\partial r_\alpha} f(r) = \int_{\partial M} d\sigma^\alpha f(r) \tag{92}$$

where ∂M is the boundary of M (which is the sphere $S^3(R)$ in our case) and $\int d\sigma^\alpha$ is the α-component of the surface integral. We get ($R \to \infty$)

$$\Delta_1(a) = \frac{e^2 a_\alpha}{(2\pi)^4} \int_{S^3(R)} d\sigma^\alpha\, \mathrm{tr}\, \gamma^\mu \gamma_5 \frac{1}{\not{r}+\not{p}}\gamma^\rho \frac{1}{\not{r}}\gamma^\sigma \frac{1}{\not{r}-\not{q}}$$
$$= \frac{e^2 a_\alpha}{(2\pi)^4} \int d\Omega^\alpha\, \mathrm{tr}\, [\gamma^\mu \gamma_5 \gamma^\nu \gamma^\rho \gamma^\delta \gamma^\sigma \gamma^\beta] R_\nu R_\delta R_\beta / R^3$$
$$= -\frac{e^2 a_\alpha}{(2\pi)^4} \int d\Omega^\alpha 4\varepsilon^{\mu\nu\rho\sigma} \frac{R_\nu}{R}\ .$$

Now we introduce polar coordinates and let the north pole point into the α-direction. Since $\int d\Omega^\alpha R_\nu/R$ is zero if $\alpha \neq \nu$ we get

$$\Delta_1(a) = -\frac{e^2 a_\nu}{\pi^3}\varepsilon^{\mu\nu\rho\sigma} \int_0^\pi d\theta \sin^2\theta \cos\theta \cdot \cos\theta = -\frac{e^2 a_\nu}{8\pi^2}\varepsilon^{\mu\nu\rho\sigma}\ .$$

The crossed term gives the same result so that the total surface term is

$$\Delta(a) = -\frac{e^2 a_\nu}{4\pi^2}\varepsilon^{\mu\nu\rho\sigma}\ . \tag{93}$$

The vector Ward identities which enssure gauge invariance have the form

$$p_\rho T^{\mu\rho\sigma}(p,q) = 0, \qquad q_\sigma T^{\mu\rho\sigma}(p,q) = 0. \tag{94}$$

The chiral Ward identity which is connected to the chiral symmetry is

$$(p+q)_\mu T^{\mu\rho\sigma}(p,q) = 0. \tag{95}$$

Gauge invariance is one of the most fundamental principles of QCD and in fact there exists a choice of the integration variable $r+a$ that enssures (94) (namely $a = -2p$ [16], p. 122), but for any other a the gauge symmetry is spoiled by the surface term (93).

Unfortunately we need different a's to assure (94) and (95). So it is impossible to have both gauge symmetry and chiral symmetry. We choose (94) to hold and get a correction on the right hand side of (95) [17], [15]:

$$(p+q)_\mu T^{\mu\rho\sigma}(p,q) = -\frac{e^2}{2\pi}\varepsilon^{\rho\sigma\mu\nu}p_\mu q_\nu. \tag{96}$$

If this is plugged into (88) we get

$$\langle \partial_\mu j_5^\mu \rangle = \frac{\varepsilon^{\mu\nu\rho\sigma}}{8\pi^2}\langle\text{tr}\,(\partial_\mu A_\nu - \partial_\nu A_\mu)(\partial_\rho A_\sigma - \partial_\sigma A_\rho)\rangle.$$

The right hand side equals $1/8\pi^2 \cdot \langle\text{tr}\,F^*_{\mu\nu}F^{\mu\nu}\rangle$ for the following reason: The third order term in A_μ vanishes since we have invariance under charge conjugation. The fourth order term is proportional to $\varepsilon^{\mu\nu\rho\sigma}\text{tr}\,([A_\mu, A_\nu][A_\rho, A_\sigma]) = -\frac{1}{2}\varepsilon^{\mu\nu\rho\sigma}A_\mu^a A_\nu^b A_\rho^c A_\sigma^d f_{abe}f_{cde}$ where f are the structure constants of the SU(3) group. One can now use the total antisymmetry of f and the Jacobi identity to show that the last expression is zero.

Thus we have motivated the final result

$$\partial_\mu j_5^\mu = \frac{1}{8\pi^2}\text{tr}\,F^*_{\mu\nu}F^{\mu\nu}. \tag{97}$$

It can be shown that there are no other contributions to the anomaly as for example virtual gluon effects[10] [15], [18]. The expression on the right hand side of (97) has a topological interpretation: It is just twice the four-dimensional Pontryagin density.

Although the axial current is not conserved we can carry on by constructing a conserved current. Due to (49) we have

$$\text{tr}\,F^*_{\mu\nu}F^{\mu\nu} = 4\partial_\mu \varepsilon^{\mu\nu\rho\sigma}\,\text{tr}\left(\frac{1}{2}A_\nu\partial_\rho A_\sigma + \frac{1}{3}A_\nu A_\rho A_\sigma\right) \tag{98}$$

and therefore

$$\partial_\mu J_5^\mu = 0 \tag{99}$$

---

[10] The fermion loop becomes more convergent with every internal gluon line and the intergrations over the gluon lines do not contribute to the anomaly.

with

$$J_5^\mu = j_5^\mu - \frac{1}{2\pi^2}\varepsilon^{\mu\nu\rho\sigma}\,\mathrm{tr}\,\left(\frac{1}{2}A_\nu\partial_\rho A_\sigma + \frac{1}{3}A_\nu A_\rho A_\sigma\right)\;. \tag{100}$$

The conserved charge Q_5 of J_5^μ is

$$\begin{aligned}Q_5 &= \int d^3r\,\left(j_5^0 - \frac{1}{2\pi^2}\varepsilon^{ijk}\,\mathrm{tr}\,\left(\frac{1}{2}A_i\partial_j A_k + \frac{1}{3}A_i A_j A_k\right)\right)\\ &= \int d^3r\,\left(j_5^0 + 2W(A)\right)\end{aligned} \tag{101}$$

where $W(A)$ is the Chern Simons three-form which was already defined in (39).

Q_5 consists of two pieces, a gauge invariant fermion contribution coming from j_5^μ and an anomalous term constructed from the gauge potentials. This term has the immediate consequence that neither J_5^μ nor Q_5 are invariant under topological non-trivial (large) gauge transformations Ω_n (under (small) gauge transformations that are smoothly connected to unity they are still invariant). Q_5 changes by two times the winding number.

$$\Omega_n Q_5 \Omega_n^{-1} = \Omega_5 - 2n \tag{102}$$

The commutator algebra of the Hamilton operator H, Q_5 and Ω_n is

$$[H, Q] = 0\;, \qquad [H, \Omega_n] = 0\;, \qquad [\Omega_n, Q_5] = 2n\Omega_n\;. \tag{103}$$

Since the θ-angle is defined by (45)

$$\Omega_n|\theta\rangle = e^{-i\theta n}|\theta\rangle \tag{104}$$

we conclude that Q_5 acts as a shift operator for θ:

$$e^{i\frac{\theta'}{2}Q_5}\psi(\theta) = \psi(\theta + \theta')\;. \tag{105}$$

Since H and Q_5 can be diagonalised simultaneously, applying $e^{i\frac{\theta'}{2}Q_5}$ can not change the energy eigenvalue of an energy eigenstate ψ. Therefore the energy spectrum does not depend on θ. The value of the θ-angle is physically irrelevant. If on the other hand fermions are massive, Equation (97) and all successive equations acquire a mass correction and we can not argue that the θ-angle has no physical consequences.

The same result may be obtained in a functional integral formulation. If one decides to have massless fermions and to translate the θ-dependence from the wavefunctions to the Lagrangean one gets the action (48)

$$Z_\theta = \int \mathcal{D}\psi \mathcal{D}\bar\psi \mathcal{D}A_\mu^a \exp\left(i\int dx \mathcal{L}_\theta(x)\right)\;, \tag{106}$$

where

$$\mathcal{L}_\theta = \frac{1}{2g^2}\,\mathrm{tr}\,[F_{\mu\nu}F^{\mu\nu}] - \frac{\theta}{16\pi^2}\,\mathrm{tr}\,[F^*{}_{\mu\nu}F^{\mu\nu}] + i\bar\psi\left(\slashed\partial + \slashed A\right)\psi\;. \tag{107}$$

Redefining the fermionic integration variables according to the chiral transformation law (80) $\mathcal{L}_\theta$ remains unaffected, but we get a contribution from the integration measure. This contribution corresponds to the anomaly, and we get

$$Z_\theta \to Z_{\theta+2\alpha} \ . \tag{108}$$

Since we just substituted our integration variables, Z_θ does not change. Therefore $Z_\theta = Z_{\theta+2\alpha}$ has to be independent of θ. So we can conclude that in the presence of massless fermions the θ-angle is no physical parameter.

1.8 The Two-Dimensional Analogon (Schwinger Model)

Let us come back to the two dimensional example QED_{1+1} that was already discussed in Sec. 1.5.

In a two dimensional space-time, the Dirac spinors become two-component objects. The Dirac matrices may be chosen to be the Pauli matrices

$$\gamma^0 = \sigma^1 \ , \qquad \gamma^1 = i\sigma^2 \ , \qquad \gamma_5 = -i\sigma^3 \ . \tag{109}$$

It is a particular property of two dimensions that axial vectors are dual to vectors

$$\Gamma_5^\mu = i\gamma^\mu \gamma_5 = \varepsilon^\mu{}_\nu \gamma^\nu \ , \qquad \varepsilon^{01} = 1 = -\varepsilon_{01} \tag{110}$$

and therefore the axial vector current is dual to the vector current

$$j_5^\mu = \varepsilon^\mu{}_\nu j^\nu \ . \tag{111}$$

Let us consider the fermionic sector of two dimensional QED. We start to calculate the divergence of the chiral current in the same way as in the four-dimensional case. $\langle j_5^\mu(x)\rangle$ can be expanded in a power series of the background A_μ field shown in figure 2. In two dimensional QED we can not use charge con-

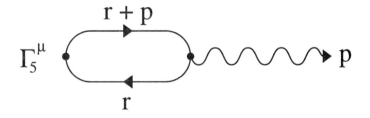

Fig. 4. Graphical representation for $T^{\mu\nu}(p)$

jugation to simplify the result since the symmetry under charge conjugation is dynamically broken. This can be concluded most easily from the existence of a constant electric field E_0 (54), which is incompatible with the symmetry under charge conjugation. However, the A_μ^2-order is already convergent enough not to produce an anomaly. Instead of the triangle graph we get the relevant contribution from $T^{\mu\nu}(p)$ shown in figure 4. We are using the duality between axial vectors and vectors to obtain

$$T^{\mu\nu}(p) = \varepsilon^\mu{}_\rho \Pi^{\rho\nu}(p) \qquad (112)$$

where $\Pi^{\rho\nu}(p)$ is the vacuum polarisation tensor. Its space-time structure is determined by the requirement of gauge invariance

$$p_\rho \Pi^{\rho\nu}(p) = p_\nu \Pi^{\rho\nu}(p) = 0 \qquad (113)$$

to be of the form

$$\Pi^{\rho\nu}(p) \propto g^{\rho\nu} - \frac{p^\rho p^\nu}{p^2} \ . \qquad (114)$$

Therefore we have

$$p_\nu T^{\mu\nu}(p) = 0 \ , \qquad (115)$$

but the Ward identity related to the chiral symmetry,

$$p_\mu T^{\mu\nu}(p) \propto p_\mu \varepsilon^{\mu\nu} \ , \qquad (116)$$

does not vanish.

So we regain the result that gauge symmetry can be maintained, while the axial symmetry is broken on the quantum level.

The result of a detailed calculation is [10], [19]

$$\partial_\mu j_5^\mu = -\frac{1}{2\pi}\varepsilon^{\mu\nu} F_{\mu\nu} = -\frac{1}{\pi} F^* \ , \qquad (117)$$

where we have once more absorbed the electromagnetic charge e in the A_μ field. The anomaly is now given by twice the two dimensional Pontryagin density.

Therefore it is again possible to define a conserved current J_5^μ and its time independent charge Q_5

$$J_5^\mu = j_5^\mu + \frac{1}{\pi}\varepsilon^{\mu\nu} A_\nu \ , \qquad (118)$$

$$Q_5 = \int dx \left(j_5^0(x) + \frac{1}{\pi} A_1(x) \right)$$
$$= \int dx\, j_5^0(x) + 2W(A) \ . \qquad (119)$$

In the Feynman path integral approach a chiral re-definition (80) of the fermionic integration variables amounts, due to the measure, to a new term in the Lagrangean

$$\int \mathcal{D}\psi \mathcal{D}\bar\psi \mathcal{D} A_\mu e^{i \int \mathcal{L}} \to \int \mathcal{D}\psi \mathcal{D}\bar\psi \mathcal{D} A_\mu e^{i \int \left(\mathcal{L} + \frac{\alpha}{2\pi} F^*\right)} \qquad (120)$$

which coincides with (65). Thus QED with massless fermions in 1+1 dimensions (the Schwinger model) has no physically relevant θ-angle [19], [20].

Let us close this section with a remark that is specific to a two dimensional theory. If we contract the gauge field equation

$$\partial_\mu F^{\mu\nu} = e^2 j^\nu \qquad (121)$$

with $\varepsilon_{\nu\rho}$ and use the antisymmetry of $F_{\mu\nu}$ we get

$$\partial^\mu F^* = e^2 j_5^\mu . \qquad (122)$$

The divergence of this equation yields

$$\Box F^* = e^2 \partial_\mu j_5^\mu = -\frac{e^2}{\pi} F^* . \qquad (123)$$

Thus the gauge field acquires the topological mass $m^2 = e^2/\pi$.

Whereas in three-dimensional space-time there exists another topological mechanism for vector meson mass generation (see below), no similarly elegant result has yet been established in four dimensions.

1.9 Conclusions of the First Part

(i) For a long time it appeared that QCD possesses too much symmetry. An additional chiral $U(1)$ symmetry would predict that there would be a particle degenerate with the pion, but no such particle exists [21]. Now we have recognised that the chiral symmetry is broken by an anomaly and the $U(1)$ problem has dissolved [13].

(ii) If the theory includes massless fermions the θ-angle is unphysical. But physical fermions are not massless and the θ-angle is supposed to remain observable. For $\theta \neq 0$ CP-invariance is violated, but in QCD the experiments require that $\theta = 0$ and CP is not violated (measurements of the electric dipole moment of the neutron give $\theta \leq 10^{-9}$ [22]).

No principle is known that insures the vanishing of θ. In fact the situation is even more complicated: If we suppose that the fermion masses arise from spontaneous symmetry breaking then we would expect that the fermion mass matrix in the QCD Lagrangean would point in an arbitrary CP direction $\bar\psi M_1 \psi + \bar\psi \gamma_5 M_2 \psi$. One can remove the M_2-term by a chiral transformation. But this induces, due to the anomaly, a $\mathrm{tr}[F^*{}_{\mu\nu} F^{\mu\nu}]$-term giving rise to a θ-angle. This angle has to be canceled by the "initial" θ-angle in the Lagrangean in order not to yield CP-violating effects[11].

This problem is not unlike that of the cosmological constant which is a parameter that in principle is present, but experiments force it to be zero.

---

[11] Or there exists a reason why even in the presence of massive fermions the θ-angle is unphysical and not CP-violating.

(iii) In the electroweak sector of the standard model couplings to $\gamma^\mu(1 - i\gamma_5)$ are present due to the coupling of only left-handed fermions to the weak charged currents. The requirement of renormalisability forces the theory to avoid the anomalies in the gauge current (anomalies may not occour in subdiagrams where the axial current couples to internal lines). This is only possible if the quarks and leptons balance in number. In particular the existence of a top-quark is demanded.

(iv) In the standard model the baryon number current acquires an anomaly [24]. The decay rate is controled by $\mathrm{tr}[F^*_{\mu\nu} F^{\mu\nu}]$. There are two mechanisms for baryon decay known:

The first involves tunnelling. The tunnelling rate is given by the exponential of the instanton action (in a semiclassical description). But $\exp(-$ instanton action$) = \exp(-8\pi^2/g^2)$ is a negligible small number ($\approx 10^{-122}$ year^{-1}) [13].

The second mechanism is connected to 't Hooft-Polyakov monopoles [25]. The magnitude of this effect is still controversial (but it seems to suffice) and moreover an experimental evidence for monopoles is still missing.

(v) The hypothesis of partial conservation of flavour $SU(2)$ axial vector currents (PCAC) implies, in the absence of anomalies, that a massless neutral pion can not decay into two photons [23]. But the physical pion does decay with a width of about 7.9 eV. This large number can only be understood with the axial anomaly [17], [15], [18]. Moreover one gets the result that the

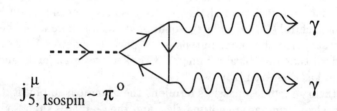

Fig. 5. *Flavour neutral axial current*

width depends on the number of quark colors. The best agreement with the experiment is achieved for $N = 3$ colors. The remaining discrepancy of about 10% can be understood as an effect due to the non-zero pion mass.

Therefore the anomaly allows an experimental determination of the number of colors.

2 High-Temperature Quantum Chromodynamics

In Section 1 we were discussing more or less settled physics, i.e. work that has been done during the eighties. Now we would like to come to talk about some current research in QCD. In this section, we are going to show you the connections between QCD at high temperature (QCD well in the deconfined, chirally symmetric region) and a three-dimensional topolgical field theory: the nonabelian Chern-Simons (CS) theory. More explicitly, we want to show you that the generating functional of the so-called hard thermal loops in QCD is the eikonal of the nonabelian CS theory. These connections have been established recently by several people [26], [27], [28]. They are relevant for the nonabelian generalization of the Kubo formula as well as for a gauge-invariant description of Landau damping in the quark-gluon plasma at high temperature.

First of all, we would like to give you a short introduction to thermal field theory. For details, see, for example, [29], [30].

2.1 Temperature Green Functions

The objects of study in a field theory at finite temperature are the temperature n-point correlation (or Green) functions

$$G_n(x_1, \ldots, x_n) := \langle \phi(x_1) \ldots \phi(x_n) \rangle \tag{124}$$

where the x_i are elements of Minkowski space, and the $\phi(x_i)$ are the generic fields of the theory in the Heisenberg picture. The angle brackets denote thermal average within the canonical ensemble

$$\langle \ldots \rangle := \frac{\operatorname{tr}(e^{-\beta H} \ldots)}{\operatorname{tr} e^{-\beta H}} . \tag{125}$$

Here, H is the Hamiltonian of the theory, and β^{-1} represents the inverse temperature in natural units that we are going to use for the rest of the talk.

Depending on the boundary conditions chosen to solve the equations of motion, one defines *various* Green functions. For example, $\langle T\phi(x)\phi(y)\rangle$ gives the *time-ordered* two-point function, whereas $\theta(x^0 - y^0)\langle[\phi(x), \phi(y)]\rangle$ defines the *retarded commutator* two-point function.

The set of all these n-point Green functions, e.g. in momentum space representation,

$$G_n(p_1, \ldots, p_n) := \int d^4x_1 \ldots d^4x_n e^{i(p_1 x_1 + \ldots p_n x_n)} G_n(x_1, \ldots, x_n) \tag{126}$$

with *real* p_i and *real* x_i, contains all the physical information about the system at finite temperature. But, as a matter of fact, perturbation theory within this description is rather difficult. A simpler perturbation theory can, however, be established on accomplishing the following *unphysical continuation*: one allows the time arguments x_i^0 to be complex valued. For Bose fields, it can be shown that

— for analyticity reasons of the n-point functions — they have to be periodic in the imaginary time direction

$$\phi(x^0, \mathbf{x}) = \phi(x^0 - i\beta, \mathbf{x}). \tag{127}$$

Similarly, fermionic fields $\psi(x^0, \mathbf{x})$ have to obey antiperiodic boundary conditions:

$$\psi(x^0, \mathbf{x}) = -\psi(x^0 - i\beta, \mathbf{x}). \tag{128}$$

Note that these boundary conditions are the essential differences between field theory at zero and field theory at finite temperature; the equations of motions do not differ except for a thermal average, of course, in the latter case. This extension to complex values of x^0 is certainly not unique. In the so-called

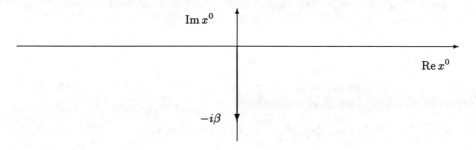

Fig. 6. *Time contour in the ITF*

imaginary-time formalism (ITF) one restricts x^0 to the imaginary axis in the complex x^0-plane, i.e. $x^0 \in [0, -i\beta]$ (cf. Fig.2.1).

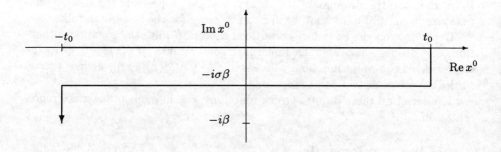

Fig. 7. *Time contour in the RTF with $\sigma = 1/2$*

This can — for Bose fields — be interpreted as a transition from the Minkowski space-time manifold $\mathbb{R}^3 \times \mathbb{R}$ to the new space-time manifold $\mathbb{R}^3 \times S^1$.

Besides the ITF scheme, another popular choice for the complex time-path contour is shown in Fig. 2.1. This is just one of infinitely many possibilities (the choice depends on the parameter σ) of setting up the *real time formalism* (RTF) using the time-path contour method. Choosing $\sigma = 1/2$ provides equivalence with yet another formulation of field theory at finite temperature, called *thermo field dynamics* [31], [32]. In actual calculations, one always considers the limit $t_0 \to \infty$. The advantage of the RTF over the ITF is that perturbation theory can be defined with Green functions depending solely on *real* time arguments. Thus one does not have the problem of a backward continuation from purely imaginary times to purely real, hence physical, time arguments.

In turn, perturbation theory is a bit more cumbersome as, for example, the RTF two-point function is a 2×2 matrix.

2.2 Imaginary-Time Formalism

In the imaginary-time formalism, perturbation theory corresponds to the well-kwown Dyson-Feynman series with the integration over p_0 replaced by an infinite sum

$$\int \frac{dp_0}{2\pi} \to iT \sum_{n \in \mathbb{Z}} . \tag{129}$$

The usual time-ordering along the real x^0-axis is converted into an imaginary-time-ordering down the imaginary x^0-axis. That is, later times are positioned *below* earlier times (cf. Fig. 2.1). Furthermore, all the Green functions are unique, because the inverse d'Alembertian $\Box^{-1}$ is unique on $\mathbb{R}^3 \times S^1$.

(Anti-)periodicity in position space on the interval $[0, -i\beta]$ provides for discrete imaginary energies $p_0 = 2\pi i n T$ (for bosons) and $p_0 = 2\pi i T(n + \frac{1}{2})$ for fermions, $n \in \mathbb{Z}$, in momentum space. These discrete energies are (proportional to) the so-called *Matsubara frequencies*.

At this point, it is interesting to take a look at the high-temperature limit. In position space, the time interval $[0, -i\beta]$ shrinks down to a point when $T \to \infty$, since then $\beta = 1/T \to 0$. Hence we lose the time dimension and end up with a three-dimenional field theory:

$$\mathbb{R}^3 \times S^1 \xrightarrow{T \to \infty} \mathbb{R}^3 . \tag{130}$$

In momentum space, the same result can be deduced by looking at some generic perturbation theoretic diagram. Let the boson propagator have the form

$$D(p) = \frac{i}{p_0^2 - \mathbf{p}^2 - m^2}, \quad \text{where} \quad p_0 = 2\pi i n T, \tag{131}$$

while a fermion propagator be

$$S(p) = \frac{i}{\gamma_0 p_0 - \boldsymbol{\gamma} \cdot \mathbf{p} - m}, \quad \text{where} \quad p_0 = 2\pi(n + \tfrac{1}{2})iT. \tag{132}$$

So our diagram might be something like

$$iT \sum_n \int \frac{d^3p}{(2\pi)^3} e \frac{i}{-4\pi^2 n^2 T^2 - \mathbf{p}^2 - m^2} e \ldots, \tag{133}$$

where e denotes the coupling constant. In the limit $T \to \infty$, all modes with $n \neq 0$ decouple — they behave like very heavy particles. Only the zero mode survives, and so we are left with

$$\int \frac{d^3p}{(2\pi)^3} e\sqrt{T} \frac{1}{\mathbf{p}^2 + m^2} e\sqrt{T} \ldots. \tag{134}$$

This is exactly what one would find in a field theory on a Eulidean space of one dimension less. Moreover, fermion contributions are obviously subdominant since the energy modes in the fermion propagator never vanish. Taking the infinite temperature limit in this way means, in the end, setting external $p_0 = 2\pi n i T \stackrel{n=0}{=} 0$. A more detailed treatment must, however, allow for a high-temperature limit with fixed, nonvanishing external p_0 in order to be able to continue back to real energies. But, even in the case p_0 is kept finite one has a problem. Namely, does one try to continue backward

$$2\pi n T \to -i p_0, \tag{135}$$

one immediately notices that this continuation is not unique. I.e., from a single Euclidean Green function one can obtain *several* Minkowski Green functions. Which one to take depends on the physical setting.

As a rule, the ITF represents the natural scheme for calculating static quantities like the effective potential.

2.3 Hard Thermal Loops

By transferring the QCD Feynman rules for $T = 0$ to finite temperature in a naive way one gets a confusing infra-red limit: on mass-shell both the sign and the magnitude of the gluon damping rate appear to be gauge dependent. Braaten and Pisarski [33] have argued that whenever a quantity is calculated perturbatively in a hot nonabelian gauge theory, sooner or later an infinite subset of diagrams nominally of higher order in the loop expansion contribute to the same order in the coupling constant g. These higher-loop diagrams have to be isolated and resummed into an effective expansion which includes all effects to leading order in g. This resummation technique is necessary to get, even at one loop, gauge invariant results.

More explicitly, hard thermal loops are the ones with exceptional (soft) external momenta

$$\text{both } p_0 \text{ and } |\mathbf{p}| \text{ of order } gT \tag{136}$$

and large (hard) internal momenta

$$k_0 \text{ and/or } |\mathbf{k}| \text{ of order } T. \tag{137}$$

$\Pi_2(p) =$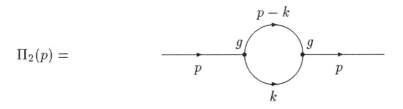

Fig. 8. *The one-loop self energy contribution $\Pi_2(p)$*

$\Pi_4(p) =$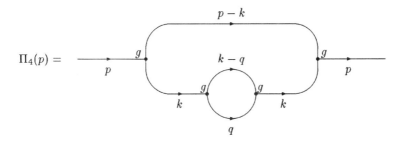

Fig. 9. *The two-loop self energy contribution $\Pi_4(p)$*

The need for resummation can be seen from a simple example. Look at the one- and two-loop contributions to the gluonic self energy depicted in Figs. 8 and 9. Let us write them as

$$\Pi_2(p) \equiv \int d^4k \, \Pi_2(k,p) \tag{138}$$

$$\Pi_4(p) \equiv \int d^4k \, \Pi_4(k,p) \, . \tag{139}$$

One can then easily derive for the following quotient ($D(k)$ symbolizes the free gluon propagator)

$$\frac{\Pi_4(k,p)}{\Pi_2(k,p)} = {}^{\prime}\Pi_2(k) D(k){}^{\prime} \, . \tag{140}$$

For small k, $\Pi_2(k)$ is known to behave like $g^2 T^2$. Hence

$$\frac{\Pi_4(k,p)}{\Pi_2(k,p)} \stackrel{\text{small } k}{\simeq} \frac{g^2 T^2}{k^2} \, . \tag{141}$$

Hence, for soft internal momentum $k \sim gT$, the two contributions are of the same order in g. Stated in a slightly different way, the fourth order diagram contains second order contributions.

2.4 The Kubo Formula

A recent application of hard thermal loops is the generalization of Kubo's formula of linear response theory to nonabelian gauge theories. In this subsection, we shall follow mainly reference [34].

Before tackling the nonabelian case, let me first remind you of Kubo's formula within quantum electrodynamics.

The behavior of electromagnetic fields in a plasma of charged particles is described by the polarization tensor $\Pi^{\mu\nu}(x,y)$ which is the two-point current correlation function

$$\Pi^{\mu\nu}(x,y) \equiv \int \frac{d^4k}{(2\pi)^4} e^{-ik(x-y)} \Pi^{\mu\nu}(k) = -i\langle j^\mu(x) j^\nu(y)\rangle. \tag{142}$$

Perturbatively, this is a one-charged-particle-loop diagram with two external photon lines.

The real part of this tensor describes phenomena such as Debyescreening and propagation of plasma waves; the imaginary part describes the damping of fields in the plasma (Landau damping). If one integrates out the charged fields in a functional integral for the theory, the polarization tensor naturally emerges as the thermal average of the time-ordered product of two currents. However, there are situations where the response of the plasma to the electromagnetic field is described as the average of the retarded commutator of currents. To see this in the case of QED with fermion current $J^\mu(x) = e\bar{\psi}(x)\gamma^\mu\psi(x)$ and interaction Lagrangean $\mathcal{L}_{\text{int}} = -J^\mu A_\mu$, one calculates J^μ in an expansion in small gauge fields A_ν. The equation of motion for this theory is

$$\partial_\nu F^{\nu\mu}(x) = J^\mu(x). \tag{143}$$

$J^\mu(x)$ is related to the scattering operator $S[A] = T\exp[-i\int d^4x\, A_\mu(x) j^\mu(x)]$ (here $j^\mu(x) \equiv e\bar{\psi}_I(x)\gamma^\mu\psi_I(x)$ and the subscript I denotes the interaction picture) in the following way:

$$J^\mu(x) = iS^{-1}\frac{\delta S[A]}{\delta A_\mu(x)}. \tag{144}$$

The rhs of (144) is ready for an epansion in A. The result up to linear order is

$$J^\mu(x) = j^\mu(x) - i\int d^4y\, \theta(x^0 - y^0)[j^\mu(x), j^\nu(x)]A_\nu(y) + \mathcal{O}(A^2). \tag{145}$$

Using this in (143) and taking the thermal average with the unperturbed density matrix $e^{-H_0/T}$, one arrives at the *Kubo formula*

$$\partial_\nu F^{\nu\mu}(x) = \int d^4y\, \Pi_R^{\mu\nu}(x,y) A_\nu(y) \tag{146}$$

where $\Pi_R^{\mu\nu}(x,y) = -i\theta(x^0 - y^0)\langle[j^\mu(x), j^\nu(x)]\rangle$. Hence, the average of the retarded commutator is the appropriate function for the situation where we perturb the plasma by the field and ask how the field evolves.

Now, we discuss the relationship between the time-ordered and the retarded response functions, $\Pi_T^{\mu\nu}$ and $\Pi_R^{\mu\nu}$. The real part is the same for both of them

$$\operatorname{Re} \Pi_T^{\mu\nu} = \operatorname{Re} \Pi_R^{\mu\nu}. \tag{147}$$

This has long been familiar, see e.g. [35]. Here we shall concentrate on their imaginary parts. A large-T calculation to one-loop order yields for their imaginary parts

$$\operatorname{Im} \Pi_R^{\mu\nu} \simeq \frac{k_0 T^2}{12} P^{\mu\nu} \tag{148}$$

as well as

$$\operatorname{Im} \Pi_T^{\mu\nu} = \operatorname{Im} \Pi_R^{\mu\nu} + \frac{T^3}{6} P^{\mu\nu}, \tag{149}$$

where

$$P^{\mu\nu} = -k^2 \theta(-k^2) \frac{6\pi}{|\mathbf{k}^2|} \left(\frac{1}{3} P_1^{\mu\nu} + \frac{1}{2} P_2^{\mu\nu} \right), \tag{150}$$

$$P_1^{\mu\nu} = g^{\mu\nu} - \frac{k^\mu k^\nu}{k^2}, \tag{151}$$

$$P_2^{\mu 0} = P_2^{0\nu} = 0, \quad P_2^{ij} = \delta^{ij} - \frac{k^i k^j}{\mathbf{k}^2}. \tag{152}$$

This relationship between $\Pi_T^{\mu\nu}$ and $\Pi_R^{\mu\nu}$ can be understood in the following way. $\Pi_R^{\mu\nu}$, being retarded, obeys a spectral representation of the form [36]

$$\Pi_R^{\mu\nu} = \Pi_{\text{sub}}^{\mu\nu} + \int dk_0' \frac{\rho^{\mu\nu}(k_0', \mathbf{k})}{k_0' - k_0 - i\epsilon} \tag{153}$$

for some spectral function $\rho^{\mu\nu}(k)$. $\Pi_{\text{sub}}^{\mu\nu}$ is a 'subtraction term' that can arise in the real part of $\Pi_R^{\mu\nu}$. For $\Pi_T^{\mu\nu}$, we then have [36]

$$\Pi_T^{\mu\nu} = \Pi_{\text{sub}}^{\mu\nu} + \int dk_0' \frac{\rho^{\mu\nu}(k_0', \mathbf{k})}{k_0' - k_0 - i\epsilon} + 2\pi i f(k_0) \rho^{\mu\nu}(k_0, \mathbf{k}), \tag{154}$$

where

$$f(k_0) = \frac{1}{e^{k_0/T} - 1}. \tag{155}$$

The bosonic distribution function $f(k_0)$ appears because $\Pi_T^{\mu\nu}$ is ultimately part of the bosonic (i.e. photon) propagator, and also because it is given by the thermal average of the T-product of two bosonic operators, viz. the two currents j_μ and j_ν. The essence of our results (148) and (149) is that the high-temperature spectral function is

$$\rho^{\mu\nu}(k) = \frac{k_0 T^2}{12\pi} P^{\mu\nu} \tag{156}$$

and the difference in the high-temperature behavior between $\operatorname{Im} \Pi_R^{\mu\nu}$ and $\operatorname{Im} \Pi_T^{\mu\nu}$ ($\mathcal{O}(T^2)$ vs. $\mathcal{O}(T^3)$) is attributed to the presence in the latter of $2\pi f(k_0) \rho^{\mu\nu}$, which according to (155) and (156) tends to $\frac{1}{6} T^3 P^{\mu\nu}$.

Our result for the retarded function $\Pi_R^{\mu\nu}$ agrees with various previous calculations [35]. It is noteworthy, that these early calculations in the Soviet literature, based on the Boltzmann and Vlasov transport equations of kinetic theory, are here regained in quantum field theory at one-loop order.

Another correlation function that is frequently considered is the imaginary-time one. It too is given by a dispersive integral

$$\Pi_{\text{im-t}}^{\mu\nu} = \Pi_{\text{sub}}^{\mu\nu} + \int dk_0' \frac{\rho^{\mu\nu}(k_0', \mathbf{k})}{k_0' - \omega_n}, \quad \omega_n = 2\pi i n T. \tag{157}$$

Because the external energy ω_n is temperature dependent in imaginary time, it makes sense to speak of high-temperature behavior only for the $n = 0$ mode, effectively reducing dimensionality to three, where the spectral function enforces an $\mathcal{O}(T^2)$ large-T behavior.

In QED, the one-loop calculations at finite temperature are useful since higher-order contributions are down by the coupling e. This is related to the following consideration of the effective action $\Gamma_{\text{high-}T}[A]$ that produces the Kubo formula as the corresponding equation of motion

$$\Gamma_{\text{high-}T}[A] = -\frac{1}{4} \int d^4x F^{\mu\nu}(x) F_{\mu\nu}(x) - \frac{1}{2} \int d^4x\, d^4y A_\mu(x) \Pi_R^{\mu\nu}(x,y) A_\nu(y). \tag{158}$$

The important fact is that the expression above is *gauge invariant*. The polarization tensor of QCD ($= SU(N)$ gauge theory with N_F flavors of fermions in the fundamental representation) at finite temperature is related to the one of QED simply by factors

$$\Pi_{ab}^{\mu\nu} = (N + \frac{1}{2} N_F) \delta_{ab}\, \Pi_{\text{QED}}^{\mu\nu}. \tag{159}$$

But, in the nonabelian case, (158) is no longer gauge invariant. The reason is, as one might already expect from the foregoing discussion of hard thermal loops, that higher-order contributions in the coupling must be taken into account. Hence, the task is to find the correct effective action $\Gamma_{\text{high-}T}[A]$ of QCD giving us the generalized Kubo formula.

2.5 Analysis of Hard Thermal Loops in QCD

One way of analyzing hard thermal loop contributions is the calculation of the corresponding Feynman diagrams (e.g. [37]). Another one is to use a high-temperature action and to require gauge invariance for it. The gauge invariance condition relates the high-T QCD to a three-dimensional Yang-Mills theory with topological mass term, a theory worked out about ten years ago [38]. The high-temperature action is deduced in different publications (e.g. [33], [37]):

$$\Gamma_{\text{high-}T}[A] = \frac{1}{2} \int d^4x \operatorname{tr} F^{\mu\nu} F_{\mu\nu} + \left(N + \frac{1}{2} N_F\right) \frac{T^2}{12\pi} \Gamma[A]. \tag{160}$$

In the following, light-like vectors are used

$$Q_\pm^\mu := \frac{1}{\sqrt{2}}(1, \pm \hat{q}), \quad \hat{q}^2 = 1 \tag{161}$$

$$A_\pm := Q_\pm^\mu A_\mu = (A_0 \pm \hat{q}^i A_i) \tag{162}$$

and an angular integration $\int d\Omega_{\hat{q}}$ over the directions of the unit vector $\hat{q}$. The temperature-independent term $\Gamma[A]$ has the form:

$$\Gamma[A] = 2\pi \int d^4x\, A_0^a A_0^a + \int d\Omega_{\hat{q}}\, W(A_+). \tag{163}$$

Gauge invariance for the action requires gauge invariance for $\Gamma[A]$:

$$\delta\Gamma[A] = \delta\left[2\pi \int d^4x\, A_0^a A_0^a\right] + \delta\left[\int d\Omega_{\hat{q}}\, W(A_+)\right] = 0 \tag{164}$$

$$\to (\partial_+ + \partial_-) A_+^a + \partial_+ \frac{\delta W(A)}{\delta A_+^a} + f^{abc} A_+^b \frac{\delta W}{\delta A_+^c} = 0 \tag{165}$$

$$\Rightarrow \partial_+ \frac{\delta}{\delta A_+^a}\left[W(A_+) + \frac{1}{2}\int d^4x\, A_+^{a'} A_+^{a'}\right] +$$
$$+ f^{abc} A_+^b \frac{\delta}{\delta A_+^c}\left[W(A_+) + \frac{1}{2}\int d^4x\, A_+^{a'} A_+^{a'}\right] - (-)\partial_- A_+^a = 0. \tag{166}$$

Calling the term in the square bracket S, the gauge invariance condition gets the form

$$\partial_1 \frac{\delta}{\delta A_1^a} S - \partial_2 A_1^a + f^{abc} A_1^b \frac{\delta}{\delta A_1^c} S = 0. \tag{167}$$

S is an integrated functional of the fields, so we set by analogy with Hamilton-Jacobi theory

$$A_2^a = \frac{\delta}{\delta A_1^a} S, \tag{168}$$

what gives for gauge invariance condition

$$\partial_1 A_2^a - \partial_2 A_1^a + f^{abc} A_1^b A_2^c = 0. \tag{169}$$

With this constraint for the A-fields we can relate the high-temperature QCD to a topic of topological field theory the Chern Simons theory. In the eighties [38] the CS–Lagrangean

$$\Omega(A) = -\frac{1}{8\pi^2} \epsilon^{ijk} \text{tr}\left(\partial_i A_j A_k + \frac{2}{3} A_i A_j A_k\right) \tag{170}$$

was used as topological mass term for three-dimensional Yang-Mills theories

$$\mathcal{L} = \frac{1}{2} \text{tr}\, F^{\mu\nu} F_{\mu\nu} + 8\pi^2 m \Omega(A) \tag{171}$$

with equations of motion

$$\mathcal{D}_\mu F^{\mu\nu} + \frac{m}{2}\epsilon^{\nu\alpha\beta}F_{\alpha\beta} = 0 \tag{172}$$

where m is a gauge invariant mass which, in nonabelian theories, is quantized so that

$$e^{i8\pi^2 m} = 1 \Longrightarrow m = n/4\pi. \tag{173}$$

So we have a topological massive gauge theory with multi-valued action. But with the quantization condition the phase exponential of the action remains gauge invariant.

In order to see that this theory resembles much of the hard thermal loop analysis in QCD, we must relate $\Gamma[A]$ to the CS functional $\Omega[A]$. This can be done by the constraint (169), which arises also in CS theory.

2.6 Pure Chern-Simons Theory

$$\mathcal{L}_{\text{CS}} = 8\pi^2 k\Omega[A] \tag{174}$$

because $\mathcal{L}$ is a volume form, the corresponding action (here integration in 2+1 dimensions) is independent of a metric. So we are dealing with a topological field theory, a framework which is used in mathematics to investigate the topology of low-dimensional manifolds [39].

The equations of motion are

$$\epsilon^{\alpha\mu\nu}F_{\mu\nu} = 0. \tag{175}$$

For doing canonical quantization we choose the condition $A_0 = 0$, then we have

$$\mathcal{L}_{CS} = \frac{k}{2}\epsilon^{ij}\dot{A}_i^a A_j^a. \tag{176}$$

Now we can choose phase space variables by a method for first order Lagrangeans of the form $\mathcal{L} = \omega^{ij}\dot{q}_i p_j$, so we have $A_1^a = q^a$, $A_2^a = p^a$. By Legendre transformation we get a vanishing Hamiltonian

$$H = 0, \tag{177}$$

which causes trivial equations of motion

$$\dot{A}_i = 0. \tag{178}$$

The 0-component equation of motion does not involve a time derivative. It is merely the Gauß' law constraint

$$\epsilon^{ij}F_{ij} = 0 \tag{179}$$

giving the generator for gauge transformations

$$G^a(\mathbf{x}) = -\frac{k}{2}\epsilon^{ij}F_{ij}^a(\mathbf{x}). \tag{180}$$

We now implement the constraint as a relation for the quantum states, what means **quantization before solving of constraints** [40].

The antisymmetric matrix in the Lagrangean determines the symplectic structure of our theory and establishes the phase space commutation relations

$$[A_i^a(\mathbf{x}), A_j^b(\mathbf{y})] = \frac{i}{k}\epsilon_{ij}\delta^{ab}\delta(\mathbf{x}-\mathbf{y}). \tag{181}$$

Since $H = 0$ all the dynamics is in the constraint

$$G^a(\mathbf{x})|\Psi\rangle = 0, \tag{182}$$

where $|\Psi\rangle$ are the physical states and

$$i\left[G^a(\mathbf{x}), G^b(\mathbf{y})\right] = f^{abc}G^c(\mathbf{x})\delta(\mathbf{x}-\mathbf{y}) \tag{183}$$

the algebra of the constraints, which follows the Lie algebra of the gauge transfomation group.

For the realization of the quantum theory we have to get an irreducible representation of an algebra of observables, which consists of functions on the phase space. The irreducibility is obtained by choosing a polarization, what means stating what is p and what is q and what is the argument of the wave functions (there is not a single unique polarization, but quantization should be independent of which one is taken). Here we choose Cartesian polarization

$$A_1^a \equiv \phi^a, \quad A_2^a = \frac{1}{ik}\frac{\delta}{\delta\phi^a} \tag{184}$$

$$|\Psi\rangle \leftrightarrow \Psi(\phi) \tag{185}$$

the phase space variables are represented as

$$A_1^a(\mathbf{x})|\Psi\rangle \leftrightarrow \phi^a(\mathbf{x})\Psi(\phi) \tag{186}$$

$$A_2^a(\mathbf{x})|\Psi\rangle \leftrightarrow \frac{1}{ik}\frac{\delta}{\delta\phi^a(\mathbf{x})}\Psi(\phi). \tag{187}$$

The constraint implies an equation for the physical states $\Psi(\phi)$

$$G^a(\mathbf{x})|\Psi\rangle = 0 \iff$$
$$\left(\partial_1\frac{\delta}{\delta\phi^a(\mathbf{x})} + f^{abc}\phi^b(\mathbf{x})\frac{\delta}{\delta\phi_c(\mathbf{x})} - ik\partial_2\phi^a(\mathbf{x})\right)\Psi(\phi) = 0. \tag{188}$$

For finding a solution of this equation we use the WKB-method. Our Lagrangean has the form

$$\mathcal{L}_{CS} = \frac{k}{2}\epsilon^{ij}\dot{A}_i^a A_j^a - H_{CS}$$
$$= \frac{k}{2}\dot{A}_1^a A_2^a - \frac{k}{2}\dot{A}_2^a A_1^a \tag{189}$$

$$\Rightarrow kA_2^a \dot{A}_1^a \equiv p\dot{q}. \tag{190}$$

The WKB-eikonal of ordinary quantum mechanics is defined as

$$\psi(q) = e^{i \int^q dq' p(q')}. \tag{191}$$

Analogously, we state for the CS–WKB-eikonal

$$\Psi(\phi) = e^{i \int^{\phi^a} \mathcal{D}A_1 k A_2^a(A_1)}. \tag{192}$$

To get $A_2^a(A_1)$, we use the zero curvature condition for the A-fields

$$\partial_1 A_2^a + f^{abc} A_1^b A_2^c = \partial_2 A_1^a. \tag{193}$$

Now we take a solution for the Gauß' law constraint (188) of the form

$$\Psi(\phi) = e^{iW(\phi)}. \tag{194}$$

In the polarization chosen above we get

$$\left(\partial_1 \frac{\delta}{\delta \phi^a(\mathbf{x})} + f^{abc} \phi^b(\mathbf{x}) \frac{\delta}{\delta \phi_c(\mathbf{x})} - ik\partial_2 \phi^a(\mathbf{x})\right) \Psi(\phi) = 0 \tag{195}$$

$$\Rightarrow \partial_1 \frac{\delta W(\phi)}{\delta \phi^a(\mathbf{x})} + f^{abc} \phi^b(\mathbf{x}) \frac{\delta W(\phi)}{\delta \phi^c(\mathbf{x})} = k\partial_2 \phi^a(\mathbf{x}). \tag{196}$$

Comparing this with (193) gives

$$\frac{\delta W(\phi)}{\delta \phi^a(\mathbf{x})} = k A_2^a(\mathbf{x}). \tag{197}$$

By identifying S from (167) with W and ϕ with A_1 and comparing the constraints (169) of the hard thermal loop analysis and of the CS-theory (193) we make the conclusion that the hard thermal loop generating function is given by the WKB-eikonal of CS-theory [41].

What still remains to do, is to construct the phase W that means solving the 'quantum' constraint. Questions of representation theory of symmetries on quantum states arise here. These are nontrivial and represent a source of anomalies. For the solution of the constraint we use a two-step strategy

(i) determine $e^{i \int_{\mathbf{x}} \lambda^a(\mathbf{x}) G^a(\mathbf{x})} \Psi(\phi)$
(ii) demand $e^{i \int_{\mathbf{x}} \lambda^a(\mathbf{x}) G^a(\mathbf{x})} \Psi(\phi) = \Psi(\phi) \Longleftarrow$ Gauß' law.

Whereas the Gauß' law represents the infinitesimal action of the Lie algebra on the states, (i) and (ii) is the action of the Lie group (λ^a are the gauge parameters). For the exponent we get

$$\int_{\mathbf{x}} \lambda^a G^a = i \int_{\mathbf{x}} \lambda^a \left(\partial_1 \frac{\delta}{\delta \phi^a} + f^{abc} \phi^b \frac{\delta}{\delta \phi^c}\right) - k \int_{\mathbf{x}} \phi^a \partial_2 \lambda^a. \tag{198}$$

We define

$$G \equiv G_\phi + 2k \int_{\mathbf{x}} \mathrm{tr}\, \phi \partial_2 \lambda, \tag{199}$$

where G_ϕ should only transform the argument

$$e^{iG_\phi}\Psi(\phi) = \Psi(\phi^g) \tag{200}$$

with $g = e^\lambda$ an element of the gauge group and

$$\phi^g \equiv g^{-1}\phi g + g^{-1}\partial_1 g. \tag{201}$$

Under gauge transformation the wave functional picks up a phase

$$\Rightarrow e^{iG}\Psi(\phi) = e^{iG}e^{-iG_\phi}\Psi(\phi^g)$$
$$e^{iG}e^{-iG_\phi} = e^{-2\pi i\alpha_1(\phi;g)}. \tag{202}$$

for α_1 one gets [40]

$$\alpha_1(\phi;g) = -\frac{k}{2\pi}\int_\mathbf{x} \text{tr}\left(2\phi\partial_2 gg^{-1} + g^{-1}\partial_1 gg^{-1}\partial_2 g\right) + 4\pi k\int_\mathbf{x} \omega^0(g). \tag{203}$$

The ω arising in α_1 is a total derivative and has the form

$$\partial^\mu \omega_\mu := \omega(g) := \frac{1}{24\pi^2}\epsilon^{\alpha\beta\gamma}\text{tr}\left(g^{-1}\partial_\alpha gg^{-1}\partial_\beta gg^{-1}\partial_\gamma g\right) \tag{204}$$

the ω-term in α_1 represents the winding number of the gauge transformation g, so $\int_\mathbf{x} \omega^0(g)$ is multi-valued, but this is innocuous when CS-quantization condition

$$4\pi k = \text{integer} \tag{205}$$

is fulfilled.

Conclusion

From the quantum mechanical transformation law

$$e^{iG}\Psi(\phi) = e^{-2\pi i\alpha_1(\phi;g)}\Psi(\phi^g)$$
$$= \Psi(\phi) \Longleftarrow \text{Gauß' law} \tag{206}$$
$$\Psi(\phi^g) = e^{2\pi i\alpha_1(\phi;g)}\Psi(\phi) \tag{207}$$
$$|\Psi(\phi^g)|^2 = |\Psi(\phi)|^2 \tag{208}$$

follows that α_1 fulfills the cocycle condition

$$\alpha_1(\phi;g) = \alpha_1(\phi;g\bar{g}) - \alpha_1(\phi^g;\bar{g}) \tag{209}$$

and so is a 1-cocycle [42]. Such objects arise in quantum mechanics, if a symmetry transformation is represented not only by shifting the argument of the wave functions, but also giving them a phase (e.g. quantum mechanical representation of Galileo boosts). The response of the action to this implementation of gauge symmetry is a change by a total derivative

$$L(A^g) - L(A) = \frac{d}{dt}2\pi\alpha_1, \tag{210}$$

what indicates a residual symmetry of the theory.

Solution

This is the explicit construction of states obeying (207). To this end, we write

$$\Psi(\phi) = e^{2\pi i \alpha_0(\phi)} \psi(\phi) \tag{211}$$

and seek a quantity $\alpha_0(\phi)$ called a 0-cochain that satisfies

$$\alpha_0(\phi^g) - \alpha_0(\phi) = \alpha_1(\phi; g). \tag{212}$$

Then (211) solves (207) with gauge invariant $\psi(\phi)$

$$\psi(\phi^g) = \psi(\phi). \tag{213}$$

If (212) holds then the 1-cocycle α_1 is trivial — it is a coboundary. It is possible [40], to construct such an α_0 which trivializes α_1

$$\alpha_0(\phi) = 4\pi k \int_{\mathbf{X}} \omega^0(h) - \frac{k}{2\pi} \int_{\mathbf{X}} \text{tr}\,(\phi h^{-1} \partial_2 h), \tag{214}$$

where h is defined by

$$\phi \equiv h^{-1} \partial_1 h. \tag{215}$$

The wave functional is single-valued provided $4\pi k =$ integer.

The Hilbert space is one-dimensional when no gauge invariant functionals of ϕ can be constructed (e.g. physical plane). For that the explicit physical states are given by

$$\Psi(\phi) = N e^{2\pi i \alpha_0(\phi)} \tag{216}$$
$$\text{with } A_i \Psi(\phi) = h^{-1} \partial_i h \Psi(\phi) \ . \tag{217}$$

References

[1] R. Jackiw, Rev. Mod. Phys. **52** (1980), 661;
    extended version in Relativity, Groups and Topology, Les Houches Session XL (1983), Vol. II, eds. B.S. DeWitt, R. Stora, Elsevier Science Publ. (1984)
[2] T. Muta, *Foundations of quantum chromodynamics*, World Scientific 1987
[3] J.D. Bjorken, S.D. Drell, *Relativistic quantum fields*,McGraw–Hill 1965
[4] P.A.M. Dirac, *The principles of quantum mechanics*, Claredon Press 1930
[5] M. Nakahara, *Geometry, topology, and physics*, Adam Hilger 1990
[6] R. Jackiw, C. Rebbi, Phys. Rev. Lett. **37** (1976) 172
[7] C.G. Callan, R.F. Dashen, D.J. Gross, Phys. Lett. **63B** (1976) 334
[8] S. Coleman, R. Jackiw, L. Susskind, Ann. Phys. (NY) **93** (1975) 267
[9] S. Coleman, Ann. Phys. (NY) **101** (1976) 239
[10] N.S. Manton, Ann. Phys. (NY) **159** (1985) 220
[11] K. Johnson, Phys. Lett. **5** (1963) 253
[12] A.A. Belavin, A.M. Polyakov, A.S. Schwartz, Y.S. Tyupkin, Phys. Lett. **59B** (1975) 85
[13] G. 't Hooft, Phys. Rev. Lett. **37** (1976) 8;
     Phys. Rev. **D12** (1978) 3432

[14] A. M. Polyakov, Gauge Fields and Strings, Harwood Academic Publishers GmbH, Chur, 1987
[15] S. Adler, Phys. Rev. **177** (1969) 2426
[16] R. Jackiw, "Field Theoretic Investigations in Current Algebra", in Current Algebra and Anomalies, editors: S. B. Treiman, R. Jackiw, B. Zumino, E. Witten, World Scientific, 1985
[17] J. Bell and R. Jackiw, Nuovo Cim. **60A** (1969), 47
[18] W. Bardeen, S. Elitzur, Y. Frishman and E. Rabinovici, Nucl. Phys. **B218** (1983) 445
[19] J. Schwinger, Phys. Rev. **128** (1962) 2425
[20] R. Jackiw, in Laws of Hadronic Matter, ed. A. Zichichi (Academic New York) (1975)
[21] S. Glashow, R. Jackiw and S. S. Shei, Phys. Rev. **187** (1969) 1916; S. Weinberg, Phys. Rev. **D11** (1975) 3583
[22] R. Crewther, P. DiVecchia, G. Veneziano and E. Witten, Phys. Lett. **88B** (1979) 123
[23] D. Sutherland, Nucl. Phys. **B2** (1967) 433;
M. Veltman, Proc. Roy. Soc. **A301** (1967) 107
[24] L. H. Ryder, Quantum Field Theory, Cambridge University
[25] N. Christ and R. Jackiw, Phys. Lett. **91B** (1980) 228
[26] R. Efraty, V.P. Nair, Phys. Rev. Lett. **68** (1992) 2891
[27] V.P. Nair, Phys. Rev. **D48** (1993) 3432
[28] J.-P. Blaizot, E. Iancu, Nucl. Phys. **B390** (1993), 589; Phys. Rev. Lett. **70** (1993) 3376
[29] J. Kapusta, *Finite temperature field theory*, Cambridge University Press (1989)
[30] N.P. Landsman, C.G. van Weert, Phys. Rep. **145** (1987) 141
[31] H. Umezawa, H. Matsumoto, M. Tachiki, *Thermo field dynamics and condensed states*, North Holland, Amsterdam (1982)
[32] I. Ojima, Ann. Phys. (NY) **137** (1981) 1
[33] E. Braaten, R.D. Pisarski, Nucl. Phys. **B337** (1990) 569;
Nucl. Phys. **B339** (1990) 310;
Phys. Rev. **D45** 1827
[34] R. Jackiw, V.P. Nair, Phys. Rev. **D48** (1993), 4991
[35] E.M. Lifshitz, L.P. Pitaevski, *Physical kinetics*, Pergamon Press, Oxford (1980)
V.P. Silin, Sov. Phys. JETP **11** (1960) 1136
[36] L. Dolan, R. Jackiw, Phys. Rev. **D9** (1974) 3320
[37] J.C. Taylor, S.M.H. Wong, Nucl. Phys. **B346** (1990) 115
[38] R. Jackiw, in *Gauge Theories of the Eighties*, Proc. Arctic School of Physics 1982, eds. R. Raitio and J. Lindfors, Lecture Notes in Physics **181**, Springer, Berlin (1983)
S. Templeton, R. Jackiw, Phys. Rev. **D23** (1981) 2291
J. Schonfeld, Nucl. Phys. **B185** (1981) 157
S. Deser, S. Templeton, R. Jackiw, Phys. Rev. Lett. **48** (1982), 975; Ann. Phys. **140** (1982) 372
[39] E. Witten, Commun. Math. Phys. **117** (1989) 353
[40] G.V. Dunne, R. Jackiw, C.A. Trugenberger, Ann. Phys. (NY) **194** (1989) 197
[41] R. Jackiw, *Physics, Geometry and Topology*, Plenum Press (1990), H. Lee ed.
[42] R. Jackiw, in: *Conformal field theory, anomalies and superstrings*, Proc. of the first Asia Pacific workshop on high energy physics, Singapore 1987, eds. C.K. Chew et. al., World Scientific, Singapore (1988)

Semiclassical Aspects of Quantum Field Theories*

L. O'Raifeartaigh[1];
Notes by M. Engelhardt[2] and S. Lenz[2]

[1] Dublin Institute for Advanced Studies, School of Theoretical Physics, 10 Burlington Road, Dublin 4, Ireland
[2] Institute for Theoretical Physics III, University of Erlangen–Nürnberg, Staudtstr. 7, 91058 Erlangen, Germany

1 Introduction

A quantum field theory is said to have an anomaly if a symmetry of the classical system is lost by quantization. A well known example is the axial anomaly in QED, where both the vector current and the axial vector current are conserved in the classical theory with massless fermions. After quantization, one of these symmetries is lost, which is generally attributed to the necessity of regularizing ultraviolet infinities. However, the concept of anomalies is more general and fundamental, as will be shown in the second part of this lecture.

Semiclassical methods are ideally suited for studying anomalies, as all known anomalies already occur on one loop level, i.e. to order $\hbar$, and in most cases higher order corrections do not modify the result. An example for this is again the axial anomaly in QED, where the leading graph to the neutral pion decay is anomalous. From this one loop graph the anomalous contribution to the axial current can be derived. Another advantage of the semiclassical approach to quantum field theories is that it is nonperturbative, as it is an expansion in the number of loops and not in powers of the coupling constant. Since non trivial classical solutions of field theories such as solitons and instantons are known, semiclassical expansion around these solutions can be used in order to gain nonperturbative results for quantum field theories. Important examples are 't Hoofts baryon decay via instantons and the formation of θ vacua by tunneling between classical vacua in Yang Mills theories. Another application of semiclassical methods is to check whether spontaneous symmetry breakdown, which is in general derived on a classical level, is stable under quantum corrections.

In the first section of this report, semiclassical methods will be reviewed. The starting point is standard time dependent WKB, where all the concepts can be developed without the technical complications of quantum field theories. It will be shown that the semiclassical term (WKB wave function in QM and the

---

* Lectures presented at the workshop "TOPICS in Field Theory" organised by the Graduiertenkolleg Erlangen–Regensburg, held on October 10th–12th, 1993 in Kloster Banz, Germany

1-loop contribution in field theory) is rather peculiar in the sense that it is the first quantum correction, but nevertheless can be expressed in terms of classical objects. In the following two sections the semiclassical treatment of anomalies and their physical consequences are discussed.

2 The Semi-Classical Approach

2.1 Introduction

Besides variational calculations and perturbation theory, the semiclassical approach is a standard approximation scheme in quantum mechanics. Semiclassical methods have been succesfully applied to numerous problems like e.g. high energy scattering (Glauber theory), tunneling problems (α-decay) and WKB calculations of bound state wave functions. For quantum field theories, semiclassical expansion is particularly useful, as it is a nonperturbative method.

The formulation of semiclassical approximations for quantum field theories and time dependent WKB in quantum mechanics is very similar. Therefore the next two sections review time dependent WKB to show the basic concepts. The results are then generalized to field theory and some applications are given.

2.2 The WKB Approximation

Starting from the Schrödinger equation

$$i\hbar \frac{\partial}{\partial t}\psi = -\frac{\hbar^2}{2m}\nabla^2\psi + V(x)\psi \tag{1}$$

and using the polar form of the wave function

$$\psi = \rho^{(1/2)} \exp \frac{i}{\hbar} S \tag{2}$$

one can show [1] that ρ and S obey the following (real) equations:

$$\frac{\partial S}{\partial t} + \frac{1}{2m}(\nabla S)^2 + V(x) = \frac{\hbar^2}{2m}\frac{\nabla^2 \rho^{(1/2)}}{\rho^{(1/2)}} \tag{3}$$

and

$$\hbar(\frac{\partial \rho}{\partial t} + \nabla \mathbf{j}) = 0, \quad \text{where } \mathbf{j} = \frac{1}{m}\rho(\nabla). \tag{4}$$

Equation (3) is the classical Hamilton Jacobi equation with a quantum correction term of the order $\hbar^2$. Equation (4) is the continuity equation and is independent of $\hbar$.

In semiclassical approximation (i.e. to order $\hbar$) the two equations decouple. S is obtained by solving the classical Hamilton Jacobi equation. The continuity equation can be solved for ρ once S is known. From the fact that quantum corrections in (3) are of the order $\hbar^2$, one sees that also the $\hbar$ term in the semiclassical wavefunction is essentially classical. One finds

$$\rho = \det(-\frac{\partial^2 S}{\partial x_i \partial a_j}) \tag{5}$$

if one imposes the boundary condition that the particle is located at position a at time $t = 0$. This equation is particularly easy to prove in one dimension. If one differentiates (3) once with respect to the initial point a and once with respect to x one finds

$$\partial_t(\frac{\partial^2 S}{\partial x \partial a}) + \frac{1}{m}\partial_x(\frac{\partial^2 S}{\partial x \partial a}\frac{\partial S}{\partial x}) = 0$$

which is just the continuity equation with $\rho \propto \partial_a \partial_x S$. The normalization is fixed by the requirement that

$$\psi(x,a,t)|_{t=0} = \delta(x-a) \tag{6}$$

A proof of (5) for more than one dimension is given in appendix A.

The WKB wavefunction is exact, if the action S is a quadratic functional. Three examples for this are:

– the free particle of mass m in d dimensions:

$$\psi^{free}(x,a,t) = \left(\frac{m}{2\pi i \hbar t}\right)^{(d/2)} \exp\left[\frac{i}{\hbar}\frac{m}{2t}(x-a)^2\right] \tag{7}$$

– a particle of mass m bound in a harmonic oscillator potential of frequency ω:

$$\psi^{ho}(x,a,t) = \left(\frac{m\omega}{2\pi i \hbar \sin(\omega t)}\right)^{(d/2)} \exp\left[\frac{i}{\hbar}\frac{m\omega}{2\sin(\omega t)}((x^2+a^2)\cos(\omega t) - 2xa)\right] \tag{8}$$

– a particle in three dimensions in a constant magnetic field $\mathbf{B} = B\mathbf{e}_3$

$$\psi^B(x,a,t) = \left(\frac{m}{2\pi i \hbar t}\right)^{1/2} \frac{\sigma}{2\pi i \hbar \sinh \sigma t} \exp\left[\frac{im}{\hbar 2}[\frac{\sigma}{\sinh \sigma t}(\cosh \sigma t \sum_{i=1}^{2}(x-a)_i^2\right.$$
$$\left.+(x \times a)_3 \sinh \sigma t) + (x-a)_3^2]\right] \tag{9}$$

$$\sigma = \frac{eB}{2cm} \ . \tag{10}$$

The classical meaning of (5) can be understood if one remembers that the derivative of the classical action with respect to the initial point of the trajectory

$$p_i := -\frac{\partial S}{\partial a_i} \tag{11}$$

is the initial momentum of the particle [2]. Therefore the inverse determinant

$$\rho^{-1} = \det(\frac{\partial}{\partial x_i} p_j)^{-1} = \det(\frac{\partial}{\partial p_i} x_j)$$

converts any small distribution in initial momentum $\Delta^3 p(0)$ into a final distribution in position $\Delta^3 x$ according to

$$\Delta^3 x(t) = \rho^{-1}(t) \Delta^3 p(0). \tag{12}$$

In other words: ρ^{-1} measures the spread of the endpoints of a trajectory, if the initial momentum is varied.

2.3 Functional Integrals

The form of the WKB wavefunction can also be derived from a functional integral approach. Starting from the path integral for the quantum mechanical wavefunction

$$\psi(x, a, t) = N^{-1} \int_a^x d(y) \exp(\frac{i}{\hbar} S) \tag{13}$$

$$S = \int_0^t d\tau (\frac{m}{2}(\dot{y}, \dot{y}) - V(y(\tau))) \tag{14}$$

one expands the integration paths around the classical solution y_c

$$y(\tau) = y_c(\tau) + \hbar^{(1/2)} \eta(\tau)$$

keeping only second order terms in η. The wavefunction is

$$\psi(x, a, t) = \exp(\frac{i}{\hbar} S_c) N^{-1} \int d[\eta] \exp(i \frac{m}{2} \tilde{S}(\eta)) \tag{15}$$

where $\tilde{S}$ is a quadratic functional

$$S(\tilde{\eta}, t) = \int_0^t d\tau ((\dot{\eta}, \dot{\eta}) - (\eta, U(\tau)\eta)) \tag{16}$$

$$U_{ij} = \frac{1}{m} \frac{\partial^2}{\partial y_i \partial y_j} V(y). \tag{17}$$

N is a potential independent normalization factor.

Note that only periodic orbits which start and end at $x = 0$ contribute to (15). The integral term measures the quantum mechanical fluctuations around the classical path. For quadratic actions, the expansion in η terminates and ψ is the exact wave function. It will shown in the following, that

$$N^{-1} \int d[\eta] \exp i\tilde{S}[\eta] = \rho^{1/2} \tag{18}$$

which means that the WKB approximation and stationary path approximation in the path integral are equivalent. This result implies that the quantum fluctuations in (15) can be expressed through purely classical objects. The factor ρ in the WKB wavefunction is connected with the quantum fluctuations around the classical path. Nevertheless it is a purely classical object, as we saw in the last section. From equation (18) one gets:

$$\frac{\rho^{1/2}}{\rho_0^{1/2}} = \frac{\int d(\eta) \exp(-i\frac{m}{2} \int d\tau(\eta, (\partial_t^2 + U(\tau))\eta))}{\int d(\eta) \exp(-i\frac{m}{2} \int d\tau(\eta, \partial_t^2 \eta))} \qquad (19)$$

$$= \frac{\det(\partial_t^2 + U(\tau))^{-1/2}}{\det(\partial_t^2)^{-1/2}} \qquad (20)$$

$$= \frac{(\prod \epsilon_r)^{-1/2}}{(\prod \epsilon_r^0)^{-1/2}} \qquad (21)$$

where the generalization of the finite dimensional formula

$$\int d^n x \exp(-(x, Ax)) = \pi^{n/2} (\det A)^{-1/2}$$

has been used. The spectrum of the operator $\partial_t^2 + U(t)$ is discrete if one imposes the boundary conditions $\eta(0) = \eta(t) = 0$ on a finite interval $[0,t]$. We now have to prove that

$$\frac{\det(\partial_t^2 + U(\tau))^{-1}}{\det(\partial_t^2)^{-1}} = \det(-\frac{\partial^2 S}{\partial x \partial a})/\det(-\frac{\partial^2 S_0}{\partial x \partial a}). \qquad (22)$$

One method to show this is the following: Consider the related differential equation

$$[\partial_t^2 + U(t)]_{ij} f_j(t,z) = z f_i(t,z) \qquad (23)$$

with the boundary conditions

$$f_i(0,z) = 0 \qquad \partial_t f_i(t,z)|_{t=0} = c_i \qquad (24)$$

where z is any complex number and c_i any real nonzero vector. Obviously the f_i are entire functions of z. Furthermore the f_i are eigenvectors of $\partial_t^2 + U(t)$ whenever z is such that $f_i(t,z) = 0$. These eigenvectors satisfy the boundary condition $f_i(0,z) = f_i(t,z) = 0$ of (21). Since the converse is also true, the determinant of the linearly independent solutions of (23) must be of the form

$$\det f_i^{(k)}(t,z) = g(t,z) \prod_r (\epsilon_r - z)$$

where $g(t,z)$ is an entire function without zeros. This is also true for $U = 0$ therefore

$$G(t,z) := \frac{g_0(t,z)}{g(t,z)} = \frac{\det f_{0i}^{(k)}}{\det f_i^{(k)}} \prod_k \frac{\epsilon_r - z}{\epsilon_r^0 - z} \qquad (25)$$

is an entire function. If the potential U is smooth on the interval $[0,t]$, (23) and its solutions become independent of U at large $|z|$. If the same bondary conditions (24) are used for free and interacting solution, these functions must approach each other at $|z| \to \infty$ and the determinant factor must tend to unity. Similarly, if one assumes that the spectrum ϵ_r is such that one can interchange the limits of forming the $\prod$ and $|z| \to \infty$, also the $\prod$ term tends to untity. Under these cirumstances, one sees that $G(z) \to 1$ as $|z| \to \infty$ and hence by Liouville's theorem $G(z) = 1$ for all z. Setting $z = 0$ one finds that the ratio of determinants (21) is related to the zero mode eigenvectors of $\partial_t^2 + U(t)$:

$$\frac{\det f_i^{(k)}}{\det f_{0i}^{(k)}} = \frac{\prod \epsilon_r}{\prod \epsilon_r^0} . \tag{26}$$

To complete the proof of (22), one has to relate this eigenvectors to the classical solutions of the problem. Consider two solutions of the classical equation of motion

$$m\partial_t^2 x_i^c(t) = -\frac{\partial V(x^c)}{\partial x_i^c}$$

x^c and $x^c + \delta x^c$ which differ by a variation δx^c. Then

$$m\partial_t^2 \delta x_i^c(t) = -\frac{\partial^2 V(x^c)}{\partial x_i^c \partial x_j^c} \delta x_j^c(t) = -U(x^c)_{ij} x_j^c(t).$$

Since the two paths satisfy the classical equation of motion, they can only differ in their initial conditions:

$$\delta x_i^c(t) = \frac{\partial x_i^c(t)}{\partial x_j^c(0)} \delta x_j(0) + \frac{\partial x_i^c(t)}{\partial \dot x_j^c(0)} \delta \dot x_j^c .$$

As we are looking for a solution with the boundary condition $f_i(0) = 0$, one has to set $\delta x_i(0) = 0$, thus the solutions with the required boundary conditions are

$$f_i^{(k)}(t) = \frac{\partial x_i^c(t)}{\partial \dot x_j^c(0)} c_j^{(k)} = \frac{\partial x_i^c(t)}{\partial p_j^c(0)} c_j^{(k)}, \tag{27}$$

which, together with (26), proves (22). Obviously the semiclassical contribution to the WKB wave function can be written in terms of classical quantities.

From the wave function with the boundary condition (6) the quantum mechanical partition function $Z(\beta)$ can be calculated:

$$Z(\beta) := \mathrm{Tr} e^{-\beta H} \tag{28}$$

$$= \int dx \langle x | e^{-\beta H} | x \rangle \tag{29}$$

$$= \int dx \psi(x, x, -i\beta) . \tag{30}$$

If e.g. the WKB result for the harmonic oscillator is inserted in this formula, one finds

$$Z_{HO}(\beta) = \left(\frac{m\omega}{2\pi\hbar\sinh(\omega\beta)}\right)^{(d/2)} \int dx \exp\left(-\frac{m\omega}{2\hbar\sinh(\omega\beta)}(\cosh(\omega\beta)-1)2x^2\right). \quad (31)$$

The term in front of the integral is the $O(\hbar)$ contribution, the integral is the classical action. If the integral in (31) is evaluated one finds that

$$Z_{HO}(\beta) = \left(\frac{m\omega}{2\pi\hbar\sinh(\omega t)}\right)^{d/2}\left(\frac{2\pi\hbar\sinh(\omega t)}{m\omega}\right)^{d/2}\left(\frac{1}{2(\cosh\omega\beta-1)}\right)^{d/2} \quad (32)$$

i.e., the result is $\hbar$ independent and classical and semiclassical contributions cancel. This again demonstrates the classical nature of the $O(\hbar)$ term.

2.4 Field Theory

The generalization of the discussion of the previous section to field theory is straightforward. The interesting object here is the generating functional

$$Z(j) := N^{-1}\int d[\phi(x)]\exp\frac{i}{\hbar}\left(S(\phi(x),\partial_x\phi(x)) + \int dxj(x)\phi(x)\right). \quad (33)$$

For illustration an action of the form

$$S(\phi,\partial\phi) = \int d^4x[(\partial_\mu\phi)^2 + V(\phi(x))] \quad (34)$$

is used, ϕ represents many fields. As for quantum mechanics, the generating functional is expanded around the classical solution of

$$\Box\phi_c(x) + V'(\phi_c(x)) = j(x)$$

with appropriate boundary conditions. The crucial role of the boundary conditions will be illustrated in the applications. One introduces new fields η

$$\phi(x) = \phi_c(x) + \hbar^{1/2}\eta(x)$$

and keeps only second order terms in the exponential to derive the WKB form of the generating functional

$$Z_{WKB}(j) = e^{\frac{i}{\hbar}S_c(j)}N^{-1}\int d[\eta(x)]\exp(i\int dxdx'\eta(x')\frac{\delta^2 S}{\delta\phi(x)\delta\phi(x')}\eta(x) \quad (35)$$

$$= e^{\frac{i}{\hbar}S_c(j)}\tilde{N}^{-1}\left(\det\frac{\delta^2 S}{\delta\phi(x)\delta\phi(x')}\right)^{-1/2}. \quad (36)$$

For an action of the form (34), this can be rewritten:

$$Z_{WKB}(j) = e^{\frac{i}{\hbar}S_c(j)} N^{-1} \int d[\eta(x)] \exp(\frac{i}{\hbar} \int dx \eta(-\Box + V''(\phi_c(x)))\eta \quad (37)$$

$$= e^{\frac{i}{\hbar}S_c(j)} \left(\frac{\det(-\Box + V''(\phi_c(x)))}{\det(-\Box + V''(0))} \right)^{-1/2} . \quad (38)$$

If the normalization N is properly defined. Note that ϕ_c is a functional of the source j, therefore the determinant term of the generating functional is also j dependent. Note also the similarity of this expression with the quantum mechanical form eqs.(2), (15) and (19).

In order to calculate the fraction in Z_{WKB}, a regularization procedure has to be given for the determinants. The determinant of an arbitrary operator D can be written as

$$\det D = \exp \ln \det D = \exp \text{Tr} \ln D.$$

The problem is that $\ln D$ is not trace class in general, which makes the above formula meaningless. A solution of this problem is to define the ζ-function of the operator D

$$\zeta(D, s) := \text{Tr} D^{-s}$$

which exists for large enough values of s (for the operators we are interested in). If $\ln D$ is trace class, one finds that

$$\text{Tr} \ln D = -\left.\frac{d}{ds}\zeta(D, s)\right|_{s=0} .$$

For all other operators the above equation can be used to define the lefthand side by analytic continuation of the ζ function from values of s where it exists to $s = 0$. Using the ζ function definition of the determinants, equation (38) reads

$$\ln Z_{WKB}(j) = \frac{i}{\hbar} S_c(j) + \frac{\hbar}{2i} \left[\zeta'(-\Box + V''(\phi_c)) - \zeta'(-\Box + V''(0))\right]_{s=0} \quad (39)$$

From (36) the effective action in one loop approximation can be derived. The connected generating functional $W(j)$ is defined as

$$Z(j) = e^{iW(j)}$$

and is related to the generating functional of the proper vertices. The effective action is defined as the Legendre transform of $W(j)$ i.e.

$$S_{eff}(\phi) = W(j) - \int d^4x \phi j \quad, \text{where} \quad \phi = \frac{\partial W}{\partial j} .$$

It follows from (36) that the effective action to order $\hbar$, i.e. to one loop order, is

$$S_{eff}(\phi) = S(\phi) + \frac{i\hbar}{2} \ln \det \frac{\delta^2 S}{\delta\phi(x)\delta\phi(x')}. \quad (40)$$

A pedagogical discussion of this can be found in [3], [4], [5]. Equation (40) is the central result of this section. From S_{eff} the proper vertices to one loop order can be calculated, if the determinant is regularized properly. The anomalies, which will be discussed in the next section, emerge from the $O(\hbar)$ term in the effective action.

2.5 Expansions About Non-trivial Configurations

Solitons: For systems with degenerate classical vacua there are topologically nontrivial classical solutions which can be interpreted in terms of a tunneling amplitude. A simple quantum mechanics example is described by the Langrange function

$$L = \frac{1}{2}(\dot\phi(t)^2 - (c^2 - \phi(t)^2)^2).$$

Classically there are two states of lowest energy $\phi(t) = c$ and $\phi(t) = -c$. Classical solutions with a finite action have to approach one of these solutions as $t \to \pm\infty$. The choice $\phi(-\infty) = -c$ and $\phi(\infty) = c$ is called an instanton solution. It has a nonzero but finite classical action. This can be seen by rewriting the Langrangean

$$L = \frac{1}{2}(\dot\phi - \phi^2 + c^2)^2 + \dot\phi(\phi^2 - c^2) \tag{41}$$

$$= \frac{1}{2}(\dot\phi - \phi^2 + c^2)^2 + \frac{d}{dt}(\frac{1}{3}\phi^3 - c^2\phi) \tag{42}$$

as the sum of a square, which vanishes for the classical solution, as can be seen from the equation of motion, and a total derivative, which gives a nonvanishing classical action for the soliton solution

$$S_c = \int_{-\infty}^{\infty} \frac{d}{dt}(\frac{1}{3}\phi^3 - c^2\phi) = -\frac{4}{3}c^3 =: Q.$$

Note that the explicit form of the classical solution has not been used so far. The WKB generating functional for $j = 0$ is then

$$Z_{WKB} = \exp \frac{i}{\hbar}QN^{-1}\int d[\eta] \exp i\int dt\eta(t)(-\partial_t^2 + 2c^2 - 6\phi_c(t))\eta(t).$$

In Euclidean spacetime the soliton can be interpreted as a tunneling amplitude between states with $\phi = \pm c$.

Instantons: An analogous situation is found in 4-dimensional Yang-Mills theories. The Yang-Mills-Fermion Lagrangian is

$$\mathcal{L}_{YM} = \frac{1}{2}(\mathbf{E}^2 - \mathbf{B}^2) + \bar\psi(i\not{D} + m)\psi \tag{43}$$

$$= \frac{1}{2}(\mathbf{E} - \mathbf{B})^2 + \mathbf{EB} + \bar\psi(i\not{D} + m)\psi . \tag{44}$$

The second term is the derivative of a current S^μ

$$\mathbf{EB} = \partial_\mu S^\mu \tag{45}$$

Thus, as in (44), the pure Yang-Mills Lagrangian is the sum of a divergence (which is topological and does not contribute to to the equations of motion) and a quadratic term. It follows that one class of classical solutions is when the quadratic term is zero or $\mathbf{E}_c = \mathbf{B}_c$. The important point is that this class of solutions is non-trivial because non-abelian groups can be mapped non-trivially onto the sphere at infinity and hence the integral

$$\nu := \int d^4x\, \partial_\mu S^\mu$$

does not necessarily vanish, indeed can be an arbitrary integer, and it is clear that for $\nu \neq 0$ the solution is necessarily non-trivial. From the point of view of the fermions the WKB expansion about such a non-trivial solution is

$$Z_{WKB} = \exp\frac{i}{\hbar}\nu N^{-1} \int d[\bar\psi\psi] \exp\frac{i}{\hbar} \int d^4x\, \bar\psi(i\slashed{\partial} + m + \slashed{A}_c)\psi . \tag{46}$$

The additional term $\slashed{A}_c$ in the fermion action couples the fermions to the instanton. It is responsible for t'Hoofts baryon decay via electroweak instantons [6], [7].

For future reference it may be worth noting that if we define the charge

$$Q(t) = \int_t d^3x\, S_o \tag{47}$$

we have

$$\nu = Q(\infty) - Q(-\infty) . \tag{48}$$

Thus, as in the 1-dimensional case, the instanton solutions can be interpreted as an indication that in the quantum theory there is tunnelling. Here the tunnelling takes place between different vacua $|n>$ characterized by $Q|n> = n|n>$. Furthermore if we define global gauge transformations as those that are generated by group elements G that map non-trivially onto the sphere at infinity, we have

$$GQG^{-1} = Q + \nu . \tag{49}$$

This shows that Q and hence the $|n>$-vacua are not gauge-invariant with respect to the global gauge-transformations, and that to obtain truly gauge-invariant vacua one must choose combinations of the form

$$|\theta> = \sum_n e^{in\theta}|n> . \tag{50}$$

The θ-vacua obtained in this way are the physical vacua.

Spontaneous Symmetry Breakdown Another application of (36) with $j = 0$ can be given for systems with spontaneous symmetry breakdown. If $V(\phi)$ has a minimum $\phi_0 \neq 0$, the classical ground state is $\phi(x) = \phi_0 = const$. If one expands the field configurations around these solutions one finds

$$S_c = V(\phi_0) \int d^4x = V(\phi_0)\Omega$$

and

$$Z = e^{\frac{i}{\hbar}\Omega V(\phi_0)} N^{-1} \int d[\eta] \exp\frac{i}{\hbar}(\int d^4x\eta(-\Box + V''(\phi_0))\eta + O(\eta^3)). \tag{51}$$

V'' is the true mass matrix. It can have vanishing eigenvalues which correspond to massless Goldstone modes. Higher derivatives of V can be interpreted as couplings between the fields η. If all higher derivatives are kept in (51), Z is the exact generating functional of the problem. It describes a system of interacting fields η. The Feynman diagrams of this theory can be calculated by introducing a new source term $\int dx \eta j$, which couples to the new field.

Effective Potentials As a last application the Coleman-Weinberg formula is derived with the help of the ζ function regularization. This time we take $j = const$ and $\phi_c = const$. The classical equation of motion becomes

$$V'(\phi_c) = j \tag{52}$$

and the classical action is

$$S_c = \int d^4x V(\phi_c) = \Omega V(\phi_c). \tag{53}$$

With the help of (39) one finds

$$\ln Z_{1-loop} = \frac{i}{\hbar}\Omega V(\phi_c) + \frac{\hbar}{2i}\left[\zeta'(-\Box + V''(\phi_c)) - \zeta'(-\Box + V''(0))\right]_{s=0}. \tag{54}$$

The ζ function term for a constant ϕ_c can be evaluated explicitly [3]:

$$\zeta(-\Box + V'', s) := Tr(-\Box + V'')^{-s} \tag{55}$$

$$= \int d^4k \frac{1}{(k^2 + V'')^s} \tag{56}$$

$$= \pi^2 \frac{(V'')^{2-s}}{(s-1)(s-2)}. \tag{57}$$

Differentiating this with respect to s and setting $s = 0$ yields:

$$\det(-\Box + V'') = \pi^2((\frac{1}{2}V'')^2 \ln V'' - \frac{3}{4}(V'')^2).$$

The effective potential becomes (up to an infinite, but field independent constant):

$$V_{eff}(\phi_c) = V(\phi_c) + \frac{i\hbar}{2\Omega}(\frac{1}{2}(V''(\phi_c))^2 \ln V''(\phi_c) - \frac{3}{4}(V'')^2).$$

This is the Coleman–Weinberg potential. Its relevance in the framework of spontaneous symmetry breaking is discussed in [6] (Chap. 6.4).

3 Anomalies [8]

An anomaly is said to occur in a quantum theory when a formal symmetry of the corresponding classical theory is not realized in the quantized version. Often one thinks of anomalies in terms of specific examples such as the axial anomaly in QED. This anomaly is usually attributed to the fact that regularizing the ultraviolet infinities in a gauge-invariant manner introduces a modified definition of the axial current which is gauge invariant but no longer commutes with the Hamiltonian (this will be elaborated upon in greater detail below). In other words, the conservation of the axial current, valid on the classical level, is broken in the quantum theory. This and other examples of anomalies may lead one to think that they are specific to field theories and somehow connected with the infinities that arise in such theories.

A closer look reveals, however, that anomalies have a much more fundamental nature; they can occur quite generally as a consequence of the quantization procedure itself, i.e. the representation of the Poisson bracket algebra in terms of self-adjoint operators on a Hilbert space. Particularly the need to specify the latter may in general introduce the breaking of a symmetry, as will be shown in the next section.

All known examples of anomalies already occur in the semiclassical approximation, i.e. at the one-loop level, and most anomalies are not modified by higher-loop effects (cf. in the case of the axial anomaly the Adler-Bardeen theorem, [9]). They are therefore essentially semiclassical phenomena and thus ideally suited for treatment by the semiclassical methods described in the previous sections. Typically, anomalous contributions will be contained in analogues of (40) which gives the effective action induced by quantum fluctuations, and can be evaluated e.g. by ζ-function methods, as will be elaborated upon further below.

3.1 Quantum Mechanical Examples

The simplest, trivial, example of breakdown of a symmetry because of quantization is the free particle, i.e. the Hamiltonian

$$H = \frac{p^2}{2} . \tag{58}$$

Classically, this Hamiltonian is invariant under translations in the coordinate x. Quantization occurs formally by substituting $p^2 \to -\hbar^2 \partial_x^2$, and specifying the Hilbert space on which this operator is to act. If one chooses $L_2(-\infty, \infty)$, translational symmetry is preserved, i.e.

$$e^{i\alpha p} x e^{-i\alpha p} = x + \hbar \alpha \quad \text{for all } \alpha \ . \tag{59}$$

If one, however, chooses the Hilbert space to be the space of functions with a finite period, e.g. $L_2(-\pi, \pi)$, then the full translational symmetry is lost. Only a lattice symmetry remains, i.e. the parameter α above is constrained to be an integer multiple of $2\pi/\hbar$.

A not quite so trivial example is the supersymmetric harmonic oscillator:

$$Q = \begin{pmatrix} 0 & p - i\hbar\omega x \\ p + i\hbar\omega x & 0 \end{pmatrix} \tag{60}$$

$$H = Q^2 = \begin{pmatrix} p^2 + (\hbar\omega)^2 x^2 + \hbar\omega & 0 \\ 0 & p^2 + (\hbar\omega)^2 x^2 - \hbar\omega \end{pmatrix} \ . \tag{61}$$

Formally, there is a parity symmetry:

$$Q(x) = \begin{pmatrix} 0 & 1 \\ 1 & 0 \end{pmatrix} Q(-x) \begin{pmatrix} 0 & 1 \\ 1 & 0 \end{pmatrix} \ . \tag{62}$$

However, Q is selfadjoined and thus $H \geq 0$. Furthermore there is a unique ground state with zero energy and it takes the form

$$\psi_0 = \begin{pmatrix} 0 \\ e^{-\hbar\omega x^2/2} \end{pmatrix} \ . \tag{63}$$

This breaks parity symmetry, since it is not invariant under

$$\psi_0 \to \begin{pmatrix} 0 & 1 \\ 1 & 0 \end{pmatrix} \psi_0 \ . \tag{64}$$

All excited states can be shown to occur in parity doublets. In the limit $\hbar \to 0$ the asymmetry vanishes. This demonstrates its quantum mechanical nature.

A very similar behaviour is exhibited by the Dirac operator

$$\slashed{D} = \begin{pmatrix} 0 & -D_t + \sigma \cdot D \\ -D_t - \sigma \cdot D & 0 \end{pmatrix} \quad \slashed{D}^\dagger = -\slashed{D} \tag{65}$$

(here, the chiral representation for the γ-matrices was used), which can serve to define the Hamiltonian

$$H = -\slashed{D}^2 = \begin{pmatrix} -D^2 - \sigma \cdot (B + E) & 0 \\ 0 & -D^2 - \sigma \cdot (B - E) \end{pmatrix} \ . \tag{66}$$

Classically, and formally, one has time reversal symmetry:

$$D(t, A_0) = \gamma_0 D(-t, -A_0) \gamma_0^{-1} \quad \text{with } \gamma_0 = \begin{pmatrix} 0 & 1 \\ 1 & 0 \end{pmatrix} \ . \tag{67}$$

However, since $\rlap{/}{D}^\dagger = -\rlap{/}{D}$ we have $H \geq 0$. Furthermore the ground states take the form

$$\psi_0 = \begin{pmatrix} 0 \\ \phi_0 \end{pmatrix} \tag{68}$$

and therefore break the symmetry. This is in fact the origin of the axial anomaly, since the multiplicity of ground states, i.e. of zero modes of $\rlap{/}{D}$, is given by the Atiyah-Singer index

$$N_{\psi_0} \propto \int {}^*F^{\mu\nu} F_{\mu\nu} \tag{69}$$

which is precisely the additional anomalous term entering the axial charge of QED[1]. Thus it becomes clear that also the axial anomaly is a general consequence of the structure of the Dirac operator in conjunction with the requirement of anti-hermiticity of $\rlap{/}{D}$, or equivalently the positivity of the Hamiltonian. It occurs independently of second quantization, which ultimately leads to the construction of the Dirac sea, the need to regularize ultraviolet divergences, etc., i.e. the path by which the anomaly is usually derived.

3.2 Anomalies in Quantum Field Theory

Despite this fundamental origin of anomalies, in physical applications they occur indeed most prominently in the framework of quantum field theory. There, one may in particular ask how they arise systematically in the path integral formalism, where the fundamental object is the generating functional

$$Z[J] = \int [d\phi] e^{S(\phi, \partial\phi) + J \cdot \phi} . \tag{70}$$

Since the action in the exponent is classical, it respects all symmetries present at the classical level. Quantization is introduced in this framework by the sum over histories. Therefore, the breaking of symmetries can only occur in the integration measure $[d\phi]$, as was first realized by Fujikawa [10]. Usually, one would choose $[d\phi]$ to respect the symmetries of the classical action S; however, under certain circumstances, this may not be possible, in particular when there is more than one symmetry. Examples of this are massless QED, which contains Poincaré, gauge, and chiral symmetry, and also general relativistic theories which contain general coordinate and Weyl invariance. These examples will be treated in detail below. Anomalies occur when there is no measure which respects all the classical symmetries. In this case one must decide which of the competing symmetries is to be regarded as more fundamental and determine the measure such that it respects those. If any of the remaining symmetries are broken by the measure there is an anomaly.

---

[1] Note that analogous considerations enter e.g. Levinson's theorem, in which one counts the number of bound states to obtain the difference in scattering phase shifts at zero energy and infinite energy, respectively; also, in the Aharonov-Bohm effect, the winding number of an electron path around the solenoid enters the phase acquired by the electron wave function in the detection plane.

Poincaré vs. Chiral Symmetry

The action

$$S = \int \bar{\psi}(\slashed{\partial} + ie\slashed{A})\psi \tag{71}$$

is Poincaré invariant and chirally invariant. If one, however, wanted to introduce a mass term, it would break one of these symmetries: The combination $m\psi^\dagger\psi$ is only chirally invariant, whereas $m\bar{\psi}\psi$ is only Poincaré invariant. Since Poincaré invariance is regarded as the more fundamental symmetry, one usually introduces the former. In precisely the same spirit one chooses $[d\bar{\psi}][d\psi]$ as the path integral measure, and not $[d\psi^\dagger][d\psi]$ and thus violates chiral invariance.

General Relativity

The presence of a gravitational field can, according to General Relativity, be expressed as a curvature in the space-time [3]. Thus the actions of the scalar, the gauge, and the Dirac field in four dimensions, and the string action in two dimensions are generalized, respectively, to

$$S_\phi = \int d^4x \, \frac{1}{\sqrt{g}} \left(g^{\alpha\beta} \partial_\alpha \phi \partial_\beta \phi + \frac{R}{6}\phi^2 \right) \tag{72}$$

$$S_A = \int d^4x \, \frac{1}{\sqrt{g}} g^{\mu\nu} g^{\alpha\beta} F_{\mu\alpha} F_{\nu\beta} \tag{73}$$

$$S_F = \int d^4x \, \frac{1}{\sqrt{g}} \bar{\psi} h_a^\mu \gamma^a D_\mu \psi \tag{74}$$

$$S_X = \int d^2x \, \frac{1}{\sqrt{g}} g^{\alpha\beta} \partial_\alpha X^d \partial_\beta X^d \tag{75}$$

where the transformation matrix between flat and curved coordinates (the local vierbein)

$$h_a^\mu(x) = \frac{\partial \xi_a}{\partial x_\mu}, \tag{76}$$

the corresponding metric tensor

$$g^{\mu\nu}(x) = \eta^{ab} h_a^\mu(x) h_b^\nu(x), \tag{77}$$

and its determinant $g = \det g^{\mu\nu}$ have been introduced, and η^{ab} is the metric tensor of Special Relativity (i.e. flat space-time). These actions are constructed so as to be general coordinate invariant, i.e. invariant under an arbitrary reparametrization of space-time. They however all have an additional invariance, the Weyl invariance, given by the combined transformations

$$g^{\mu\nu}(x) \to e^{2\alpha(x)} g^{\mu\nu}(x) \tag{78}$$

and, respectively,

$$\phi(x) \to e^{\alpha(x)} \phi(x) \tag{79}$$

$$A_\mu(x) \to A_\mu(x) \tag{80}$$

$$\psi(x) \to e^{3\alpha(x)/2} \psi(x) \tag{81}$$

$$X^d(x) \to X^d(x). \tag{82}$$

Since general coordinate invariance is regarded as the more fundamental symmetry, the path integral measure is fixed (uniquely) to respect this symmetry; this leads to the choice

$$\left[dg^{\frac{1}{4}-\frac{s}{2n}}T_s\right] = \prod_x d\left(g^{\frac{1}{4}-\frac{s}{2n}}T_s(x)\right) \tag{83}$$

for any tensor field T_s of rank s in n space-time dimensions. The case of the scalar field ($s = 0$) will be treated explicitely below. The measure (83) in general breaks Weyl invariance, leading to the Weyl anomaly (also called the conformal anomaly). The Weyl anomaly is an example of an anomaly which is not already completely given by one-loop effects (however, in all known cases, anomalies already show up at the one-loop level).

To see why the factor $g^{1/4-s/2n}$ is needed in (83) let us consider for example the case of a scalar field, with a path integral measure of the form

$$\prod_x d(g^k(x)\phi(x)) . \tag{84}$$

Under an arbitrary infinitesimal reparametrization of space-time, $x' = x + \epsilon$, the measure changes as

$$\prod_x d(g^k(x)\phi(x)) = \prod_{x'} \left|\frac{\partial x'}{\partial x}\right|^{2k} d(g^k(x')\phi(x')) \tag{85}$$

$$= \prod_{x'}(1 + 2k(\partial_\mu \epsilon^\mu))d(g^k(x')\phi(x')) \tag{86}$$

where the factor 2 in the exponent stems from the fact that $g^{\mu\nu}$ is a second-rank tensor (cf. (77)). On the other hand,

$$d(g^k(x')\phi(x')) = (1 + \epsilon^\mu \partial_\mu)d(g^k(x)\phi(x)) \tag{87}$$

so that, to first order in ϵ, the Jacobian of the transformation is

$$J = \det(1 + 2k(\partial_\mu \epsilon^\mu) + \epsilon^\mu \partial_\mu) \tag{88}$$
$$= \exp(\text{Tr}(2k(\partial_\mu \epsilon^\mu) + \epsilon^\mu \partial_\mu)) . \tag{89}$$

The trace can be evaluated as

$$\text{Tr}(2k(\partial_\mu \epsilon^\mu) + \epsilon^\mu \partial_\mu) =$$
$$= \sum_n \int dx\, \phi_n^*(x)(2k(\partial_\mu \epsilon^\mu) + \epsilon^\mu \partial_\mu)\phi_n(x) \tag{90}$$
$$= \sum_n \int dx\, \left[\frac{1}{2}\partial_\mu(\phi_n^*(x)\epsilon^\mu \phi_n(x)) + \left(2k - \frac{1}{2}\right)\phi_n^*(x)\phi_n(x)(\partial_\mu \epsilon^\mu)\right] . \tag{91}$$

The first term, being a total divergence, vanishes. However, for the second one to vanish (and thus for general coordinate invariance to be preserved), one must choose $k = 1/4$.

3.3 Quantitative Computations

Having established the source of the anomalous contributions in the path integral representation of the generating functional, the task of computing them in practice remains nevertheless formidable. In general, there are two ways to proceed, the systematics of which one can illustrate in the example of the integral

$$I = \int dx\, e^{-a^2 x^2} = \sqrt{\pi}/a \,. \tag{92}$$

Assume one is interested in the behaviour of I under a rescaling $a \to \lambda a$. One can either evaluate the integral, observe $I \propto 1/a$, and conclude $I \to I/\lambda$. On the other hand, one can rescale x such that the exponent is invariant, i.e. $x \to x/\lambda$, which leads to $dx \to dx/\lambda$, and thus to the same conclusion without having to evaluate the integral. In the context of the path integral, the first method would typically correspond to an evaluation via ζ-function regularization,

$$N \int [d\phi] e^{-\phi D \phi} = (\det D)^p = e^{-p\zeta'(0,D)} \tag{93}$$

where $\zeta(s, D) = \mathrm{Tr}(D^{-s})$, p depends on the type of fields considered and N is a field independent normalization constant. Then one can investigate the response of the calculated object under the relevant transformations, i.e. obtain $\delta\zeta'(0, D)$ as the limit $\lim_{s\to 0} \delta\zeta'(s, D)$.

The second method, on the other hand, corresponds to the approach taken by Fujikawa: Transform the field ϕ as

$$\phi(x) \to \phi'(x) = e^{i\epsilon(x)} \phi(x) \tag{94}$$

where $\epsilon(x)$ is an element of the Lie algebra generating the symmetry transformations. The action remains invariant; however, the measure acquires a Jacobian

$$[d\phi] \to [d\phi'] = \left| \frac{\delta\phi}{\delta\phi'} \right| [d\phi] = \exp(i\mathrm{Tr}\,\epsilon(x))[d\phi] \tag{95}$$

The Jacobian needs to be regularized; this is done by the Fujikawa prescription:

$$\mathrm{Tr}_D \epsilon(x) = \lim_{M \to \infty} \mathrm{Tr}(\epsilon(x) e^{-D/M^2}) \tag{96}$$

where D is precisely the operator appearing in the path integral. This is an ad hoc prescription. Its arbitrariness can be reduced somewhat by checking whether the same result is obtained using

$$\mathrm{Tr}_D \epsilon(x) = \lim_{M \to \infty} \mathrm{Tr}(\epsilon(x) f(D/M^2)) \tag{97}$$

with any $f(x)$ such that

$$f(0) = 1 \quad f(\infty) = f'(\infty) = f''(\infty) = \ldots = 0 \,. \tag{98}$$

This can be verified for most of the interesting theories. Note that (96) is nothing but a $\epsilon(x)$-weighted partition function in the limit of infinite temperature M^2:

$$\lim_{M\to\infty} \text{Tr}(\epsilon(x)e^{-D/M^2}) = \lim_{t\to 0} \text{Tr}(\epsilon(x)e^{-Dt}) \tag{99}$$

$$= \lim_{t\to 0} \sum_n \int dx\, \phi_n^*(x)\epsilon(x)e^{-Dt}\phi_n(x) \tag{100}$$

$$= \lim_{t\to 0} \sum_n \int dx\, \phi_n^*(x)\epsilon(x)G_D(x,y,it)\phi_n(y) \tag{101}$$

$$= \lim_{t\to 0} \int dx\, \epsilon(x)G_D(x,x,it) . \tag{102}$$

Thus the task of computing the 1—loop contribution to a quantum field theoretical anomaly can be reduced to the task of computing a quantum mechanical partition function.

3.4 The Axial ($U(1)$) Anomaly

This section is devoted to an evaluation of the axial anomaly of QED. This is the application initially considered by Fujikawa; here, his calculation will be discussed, and it will also become clear that the ζ-function method [8] precisely reproduces the Fujikawa prescription.

The fermionic part of the QED path integral is

$$Z = \int [d\bar\psi][d\psi] \exp(\bar\psi i\slashed{D}\psi) . \tag{103}$$

One now considers local chiral rotations,

$$\psi \to \psi' = e^{\alpha(x)\gamma_5}\psi \tag{104}$$

or, for infinitesimal transformations,

$$\delta\psi = \alpha(x)\gamma_5\psi \tag{105}$$

and evaluates the change in the path integral measure; this gives the gluonic contribution which violates axial current conservation. According to Fujikawa, the Jacobian of the transformation,

$$[d\bar\psi][d\psi] \to \exp(2\text{Tr}(\alpha(x)\gamma_5))[d\bar\psi][d\psi] \tag{106}$$

must be regularized as

$$2\text{Tr}(\alpha(x)\gamma_5) := 2\lim_{M^2\to\infty} \text{Tr}(\alpha(x)\gamma_5 \exp(-\slashed{D}^2/M^2)) . \tag{107}$$

Before proceeding with the evaluation, consider first the anomalous contribution as it would result from ζ-function regularization. There, the path integral (103) evaluates to

$$\det(i\slashed{D}) = \exp(\text{Tr}\ln i\slashed{D}) = \exp(\text{Tr}\ln(-\slashed{D}^2)/2) . \tag{108}$$

Note the analogy with the WKB formula (40) for the effective action induced by quantum fluctuations at the semiclassical level; this illustrates the semiclassical character of the anomaly. The object above is easier to evaluate if one more generally considers an arbitrary fermion mass, i.e.

$$\exp(\text{Tr}\ln(-\slashed{D}^2)/2) \to \exp(\text{Tr}\ln(-\slashed{D}^2 + m^2)/2) \,. \tag{109}$$

The trace is regularized by writing it in terms of a ζ-function:

$$\frac{1}{2}\text{Tr}\ln(-\slashed{D}^2 + m^2) = -\left.\frac{d}{ds}\zeta(s)\right|_{s=0} \tag{110}$$

with

$$\zeta(s) = \frac{1}{2}\text{Tr}(-\slashed{D}^2 + m^2)^{-s} \,. \tag{111}$$

For sufficiently large s, this trace converges. Consider now the change of the Dirac operator in the Lagrangian under an infinitesimal chiral rotation:

$$\slashed{D} \to (1+\alpha\gamma_5)\slashed{D}(1+\alpha\gamma_5) \Rightarrow \delta\slashed{D} = \{\slashed{D}, \alpha\gamma_5\} \tag{112}$$

$$m \to (1+\alpha\gamma_5)m(1+\alpha\gamma_5) \Rightarrow \delta m = 2m\alpha\gamma_5 \tag{113}$$

and therefore,

$$\delta(-\slashed{D}^2 + m^2) = -\slashed{D}^2\alpha\gamma_5 - 2\slashed{D}\alpha\gamma_5\slashed{D} - \alpha\gamma_5\slashed{D}^2 + 4m^2\alpha\gamma_5 \,. \tag{114}$$

Under the (for sufficiently large s) convergent trace, however, one can cyclically permute the operators and thus, the variation of the ζ-function can be given as

$$\delta\zeta(s) = 2(-s)\text{Tr}[(-\slashed{D}^2 + m^2)^{-s}\alpha(x)\gamma_5] \,. \tag{115}$$

Consequently, the derivative of the ζ-function in the vicinity of $s=0$ varies as

$$\delta\zeta'(s) \approx -2\text{Tr}[(-\slashed{D}^2 + m^2)^{-s}\alpha(x)\gamma_5] \tag{116}$$

since the other term is suppressed quadratically in s. This can be written in integral form:

$$\delta\zeta'(s) = \frac{-2}{\Gamma(s)}\int dt\, t^{s-1}e^{-m^2 t}\text{Tr}(\exp(-t\slashed{D}^2)\alpha(x)\gamma_5) \,. \tag{117}$$

However, because of

$$\lim_{s\to 0}\frac{1}{\Gamma(s)}\int dt\, t^{s-1}e^{-m^2 t}t^n = \lim_{s\to 0}\frac{\Gamma(n+s)}{\Gamma(s)}(m^2)^{-(n+s)} = \delta_{n0} \tag{118}$$

the integral (117) is dominated by the value of the trace at $t=0$, provided that the trace is a regular function in t (this will be verified below). Thus,

$$-\delta\zeta'(s)|_{s=0} = 2\lim_{t\to 0}\text{Tr}(\exp(-t\slashed{D}^2)\alpha(x)\gamma_5) \tag{119}$$

which is exactly the Fujikawa prescription (107) for the anomalous contribution.

The anomaly can finally be computed explicitly in the following way, where the trace runs over Dirac indices and space-time:

$$2\text{Tr}(\alpha(x)\gamma_5) := 2\lim_{t\to 0}\text{Tr}(\alpha(x)\gamma_5\exp(-t\slashed{D}^2)) \tag{120}$$

$$= 2\lim_{t\to 0}\text{Tr}[(\exp(-t\slashed{D}_+^2) - \exp(-t\slashed{D}_-^2))\alpha(x)] \tag{121}$$

where

$$\slashed{D}_\pm^2 = \partial^2 + (D^2 - \partial^2) + \sigma\cdot(B\pm E) \tag{122}$$

and the Dirac trace has been partly taken. To leading order in t, the exponentials can be separated,

$$2\text{Tr}(\alpha(x)\gamma_5) = 2\lim_{t\to 0}\text{Tr}[\alpha(x)e^{-t\partial^2}\left(e^{-t(D^2-\partial^2+\sigma\cdot(B+E))} - e^{-t(D^2-\partial^2+\sigma\cdot(B-E))}\right)]$$

$$= 2\lim_{t\to 0}\text{Tr}[\alpha(x)e^{-t\partial^2}(4t^2 E\cdot B + O(t^3))] \tag{123}$$

where the Dirac trace has now been completely taken, eliminating all terms proportional to σ-matrices. The trace over space-time indices can be performed e.g. in a basis of eigenfunctions of the (four-dimensional) position operator (Fujikawa orginally used a plane wave basis). Consider a diagonal matrix element corresponding to the position vector x_0. The electric and magnetic fields are simply evaluated at x_0; the operator $\exp(-t\partial^2)$ corresponds to the (Euclidean) "time"-evolution operator for a free particle on four-dimensional space. The wave function for this particle given that, initially, it is taken to be localized, is

$$\langle x|e^{-t\partial^2}|x_0\rangle = \frac{1}{(4\pi t)^2}\exp(-(x-x_0)^2/4t) \tag{124}$$

(cf. (7), where the mass has been set to $m=1/2$, and the direction of propagation has been taken to be the (negative) imaginary time axis). Thus, taking the diagonal matrix element yields

$$\langle x_0|4t^2\alpha E\cdot B e^{-t\partial^2}|x_0\rangle = 4t^2\alpha(x_0)E(x_0)\cdot B(x_0)\frac{1}{(4\pi t)^2} \tag{125}$$

to leading order in t. Taking the whole trace now corresponds to summing over all x_0, i.e.

$$2\text{Tr}(\alpha(x)\gamma_5) = \frac{1}{2\pi^2}\int d^4x_0\,\alpha E\cdot B = -\frac{1}{8\pi^2}\int d^4x_0\,\alpha\,{}^*F_{\mu\nu}F^{\mu\nu}. \tag{126}$$

Note that this derivation also explicitly shows that the limit $t\to 0$ in the definition (120) is indeed regular, as claimed above.

Furthermore, note that in QED_{1+1}, the same treatment yields, using

$$\slashed{D}_\pm^2 = \partial^2 + (D^2 - \partial^2) \pm E, \tag{127}$$

for the anomaly:

$$2\text{Tr}(\alpha(x)\gamma_5) = 2\lim_{t\to 0} \text{Tr}[\alpha(x)e^{-t\partial^2}(-2tE + O(t^2))] \tag{128}$$

$$= \int d^2x_0\, \alpha(x_0)(-4tE(x_0))\frac{1}{4\pi t} \tag{129}$$

$$= -\frac{1}{\pi}\int d^2x_0\, \alpha E \tag{130}$$

$$= -\frac{1}{2\pi}\int d^2x_0\, \alpha\epsilon_{\mu\nu}F^{\mu\nu} . \tag{131}$$

In order to study the physical consequences of the anomaly (cf. Sect. 4), it is important to clarify how it enters physical quantities, such as currents. The divergence of the Noether current associated with a symmetry transformation is

$$\partial^\mu J_\mu(x) = \frac{\delta L}{\delta \alpha(x)} \tag{132}$$

where $\alpha(x)$ parametrizes the symmetry transformation. Thus, if L is invariant under the transformation, the current is conserved, such as the axial current $J^5_\mu(x) = \bar\psi(x)\gamma_\mu\gamma_5\psi(x)$ in the classical theory for zero fermion mass. Including a mass term explicitly breaks chiral symmetry and one obtains

$$\partial^\mu J^5_\mu(x) = 2im\bar\psi(x)\gamma_5\psi(x) . \tag{133}$$

In the quantum theory, the additional effective term (126) is subtracted from the action and (133) is modified to[2]

$$\partial^\mu J^5_\mu(x) = 2im\bar\psi(x)\gamma_5\psi(x) + \frac{e^2}{8\pi^2}{}^*F^{\mu\nu}F_{\mu\nu} . \tag{134}$$

This relation will e.g. allow directly a computation of the $\pi^0 \to 2\gamma$ decay width.

Historically the axial anomaly was first discovered computing the triangle diagram (Fig. 2). The relationship between the anomaly defined in terms of the Fujikawa prescription and the diagram can be established in the following way. First the fermion mass is neglected, as we already know that it does not appear in the final result. From the original expression for the quantum corrections to the Dirac or Yang-Mills Lagrangian, i.e. $\text{Tr}\ln\slashed{D}$, one sees that its change under chiral rotations is

$$\delta\text{Tr}\ln\slashed{D} = 2\text{Tr}[\gamma_5\gamma_\mu\alpha(x)(\slashed{D})^{-1}] \tag{135}$$

where (112) has been used. Now $(\slashed{D})^{-1}$ in (135) is expanded in the form

$$(\slashed{D})^{-1} = (\slashed{\partial})^{-1} + (\slashed{\partial})^{-1}e\slashed{A}(\slashed{\partial})^{-1} + (\slashed{\partial})^{-1}e\slashed{A}(\slashed{\partial})^{-1}e\slashed{A}(\slashed{\partial})^{-1} + \cdots .$$

The Dirac trace cancels the two leading terms, and so

$$\delta\text{Tr}\ln\slashed{D} = 2\text{Tr}[(\slashed{\partial})^{-1}e\slashed{A}(\slashed{\partial})^{-1}e\slashed{A}(\slashed{\partial})^{-1}\gamma_5\gamma_\mu\partial^\mu\alpha + \ldots] . \tag{136}$$

This shows that the leading term in the anomaly is just the triangle graph.

---

[2] Here, the electrical charge has been pulled out of the field strength to explicitly display the dependence.

3.5 The Weyl Anomaly

In this section, the Weyl anomaly is considered in the case of a two-dimensional string theory in a gravitational field. The corresponding path integral is

$$Z = \int [d\,g^{1/4} X] \exp\left(-\int d^2x \, \frac{1}{\sqrt{g}} g^{\mu\nu} \partial_\mu X^d \partial_\nu X^d\right) \tag{137}$$

(cf. Sect. 3.2), which is constructed to be invariant under general coordinate transformations but has an anomaly with respect to Weyl invariance. General coordinate invariance in particular permits the choice

$$g^{\mu\nu}(x) = \rho(x)\eta^{\mu\nu} = e^{2\alpha(x)}\eta^{\mu\nu} \tag{138}$$

for the metric; in this frame, a Weyl transformation corresponds to $\alpha(x) \to \alpha(x) + \omega(x)$. The path integral now reads

$$Z = \int [d\,\rho^{1/2} X] \exp\left(\int d^2x \, X^d \partial^2 X^d\right) \tag{139}$$

$$= \int [dY] \exp\left(-\int d^2x \, Y^d D Y^d\right) \tag{140}$$

where

$$D = -\frac{1}{\sqrt{\rho}} \partial^2 \frac{1}{\sqrt{\rho}} \,. \tag{141}$$

Under a Weyl transformation,

$$Y \to Y' = e^{\omega(x)} Y \tag{142}$$

the change in the path integral measure is (cf. Sect. 3.3)

$$J = \exp(\mathrm{Tr}\omega(x)) \tag{143}$$

$$= \exp\left(\lim_{t\to 0} 2 \int d^2x \, \omega(x) G_D(x,x,it)\right) \tag{144}$$

where the overall factor 2 comes from the trace over string components. The kernel evaluates as

$$G_D(x,x,it) =$$
$$= \sum_n \phi_n^*(x) e^{-Dt} \phi_n(x) \tag{145}$$

$$= \frac{1}{(2\pi)^2} \int d^2k \, e^{-ikx} \exp\left(\frac{1}{\sqrt{\rho}} \partial^2 \frac{1}{\sqrt{\rho}} t\right) e^{ikx} \tag{146}$$

$$= \frac{1}{(2\pi)^2} \int d^2k \, \exp\left(\frac{1}{\sqrt{\rho}}(\partial + ik)^2 \frac{1}{\sqrt{\rho}} t\right) \tag{147}$$

$$= \frac{1}{(2\pi)^2 t} \int d^2k \, \exp\left(\frac{1}{\sqrt{\rho}}(ik + \sqrt{t}\partial)^2 \frac{1}{\sqrt{\rho}}\right) \tag{148}$$

$$= \frac{1}{(2\pi)^2 t} \int k\, dk\, d\varphi \, \exp\left(\frac{1}{\sqrt{\rho}}(-k^2 + 2i\sqrt{t}k\partial + t\partial^2)\frac{1}{\sqrt{\rho}}\right) \quad (149)$$

$$= \frac{1}{(2\pi)^2 t} \int k\, dk\, d\varphi \, e^{-k^2/\rho}$$

$$\left(1 - \frac{i\sqrt{t}}{\rho^2}(\mathbf{k}\partial\rho) + t\frac{1}{\sqrt{\rho}}\partial^2 \frac{1}{\sqrt{\rho}} - \frac{t}{2\rho^4}(\mathbf{k}\partial\rho)^2 + O(t^{3/2})\right) \quad (150)$$

$$= \frac{1}{2\pi t}\left(\frac{\rho}{2} + t\frac{\rho}{2}\frac{1}{\sqrt{\rho}}\partial^2\frac{1}{\sqrt{\rho}} - t\frac{1}{8\rho^2}(\partial\rho)^2\right) \quad (151)$$

$$= \frac{1}{t}\left(\frac{\rho}{4\pi} + \frac{t}{8\pi}(-\partial^2 \ln \rho)\right) \quad (152)$$

$$= -\frac{1}{4\pi}\partial^2\alpha + \frac{1}{4\pi t}e^{2\alpha} \,. \quad (153)$$

Note that, in contrast to the axial anomaly, the singular piece in t does not cancel. Physically, this can be interpreted as an (infinite) renormalization of the cosmological constant; the motivation for this will become clearer further below. Denoting the renormalized cosmological constant as $\kappa/2\pi$, the anomaly finally becomes

$$\ln J = \int d^2 x\, \omega(x) \frac{1}{2\pi}(-\partial^2 \alpha(x) + \kappa e^{2\alpha(x)}) \,. \quad (154)$$

Having calculated the Weyl anomaly, one can show that it is equivalent to the presence of a dynamical gravitational field, described by the so-called Liouville Lagrangian. To see this, transform the metric $g^{\mu\nu}(x)$ by a sequence of infinitesimal Weyl transformations to the flat metric $\eta^{\mu\nu}$, i.e. transform away $\alpha(x)$:

$$\alpha \to \alpha(1 - \Delta s) \to \alpha(1 - 2\Delta s) \to \ldots \to 0 \,. \quad (155)$$

At the n-th step, where $n\Delta s = s$, one has

$$\alpha_s(x) = \alpha(x)(1 - s) \quad (156)$$
$$\omega(x) = \omega\alpha_s(x) = \alpha(x)\Delta s \quad (157)$$

and the resulting anomalous contribution from this step is

$$\ln J_s = \int d^2 x\, \alpha(x) \Delta s \left(-\frac{1}{4\pi}\partial^2\alpha(x)(1-s) + \frac{\kappa}{4\pi}e^{2\alpha(x)(1-s)}\right) \,. \quad (158)$$

The accumulated contributions when $s = 1$, i.e. $\alpha_s = 0$, amount to

$$\int_0^1 ds\, \ln J_s = \frac{1}{8\pi}\int d^2 x \left(\alpha(x)(-\partial^2)\alpha(x) + \kappa e^{2\alpha(x)} - \kappa\right) \,. \quad (159)$$

The last term can be dropped, since it merely constitutes a renormalization of the energy, and thus one obtains the Liouville action

$$S_L = \int d^2 x\, \frac{1}{8\pi}((\partial\alpha)^2 + \kappa e^{2\alpha}) = \int d^2 x\, \mathcal{L}_L \,. \quad (160)$$

In covariant notation,

$$S_L \propto \int d^2x \sqrt{g}(R\Box^{-1}R + \kappa) \tag{161}$$

where R is the Riemann scalar, $R \sim \partial^2\alpha$. This is Polyakov's two-dimensional version of gravity; note that Einstein's gravity becomes trivial in two dimensions because it reduces to a total divergence. Note also that the factor $\sqrt{g}$ multiplying κ in (161) is simply the covariant measure for the space-time integral. Thus the interpretation of κ as a cosmological constant (which can be arbitrarily chosen) is justified.

Summing up, the Weyl anomaly has been substituted by dynamical terms for the gravitational field:

$$\int [d\,g^{1/4}X]\exp\left(-\int d^2x\,\frac{1}{\sqrt{g}}g^{\mu\nu}\partial_\mu X^d\partial_\nu X^d\right)$$
$$= \int [dX]\exp\left(-\int d^2x\,\frac{1}{\sqrt{g}}g^{\mu\nu}\partial_\mu X^d\partial_\nu X^d - \mathcal{L}_L\right). \tag{162}$$

4 Physical Consequences of Anomalies

4.1 Fermion Generations and the Number of Quark Colours

In the Standard Model, the requirement of renormalizability, which can be spoiled by anomalies, places restrictions on the number and types of fundamental fields. More specifically, the electroweak Lagrangian density

$$\mathcal{L} = \mathcal{L}_{Gauge} + \mathcal{L}_{Higgs} + \mathcal{L}_F \tag{163}$$

in particular contains the fermionic part

$$\mathcal{L}_F = \bar{\psi}_L\gamma^\mu(i\partial_\mu + g_2(\sigma^a/2)W^a_\mu - g_1(Y/2)B_\mu)\psi_L \tag{164}$$
$$+\bar{a}_R\gamma^\mu(i\partial_\mu + g_1(\alpha/2)B_\mu)a_R \tag{165}$$
$$+\bar{b}_R\gamma^\mu(i\partial_\mu + g_1(\beta/2)B_\mu)b_R \tag{166}$$

where

$$\alpha = 1 + Y \qquad \beta = -1 + Y \tag{167}$$
$$\psi = (a,b) = (\nu_e, e^-), (u,d), \ldots . \tag{168}$$

Since the projection operators onto the chiral components ψ_L and ψ_R contain γ_5, the theory allows for anomalous triangle graphs. These have potentially disastrous consequences for the renormalizability of the theory. For instance, the gauge field propagator contains graphs of the form shown in Fig. 4.1, the anomalous contributions of which give the propagator a longitudinal part. When renormalizing this part, the necessary counterterms would be of a form which

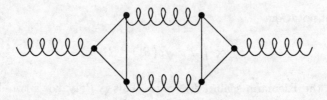

Fig. 1. *Contribution to the gluon propagator*

does not appear in the original Lagrangian [3], thus spoiling renormalizability. For a further discussion, see also [11].

The only known way out of this problem is the following: In internal triangle graphs, as above, all possible fermionic particles participate; therefore, a judicious choice of the fundamental fermion fields allows for a cancellation of the anomalous contributions. In the electroweak Lagrangian, the coupling at the vertices is proportional to the hypercharge Y or the weak isospin $\sigma^a/2$; when calculating the triangle graph, the trace with respect to the corresponding internal indices must be performed, i.e. one has contributions proportional to

$$\mathrm{tr}(Y\sigma^a\sigma^a) \quad \text{and} \quad Y^3 \tag{169}$$

(contributions containing an odd number of isospin couplings already vanish when taking the trace). Thus, to guarantee an anomaly-free theory, the sums over all particles must vanish:

$$\sum \mathrm{tr}(Y\sigma^a\sigma^a) = \sum Y^3 = 0 \ . \tag{170}$$

Since the vertices do not change the helicity, right- and left-handed particles give the same contribution; it therefore suffices to consider left-handed particles, the hypercharge assignments of which are

$$\begin{array}{ccccc} (\nu_L^e, e_L) & \bar{e}_L & (u_L, d_L) & \bar{u}_L & \bar{d}_L \\ (-1,-1) & 2 & (1/3, 1/3) & -4/3 & 2/3 \end{array} \tag{171}$$

(the brackets denote weak isospin doublets). Note also that the higher generations of fermions merely replicate this scheme. Thus one obtains

$$\sum Y^3 = 6 - 2N_C \tag{172}$$

$$\sum \mathrm{tr}(Y\sigma^a\sigma^a) = 3(-2 + 2N_C/3) \tag{173}$$

where it has been taken into account that quarks come in N_C colours, and that $\mathrm{tr}(\sigma^a\sigma^a) = 0$ for isospin singlets, and $\mathrm{tr}(\sigma^a\sigma^a) = 3$ for doublets. One must now conclude that the Standard Model is only anomaly-free if the number of quark colours is $N_C = 3$, and if leptons and quarks always come grouped into generations like the first generation treated explicitly above. In particular, the existence of the top quark is necessary to complete the third generation.

4.2 π^0-Decay

An important consequence of the anomaly is the fact that the neutral pion can decay into two photons. The pion is interpreted to be the Goldstone boson of the spontaneously broken approximate $SU(2)_{fR} \times SU(2)_{fL}$ chiral symmetry of QCD. Only the vector part of this symmetry survives; the axial current by contrast does not annihilate the vacuum, but its divergence at zero momentum creates the Goldstone boson. The assumption of PCAC is that this relation can be extrapolated to finite momentum:

$$\partial^\mu J_\mu^{5,a}(q) = f_\pi m_\pi^2 \pi^a(q) \tag{174}$$

A more thorough discussion of this can be found e.g. in [6].

From this, one can evaluate the decay of the neutral pion

$$\langle \pi^0 | 2\gamma \rangle = (m_\pi^2 f_\pi)^{-1} \langle 0 | \partial^\mu J_\mu^{5,3} | 2\gamma \rangle \tag{175}$$

(the third component of the $SU(2)$ current corresponds to the neutral pion). One can show that, for low momenta, the contribution to $\partial^\mu J_\mu^5$ from the quark mass term (cf. (133)) vanishes in the matrix element (175); neutral pion decay is due to the additional anomalous term (cf. (134)) which $\partial^\mu J_\mu^5$ acquires when coupled to the electromagnetic interaction:

$$\langle \pi^0 | 2\gamma \rangle = (m_\pi^2 f_\pi)^{-1} \langle 0 |^* F^{\mu\nu} F_{\mu\nu} | 2\gamma \rangle \frac{1}{8\pi^2} \text{tr}(Q^2 \sigma^3 / 2) \tag{176}$$

$$= \frac{1}{16\pi^2 m_\pi^2 f_\pi} \text{tr}(Q^2 \sigma^3) \epsilon_{\mu\nu\lambda\eta} k_1^\mu k_2^\nu e^\lambda e'^\eta \tag{177}$$

where k_1, k_2 are the momenta and e, e' the polarization vectors of the photons. It was already pointed out in Sect. 3.4 that this result can also be obtained by evaluating the corresponding triangle graph (Fig. 2). It is in good agreement with

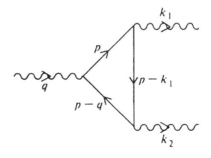

Fig. 2. *The triangle graph*

experiment if one notes that, for this low-energy effect, the only participating fermions are the proton and the neutron, and thus

$$\text{tr}(Q^2 \sigma^3) = Q_p^2 - Q_n^2 = 1 \; . \tag{178}$$

On the other hand, if one describes the decay on the quark level, one has

$$\text{tr}(Q^2 \sigma^3) = N_C(Q_u^2 - Q_d^2) = N_C/3 \tag{179}$$

which is further evidence that the number of quark colours is indeed $N_C = 3$.

4.3 Baryon–Lepton Decay

Additional consequences of the anomalous gluonic contribution to the axial current (cf. (180)) result from the fact that this contribution induces an instanton-fermion interaction in analogy to the electromagnetic coupling responsible for the decay $\pi^0 \to 2\gamma$ discussed in the preceding section. One can derive an effective fermion-fermion interaction Lagrangian in the presence of an instanton background [7]. This leads to baryon decay in the case of an electroweak instanton background (see Sect. 2.5). The amplitude for such a decay is, however, proportional to the tunneling amplitude between two distinct $|n\rangle$-vacua, i.e. the exponential of the instanton action. This gives a typical half-life of $10^{150}a$, which is negligible compared to the proton half-life derived from GUT theories, $10^{30}a$. However, at temperatures above 200 GeV, which prevailed in the early universe, direct transitions between different $|n\rangle$-vacua become possible, and thus baryon number-changing processes by this mechanism may not be negligible anymore.

4.4 The $U(1)$-Problem

The axial anomaly also plays an important role in the resolution of the so-called U(1) problem. The QCD Lagrangian has a $U(1)_{f,L} \times U(1)_{f,R}$ symmetry, of which only the vector part manifests itself, namely in the conservation of baryon number. Explicit realization of the axial part would mean the seperate conservation of left- and right-handed fermion doublets, e.g. negative parity partners of nucleons. Such particles are not found experimentally. Thus one would conclude that the $U(1)_A$ symmetry is spontaneously broken. In that case, however, one should observe a corresponding (isoscalar) Goldstone boson. Such a particle is also not found in the physical spectrum. This puzzle is known as the U(1) problem.

The puzzle is solved by the fact that, due to the axial anomaly, the $U(1)_A$ symmetry is broken explicitly since

$$\partial^\mu J_\mu^5 = 2i(m_u \bar{u}\gamma_5 u + m_d \bar{d}\gamma_5 d) + \frac{g^2}{4\pi^2}\text{tr}(^*G^{\mu\nu}G_{\mu\nu}) \tag{180}$$

and the right-hand side is not zero even in the zero quark mass limit. Thus the symmetry is broken explicitly and there is no Goldstone boson.

At first sight this explanation may seem too simplistic, because the anomaly term on the right-hand side of (180) is itself a divergence, namely $\partial^\mu S_\mu$ where S_μ is defined in (45) and thus (in the zero quark-mass limit) the total current $J_\mu^5 + S_\mu$ is conserved. However, as we have already seen, the charge Q associated with S_μ is gauge-variant. Hence the total current is gauge-variant and non-physical. What this means is that in the non-physical space of gauge-variant quantities the symmetry breakdown may be regarded as spontaneous (with the massless field S_μ-field as the gauge-variant Goldstone field). But in the physical space of gauge-invariant quantities the breakdown is explicit and there are no Goldstone fields.

One of the most interesting features of this resolution of the $U(1)$ problem is that it not only explains why the η' particle is not a Goldstone particle but, when combined with the existence of the θ-vacua of Sect. 2.5, it provides an estimate for the η' mass and for its decay into three pions. The point is that if one adds to the Lagrangian a term of the form $\theta \int F * F$ to take care of the θ-vacua, expands the Fermion Lagrangian about the instanton solutions and sums over all instantons, the term $\theta \int F * F$ and the zero-modes of the instantons combine to generate an effective potential for the η' field and this effective potential provides a mass-term (proportional to the gluonic condensate $< (F^*F)^2 >$) and a $3-\pi$ decay for the η'. This is explained in references [12], [13] and [14], though it should be mentioned that the arguments given there are somewhat qualitative and there is still some dispute [15] about the details.

A Proof of Equation (5)

One starts from (11) and differentiates it once with respect to time. Using the classical Hamilton Jacobi equation ((3) with $\hbar^2 = 0$) one gets

$$\frac{\partial}{\partial t} p_i + \frac{1}{m} \sum_k \frac{\partial}{\partial x_k} S \frac{\partial}{\partial x_k} p_i = 0 \ .$$

This equation is differentiated once more with respect to the final position x_j. With the definitions

$$M_{ij} = \frac{\partial}{\partial x_i} p_j$$

and

$$N_{ij} = \frac{\partial^2}{\partial x_i \partial x_j} S$$

one gets

$$\partial_t M + \frac{1}{m} NM + \frac{1}{m} (\boldsymbol{\nabla} S)(\boldsymbol{\nabla} M) = 0 \ .$$

This equation is multiplied from the right with M^{-1}, then one takes the trace of the matrices.

$$\mathrm{Tr}(M^{-1} \partial_t M) + \frac{1}{m} \mathrm{Tr} N + \frac{1}{m} \boldsymbol{\nabla} S \mathrm{Tr}(M^{-1} \boldsymbol{\nabla} M) \ . \tag{181}$$

For the next step the following identity is used:

$$\operatorname{Tr} M(\lambda)^{-1} \frac{\partial}{\partial \lambda} M(\lambda) = \frac{\frac{\partial}{\partial \lambda} \det M}{\det M} .$$

This can be shown easily by writing the trace and the determinant in terms of the eigenvalues of M. With this and $\operatorname{Tr} N = \nabla^2 S$, equation (181) becomes:

$$\partial_t \det M + \frac{1}{m}(\nabla^2 S \det M + \nabla S \nabla \det M) = 0 .$$

This is again the continuity equation (4). Therefore

$$\rho \propto \det M = \det(-\frac{\partial^2 S}{\partial a_i \partial x_j}).$$

The normalization is again fixed by the initial condition.

References

[1] Messiah, "Quantenmechanik", Band 1, de Gruyter Berlin (1976), p. 201
[2] L.D.Landau, "Klassische Mechanik", Akademie Verlag Berlin (1984)
[3] P.Ramond, "Field Theory: A Modern Primer", Addison-Wesley, Redwood City 1989
[4] L.H.Ryder, "Quantum Field Theory", Cambridge University Press, Cambridge (1985)
[5] L.O'Raifeartaigh and A.Wipf, "WKB properties of the Time-Dependent Schrödinger System", Foundations of Physics Vol.18, No.3, March 1988, pp. 307
[6] T.Cheng, L.Li, "Gauge Theory of Elementary Particle Physics", Clarendon Press, Oxford 1989
[7] G.'t Hooft, Phys. Rev. Lett. 37 (1976) 8
[8] L.O'Raifeartaigh, Axial Anomalies, Proceedings of the ZUOZ (PSI) Spring School 1989
[9] W.Bardeen, Phys. Rev. 184 (1973) 1848
[10] K.Fujikawa, Phys. Rev. D 21 (1980) 2848
[11] D.J.Gross, R.Jackiw, Phys. Rev. D 6 (1972) 477
[12] E. Witten, Nucl. Phys. **B156** (1979) 269
[13] G. 't Hooft, Physics Reports **C142** (1986)357
[14] D. Dyakonov and M. Eides, Sov. Phys. JETP **54(2)** (1981) 232
[15] G. Christos, Physics Reports **C116** (1984)251

Anomalies in Gauge Theories*

M.A. Shifman[1];
Notes by M. Engelhardt[2] and D. Stoll[2]

[1] Theoretical Physics Institute, University of Minnesota, Minneapolis, MN 55455, USA
[2] Institute for Theoretical Physics III, University of Erlangen–Nürnberg, Staudtstr. 7, 91058 Erlangen, Germany

1 Introduction

In this lecture we discuss anomalies in QED and QCD [1], [2]. The physical picture which we shall present for the class of chiral anomalies is believed to be universal, independent of the dimension of space used for the formulation of the theory. However, only in two dimensions a simple fermion dynamics allows us to develop the picture explicitely, and thus in the first part of the lecture we shall be concerned with two dimensional QED with massless fermions, the Schwinger model. In the second part we shall introduce still another class of anomalies, so called trace anomalies and turn to the question of the implications of the existence of anomalies on hadronic reactions. In particular we discuss the decay of a light Higgs particle into a fermion–antifermion pair and the influence of the anomaly on the branching ratio for decay into leptons vs. decay into hadrons [2].

2 The Chiral Anomaly in the Schwinger Model

We now study chiral QED in 1+1 dimensions, and in order to be able to extract the essence of the chiral anomaly we shall simplify the model sucessively. In this way we shall arrive at a picture for the anomaly which indeed is very simple to interpret, since it originates from the discussion of fermions coupled to a single quantum mechanical degree of freedom and the remaining global symmetries.

2.1 QED on a Small Circle

The starting point for these considerations is the QED Lagrangian

$$\mathcal{L} = -\frac{1}{4e_0^2} F_{\mu\nu} F^{\mu\nu} + \overline{\psi}\gamma^\mu \left(i\partial_\mu + A_\mu\right)\psi$$

---

* Lectures presented at the workshop "QCD and Hadron Structure" organised by the Graduiertenkolleg Erlangen–Regensburg, held on June 9th–11th, 1992 in Kloster Banz, Germany

with the dimensionful coupling constant e_0 (dimension of e_0 = mass) and the following choices for the (chiral) representation of the fermion field and the γ-matrices

$$\psi = \begin{pmatrix} \psi_1 \\ \psi_2 \end{pmatrix}; \quad \gamma^0 = \sigma_2, \; \gamma^1 = i\sigma_1, \; \gamma^5 = \sigma_3 \;.$$

The components of the fermion spinor are eigenstates of chirality, i.e. right (left) handed spinors are

$$\psi_R = \begin{pmatrix} \psi_1 \\ 0 \end{pmatrix}; \; \psi_L = \begin{pmatrix} 0 \\ \psi_2 \end{pmatrix}; \quad \gamma^5 \psi_R = \psi_R, \; \gamma^5 \psi_L = -\psi_L \;.$$

Although the dynamics of the fermions is already very much simplified by using a formulation of the theory in one spatial dimension only, it is nevertheless still nontrivial. This can be seen easily from the fact that a charged particle is the source for a linearly rising scalar potential (two static electric charges can not be infinitely separated in one dimension) which acts as a confining potential for a particle of opposite charge. Consequently the physically observable states will not be simply related to the fields appearing in the Lagrangian and therefore the dynamical problem can not be simple[1]. In order to simplify the dynamics further we use the fact that the kinetic energy of quantum states increases compared to the interaction energy when they are enclosed in a small box. Thus we replace the infinite interval by a circle of finite length and thus introduce a new dimensionful parameter, the length of the circle L. The advantage of having another length scale is that the interaction strength can now be measured. It then becomes possible to go to the limit where $e_0 L$ is very small and where the Coulomb interaction can be treated as a perturbation, which actually will be even neglected in our discussion.

In such a formulation on the circle it is then necessary to specify boundary conditions for the fields, which we choose for convenience in the following way

$$A_\mu(x = -\frac{L}{2}, t) = A_\mu(x = \frac{L}{2}, t); \quad \psi(x = -\frac{L}{2}, t) = -\psi(x = \frac{L}{2}, t)$$

Note that for fermions more general boundary conditions could be used due to the fact that only bilinears of fermions are important. According to these boundary conditions we have the following Fourier decomposition of the fields

$$A_\mu(x,t) = \sum_k a_\mu(k,t) e^{i2\pi k x/L} \tag{1}$$

$$\psi(x,t) = \sum_k \psi(k,t) e^{i\pi(2k+1)x/L} \;. \tag{2}$$

---

[1] An argument supporting the existence of confinement independently of the assumption of nonrelativistic motion of the fermions is the infinite difference in field energy associated with charged states compared to charge neutral states.

The gauge freedom which we did not yet take into account can then be used to gauge all Fourier modes of A_1 away except for the zero mode $a_1(k=0,t)$. This elimination is achieved by applying the following gauge transformation

$$\alpha(x,t) = -\sum_{k\neq 0} \frac{L}{2\pi i k} a_1(k,t) e^{i2\pi kx/L}$$

which is obviously not applicable for $k=0$. The result for the gauge field is

$$A_1' = A_1(x,t) - \partial_1\alpha(x,t) = a_1(k=0,t) = a_1,$$

a space independent component a_1. In other words a single quantum mechanical degree of freedom is left, since the gauge we have chosen is just the Coulomb gauge $\partial_1 A_1 = 0$. The scalar potential A_0 on the other hand can be neglected entirely in the limit of very small values for $e_0 L$ which is evident from Poisson's equation (in the Coulomb gauge)[2]

$$\partial_1^2 A_0 = -e_0 \psi^\dagger \psi \Rightarrow a_0(k,t) \propto e_0 L \ll 1.$$

Thus in this simplified version we are left with fermions only coupled to a single quantum mechanical degree of freedom $a_1(0,t)$. Since this gauge field excitation does not carry any momentum, there is no coupling of different Fourier modes of the fermion field in the Hamiltonian and the dynamics consequently is almost trivial. Nevertheless the anomaly is still present even after all these simplifications have been introduced, since it reflects only global features of the theory, as we shall see soon.

2.2 The Origin of the Chiral Anomaly

The existence of the chiral anomaly is closely related to the fact that despite fixing the gauge as we did, it is possible to perform further non–trivial gauge transformations with

$$\alpha(x,t) = \frac{2\pi}{L} nx; \quad n = \pm 1, \pm 2, \ldots.$$

These are compatible with the gauge fixing and with the boundary conditions since both $\exp[i\alpha]$ and $\partial_1 \alpha$ are periodic. Since a_1 changes by $2\pi n/L$ under these transformations, these values should in fact be identified which implies that a_1 is itself a periodic variable just like an angle [3]. Before following this line of argument further we briefly remind you where the notion of an anomaly arises. The basic observation is that for massless fermions, the case we are dealing

---

[2] Note that the simple argument can be used because of the super–renormalizability of QED in 1+1 dimensions, according to which we do not have to worry about possible renormalizations of these expressions.

with, the Lagrangian posesses two independent symmetries on the classical level. Namely it is invariant under (phase) rotations

$$\psi \to e^{i\alpha}\psi, \psi^\dagger \to \psi^\dagger e^{-i\alpha}; \quad \Rightarrow \quad \partial_\mu j^\mu = \partial_\mu \bar\psi \gamma^\mu \psi = 0$$

and it is invariant under chiral rotations

$$\psi \to e^{i\alpha\gamma^5}\psi, \psi^\dagger \to \psi^\dagger e^{-i\alpha\gamma^5}; \quad \Rightarrow \quad \partial_\mu j^{5\mu} = \partial_\mu \bar\psi \gamma^\mu \gamma^5 \psi = 0 \ .$$

The corresponding charges Q and Q_5 are conserved and they are just the sum and difference of the charges of the right handed and left handed fermions respectively. Thus the numbers of these fermions, which differ in their charge and axial charge assignments as shown in Table 1, are separately conserved on the

Table 1. *Charge and axial charge of left and right handed fermions and anti–fermions*

	ψ_L	ψ_R	$\bar\psi_L$	$\bar\psi_R$
Q	+1	+1	-1	-1
Q_5	+1	-1	-1	+1

classical level, i.e. in any Feynman diagram which does not contain loops. The anomaly then consists in the observation that the axial charge Q_5 is not conserved anymore after quantization.

In order to understand how this anomaly arises we make an additional assumption which will be justified from the result. We treat the quantum mechanical variable a_1 as a slow variable which may be considered as frozen while we are solving for the fermionic dynamics. Thus we only have to solve the Dirac equation for the fermions with a_1 fixed

$$\left[i\frac{\partial}{\partial t} + \sigma_3 \left(i\frac{\partial}{\partial x} - a_1\right)\right]\psi = 0 \ .$$

Since a_1 is considered as constant, the solution for the coefficients of the Fourier expansion (2) is

$$\psi(k,t) = \begin{pmatrix} \exp\left[-iE_{k(R)}t\right] \\ \exp\left[-iE_{k(L)}t\right] \end{pmatrix} \tag{3}$$

$$E_{k(L)} = \left(k + \frac{1}{2}\right)\frac{2\pi}{L} + a_1 \tag{4}$$

$$E_{k(R)} = -\left(k + \frac{1}{2}\right)\frac{2\pi}{L} - a_1 \tag{5}$$

and we observe that for $a_1 = 0, a_1 = 2\pi/L$, which are gauge copies, the spectra are identical. However, having filled the negative energy states for $a_1 = 0$ and changing a_1 adiabatically from $a_1 = 0$ to $a_1 = 2\pi/L$ we notice that the resulting state is no longer a vacuum state, since a hole associated with the right handed fermions and a particle associated with the left handed fermions is produced (Fig. 1). Thus we observe that the total charge is conserved in this transition, because particle and hole irrespective of their handedness have opposite charges, but the axial charge is not conserved and in fact one finds that it changes by two units. From this observation we can then reproduce the standard form of the anomaly equation

$$\Delta Q_5 = \frac{L}{\pi}\Delta a_1 \Rightarrow \dot{Q}_5 = \frac{1}{\pi}\int_{-L/2}^{L/2} dx\, \dot{A}_1(x,t) \tag{6}$$

$$\Rightarrow \int dx \left(j^{50} - \frac{1}{\pi}A_1\right) = constant \tag{7}$$

$$\Rightarrow \partial_\mu \left(j^{5\mu} - \frac{1}{\pi}\epsilon^{\mu\nu}A_\nu\right) = 0 \,. \tag{8}$$

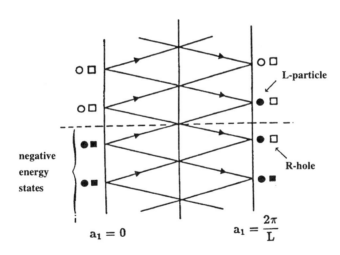

Fig. 1. *Single particle levels in the zero mode background field*

Introducing the field strength tensor into this expression we recover the form in which the anomaly is usually expressed

$$\partial_\mu j^{5\mu} = \frac{1}{2\pi}\epsilon^{\mu\nu}F_{\mu\nu} \,. \tag{9}$$

Although the anomaly equation is correctly reproduced including all factors footing on the observation that the axial charge changes by two units, the rigorous derivation of this fact is actually more subtle. Indeed, what is needed is a proper definition of the charge operators, since the existence of an infinite number of negative energy states and the necessity of filling the Dirac–sea makes them ill defined without regularization. One way of defining them is point splitting of the current operators. However, in order to maintain gauge invariance it is necessary to add an appropriate line integral over the gauge field. Thus we can define the current in the following way

$$j_\mu^{reg} = \lim_{\epsilon \to \infty} \left[\overline{\psi}(x+\epsilon,t)\gamma_\mu \exp\left[-i \int_x^{x+\epsilon} A_1(z,t)dz\right] \psi(x,t) - <0|\ldots|0> \right]$$

which results in the expression for the regularized right(left) handed charges $Q_{R(L)}$

$$Q_{R(L)} = \sum_{k_{R(L)}} e^{-i\epsilon[(2k+1)\pi/L + a_1]}$$

where the sum extends over the occupied states which are different in the vacuum for right and left handed fermions. In the vacuum defined previously one then finds

$$(Q_L)_{vac} = \frac{L}{-i2\pi\epsilon} + \frac{L}{2\pi}a_1 + O(\epsilon) ,\qquad(10)$$

$$(Q_R)_{vac} = \frac{L}{i2\pi\epsilon} - \frac{L}{2\pi}a_1 + O(\epsilon) .\qquad(11)$$

The singular contribution to the charges can be subtracted, since it is independent of the dynamics and just reflects the contribution of the infinite number of states in the Dirac–sea. One therefore finds that the regularized charge operator $Q = Q_R + Q_L$ is time independent, whereas the regularized axial charge operator $Q_5 = Q_R - Q_L$ is time dependent through the appearance of a_1 which is just the observation of the chiral anomaly. Note also that we have now two facets of the anomaly; one is level crossing at zero energy, the other comes from ultraviolet regularization which forces a regularization and allows for appearance and disappearence of energy levels with energies in the cutoff region.

We thus see that there are in fact three essential ingredients for the existence of the anomaly

(i) The existence of an infinite number of levels
(ii) Appearance of the Dirac–sea after second quantization
(iii) Level crossing under gauge transformation and the reinterpretation of the vacuum wave function associated with it.

There is however no dynamics involved in the whole derivation and we believe this to be true also in gauge theories in four dimensions such as QCD, where anomalies are known to exist. In these theories, however, the simple reduction of the dynamics no longer goes through. It is still possible to obtain the anomalous divergencies of axial currents as an effect of the ultraviolet regularization, but the connection to level crossing at zero energy can no more be obtained because of complicated and basically unknown infrared dynamics in QCD. However, an understanding of the anomaly in terms of the existence of a level flow leading to the disappearance of levels in the cutoff region is still accessible.

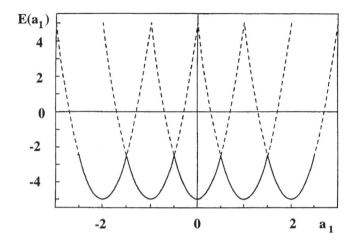

Fig. 2. Effective potential for the a_1 variable obtained from the Born–Oppenheimer approximation

2.3 Vacuum Wave Function and Remarks

Let us finally come to the justification for working with a_1 as an adiabatic variable. It is obtained by solving the dynamical equation for the gauge field variable after having determined the fermion state in the presence of a static a_1 variable. This is the usual procedure called Born–Oppenheimer-Approximation and it leads to the following effective Hamiltonian for the gauge field variable, when the expectation value of the regularized Hamiltonian in the fermionic ground state is taken

$$< 0_{ferm.}|H_{reg}|0_{ferm.}> = \frac{L}{2e_0^2}\dot{a_1}^2 + <0_{ferm.}|-\psi^\dagger(x+\epsilon,t)\left(i\frac{\partial}{\partial x}-a_1\right)$$
$$\exp\left[-i\int_x^{x+\epsilon} A_1(z,t)dz\right]\psi(x,t)|0_{ferm.}> \quad (12)$$
$$= \frac{L}{2e_0^2}\dot{a_1}^2 + \frac{L}{2\pi}\left(a_1^2 - \frac{\pi^2}{L^2}\right) + O(\epsilon) \quad (13)$$

where $L\dot{a}_1$ should be interpreted as canonical momentum associated with a_1.

The resulting Hamiltonian, the potential is shown in Fig. 2, obviously describes a harmonic oscillator in the a_1 variable which is oscillating with frequency

$\omega_A = e_0/\sqrt{\pi}$. This frequency has to be compared with the corresponding one for the fermions which according to eqs.(4,5) is $\omega_\psi \approx 1/L$ and this yields the justification for the Born–Oppenheimer–Approximation since we find

$$\frac{\omega_A}{\omega_\psi} \approx e_0 L \ll 1$$

We thus observe that filling the negative energy states of the fermions for $a_1 = 0$ leaves a_1 oscillating slightly around the value zero; filling the negative energy states for $a_1 = 2\pi n/L$ gives an oscillation of the gauge degree of freedom around this value. Note that large fluctuations around the equilibrium positions would have invalidated the Born–Oppenheimer approximation. As a consequence there exist for all n stable solutions which are mapped into each other under gauge transformations. The true ground state (because of its cluster decomposition properties [4]) is obtained as a superposition of these equivalent states for definite value of n which yields the well known θ–vacuum in chiral QED(1+1)

$$\Psi_n = \psi_A(a_1 - \frac{2\pi}{L}n) \cdot \psi_n(fermions) \tag{14}$$

$$\Psi_{Vac} = \sum_n e^{i\theta n} \Psi_n \tag{15}$$

which has the correct cluster decomposition properties. Furthermore, the chiral symmetry is broken in this vacuum resulting in a non–vanishing fermionic condensate value.

As a final remark we would like to add that the result for the anomaly may also be obtained by standard use of Feynman graph techniques. It is thus not inevitable to follow the reasoning of this lecture, where we have put the emphasis more on the infrared aspect of this anomaly, but it may also be obtained purely as an ultraviolet effect. (See, e.g., the lectures by L. O'Raifeartaigh or R. Jackiw in this volume.)

3 Anomalies in QCD and Implications

To discuss the implications of anomalies for hadronic reactions it is of course necessary to switch to four dimensions and, in particular, we shall discuss the consequences in "QCD". Since gauge theories in four dimensions have to be renormalized, the scale invariance of the classical theory gets lost after quantization and a length scale is introduced. The consequence is that not only chiral anomalies occur which may already exist in super–renormalizable theories like the Schwinger model, but, in addition, one encounters the so–called scale (or trace) anomalies.

3.1 Anomalies in QCD

To become more specific, we recall that the classical Lagrangian

$$\mathcal{L} = -\frac{1}{4}G^a_{\mu\nu}G^{a\,\mu\nu} + \sum_{q=u,d,s}\bar{q}\gamma^\mu\left(i\partial_\mu + gA_\mu\right)q$$

is invariant (for massless quarks) under

(i) Scale transformations $x \to \lambda x$

$$q(x) \to \lambda^{3/2}q(\lambda x) \tag{16}$$
$$A_\mu(x) \to \lambda A_\mu(\lambda x) \tag{17}$$

(ii) Flavor rotations generated by the group $SU(3)_V \times SU(3)_A \times U(1)_A \times U(1)_V$ which have different realizations in nature

  (a) $SU(3)_V$ is realized linearly (in the Wigner–Weyl mode, where the ground state respects the symmetry), leading to $SU(3)$ multiplets in the observed spectra of hadrons
  (b) $SU(3)_A$ is realized non–linearly (in the Nambu–Goldstone mode, where the ground state breakes the symmetry), causing the existence of the octet of nearly massless Goldstone bosons
  (c) $U(1)_A$ is afflicted with an anomaly (the latter manifests itself in a comparatively large mass of the η'–meson compared to the other pseudoscalar mesons)
  (d) $U(1)_V$ is not anomalous and is associated with the conserved baryon number

By quantization and renormalization two anomalies are induced: the chiral anomaly in the axial flavor–singlet current which has already been discussed in the previous section (in the context of QED)

$$\partial^\mu j^5_\mu = \frac{n_F \alpha_s}{8\pi}\epsilon^{\mu\nu\rho\sigma}G^a_{\mu\nu}G^a_{\rho\sigma}$$

and the scale anomaly in the trace of the energy momentum tensor $\theta_{\mu\nu}$

$$\theta^\mu_\mu = \frac{\beta(\alpha_s)}{8\alpha_s}G^{a\,\mu\nu}G^a_{\mu\nu} = -\frac{b\alpha_s}{4\pi}G^{a\,\mu\nu}G^a_{\mu\nu} + \ldots \tag{18}$$

where α_s is the strong coupling constant, $\beta(\alpha_s) = -b\alpha_s^2/2\pi$ is the renormalization group β–function and n_F stands for the number of quark flavors.

Taking into account the coupling of an electromagnetic gauge field as well as the gluon field, the scale anomaly gets an additional contribution

$$\theta^\mu_\mu = c_1 G^{a\,\mu\nu}G^a_{\mu\nu} + c_2 F^{\mu\nu}F_{\mu\nu}$$

and in addition an anomaly arises in the neutral axial isospin current, leading to the famous contribution to the $\pi^0 \to 2\gamma$ decay at low energies [5], [6]

$$\partial^\mu\left(\bar{u}\gamma_\mu\gamma^5 u - \bar{d}\gamma_\mu\gamma^5 d\right) = \frac{\alpha}{4\pi}\left(Q_u^2 - Q_d^2\right)\epsilon^{\mu\nu\rho\sigma}F_{\mu\nu}F_{\rho\sigma}$$

3.2 Decay of Light Higgs Particles

To demonstrate consequences of the existence of anomalies on reaction cross sections, we assume now that in addition to gluons and photons there exists a light Higgs particle coupled to heavy quarks. As we shall argue, the presence of the scale anomaly reverses the naive dominance of the decay of a light Higgs particle into $\mu^+\mu^-$ pairs into dominance of the decay into hadronic channels.

We assume the coupling of the Higgs field H to be similar to the coupling in the standard model

$$\mathcal{L}_{int} = -\sum_f \lambda_f \hat{H} \bar{f} f; \quad \lambda_f = \frac{m_f}{<H>}$$

with λ_f depending on the current quark (lepton) masses m_f and the non-vanishing vacuum expectation value of the Higgs field $<H>$. Note that the vacuum expectation value is not generally related to Fermi's weak coupling constant G_F. We assume furthermore that the mass of the Higgs particle is about $m_H = 1 GeV$ in order to be able to eliminate the quarks much heavier than the Higgs in low energy reactions deriving in this way an effective interaction. Using the interaction Lagrangian as given above we expect the cross section for decay of the Higgs particle into leptons $\mu^+\mu^-$ to be of the order of

$$\Gamma(H \to 2\mu) \propto \lambda_\mu^2 \propto \frac{m_\mu^2}{<H>^2}.$$

Correspondingly, naively one would expect the decay into pions to be of the order of (neglecting effects from hadronization which should not depend strongly on the quark mass)

$$\Gamma(H \to 2\pi) \propto \lambda_{u,d}^2 \propto \frac{m_{u,d}^2}{<H>^2}$$

and therefore leptonic decay should be favored by two orders of magnitude over decay into pions

$$\left.\frac{\Gamma(H \to 2\mu)}{\Gamma(H \to 2\pi)}\right|_{no\ anomaly} \propto \frac{m_\mu^2}{m_{u,d}^2} \approx 10^2.$$

Note that due to the similarity in mass of muons and pions we do not have to worry about phase space factors in order to get a first estimate. However, the presence of the anomaly reverses this result and one eventually finds that

$$\left.\frac{\Gamma(H \to 2\mu)}{\Gamma(H \to 2\pi)}\right|_{with\ anomaly} \propto \frac{m_\mu^2}{m_H^2} \approx 10^{-2}. \tag{19}$$

The reason is that the transition into virtual heavy quarks decaying in an intermediate stage to gluons has to be taken into account, as well. This then leads to a decay matrix element of the form

$$<0| -\frac{b\alpha_s}{8\pi} G^a_{\mu\nu} G^{a\,\mu\nu} |\pi^+(p_1)\pi^-(p_2)> = (p_1+p_2)^2 + O(p^4)$$

by virtue of the scale anomaly which then becomes proportional to the invariant mass squared of the Higgs meson.

3.3 The Anomaly Contribution to Higgs Decay

The assertion stated above will be proven in two steps. First we derive from general principles the form of the matrix element of the energy momentum tensor for transition from the vacuum into a two pion state. Next we derive the low energy form of the coupling of a Higgs to gluons by eliminating the heavy quarks.

From Lorentz invariance and symmetry of $\theta_{\mu\nu}$ the structure of the matrix element must be

$$M_{\mu\nu} = <\pi^+\pi^-|\theta_{\mu\nu}|0> = Ar_\mu r_\nu + Bq_\mu q_\nu + C(r_\mu q_\nu + q_\mu r_\nu) \quad (20)$$
$$+ g_{\mu\nu}\left(D_1 q^2 + D_2 r^2 + D_3 r \cdot q\right) \quad (21)$$

with $\quad r_\mu = p_\mu^{(1)} - p_\mu^{(2)}; \quad q_\mu = p_\mu^{(1)} + p_\mu^{(2)} \quad (22)$

with $p_\mu^{(1,2)}$ the four-momenta of $\pi^{+,-}$ respectively. From charge conjugation invariance of $\theta_{\mu\nu}$ and the vacuum we have $C = D_3 = 0$ and using energy momentum conservation $\partial^\mu \theta_{\mu 0} = 0$ this expression can be reduced to the form

$$q^\mu M_{\mu\nu} = r^2 D_2 q_0 + q^2 \left(Bq_0 + Cr_0 + D_1 q_0\right) \quad (23)$$
$$\Rightarrow \quad D_2 = 0; \; B = -D_1 \quad (24)$$
$$\Rightarrow \quad M_{\mu\nu} = Ar_\mu r_\nu + B\left(q_\mu q_\nu - g_{\mu\nu} q^2\right) \quad (25)$$

Using the neutrality of $\theta_{\mu\nu}$, i.e. $[\theta_{\mu\nu}, Q^5] = 0$, we find $A = -B$ and using crossing symmetry to relate the upper matrix element to the matrix element $<\pi^+|\theta_{\mu\nu}|\pi^+> = 2p_\mu p_\nu$ we find the normalization condition that $A = 1/2$.

Having determined in this way the form of the matrix element, we derive an effective Lagrangian for the interaction of the light Higgs particle. For this purpose we have to calculate the matrix element for coupling to heavy quarks shown in Fig. 3a. As indicated in the figure, the virtual decay of the Higgs particle

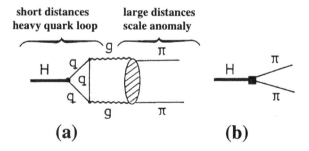

Fig. 3. (a) Decay of a Higgs particle into pions via the scale anomaly; (b) effective coupling of the Higgs particle to pions

into heavier quarks is a short distance effect and leads to an effective coupling to gluons in a first step (Fig. 4). The gluonic state decays at long distance into pions thus causing an effective coupling of the Higgs particle to pions (Fig. 3b). Therefore one finds from elimination of the heavy quarks the effective interaction (see appendix)

$$\mathcal{L}_{eff} = \frac{\alpha_s}{12\pi <H>}\hat{H}G^a_{\mu\nu}G^{a\,\mu\nu} \qquad (26)$$

and using the relation between the gluon fields and the energy momentum tensor provided by the scale anomaly (18) as well as the result for the matrix element

$$<\pi^+\pi^-|\theta^\mu_\mu|0> = q^2 = M_H^2 \qquad (27)$$

we find the effective coupling to pions to be of the form

$$\mathcal{L}_{eff} \propto \frac{2M_H^2}{3b<H>}\hat{H}\hat{\pi}^+\hat{\pi}^-$$

where the energy momentum tensor known from the matrix element (27) has been replaced by $\theta^\mu_\mu \to M_H^2\hat{\pi}^+\hat{\pi}^-$. Therefore the quark masses in the aforementioned ratio between decay into muons and into pions have to be replaced by the Higgs mass which reverses the result of the branching ratio as mentioned in (19).

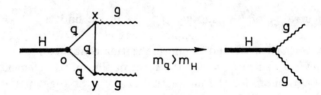

Fig. 4. *Heavy quark loop leading to an effective coupling of Higgs particles to gluons*

Although we have been discussing in this lecture only the effect of the scale anomaly on the decay of a light Higgs particle which is of course still a hypothetical one, we would like to mention that there are indeed effects in pure QCD to be observed, as well. There are e.g. the S–D–wave contributions to the decay $\Psi' \to J/\psi + 2\pi$ where experimentally found small D–wave contributions can be attributed to the scale anomaly.

A The Effective Interaction Between Higgs Particles and Gluons

In this appendix we want to demonstrate how the heavy quark mass in the coupling of the Higgs particle to quarks is cancelled when deriving an effective coupling to gluons and how the field strength tensor arises. We work in the Fock–Schwinger gauge $x^\lambda A^a_\lambda = 0$ which allows us to express the gauge field by the field strength tensor in the following way. We start from the expression of the field strength tensor

$$x^\mu G^a_{\mu\nu} = x^\mu \left(\partial_\mu A^a_\nu - \partial_\nu A^a_\mu + gf^{abc}a^b_\mu A^c_\nu\right) \tag{28}$$

$$= x^\mu - \partial_\nu \left(x^\mu A^a_\mu\right) + A^a_\nu \tag{29}$$

$$= x^\mu \partial_\mu A^a_\nu + A^a_\nu \tag{30}$$

and then rescale the $x^\mu \to \alpha x^\mu$ in order to use

$$\frac{d}{d\alpha}\left(\alpha A^a_\mu(\alpha x)\right) = A^a_\mu(\alpha x) + x^\rho \partial_\rho A^a_\mu(\alpha x) = \alpha x^\rho G^a_{\rho\mu}(\alpha x) \tag{31}$$

$$\Rightarrow A^a_\lambda(x) = \int_0^1 d\alpha\, \alpha G^a_{\rho\lambda}(\alpha x)x^\rho \tag{32}$$

$$= x^\rho \left[\frac{1}{2} + \frac{1}{3}x^\sigma D_\sigma + \ldots\right] G^a_{\rho\lambda}(0) \ . \tag{33}$$

Using this relation, we can calculate the loop shown in Fig. 4 for heavy quarks which propagate only over distances of order $1/m_q$ for which the expansion (33) is valid. The matrix element for $H \to 2g$ then becomes

$$M(H \to 2g) = \lambda\alpha_s \int d^4x\, d^4y\, A^a_\mu(x)A^a_\nu(y) Tr\left\{S(x)\gamma^\mu S(y-x)\gamma^\nu S(-y)\right\}$$

$$\approx \frac{\lambda\alpha_s}{4} G^a_{\mu\rho}(0)G^a_{\nu\sigma}(0) \int d^4x\, d^4y\, x^\rho y^\sigma Tr\left\{S(x)\gamma^\mu S(y-x)\gamma^\nu S(-y)\right\}$$

$$= \frac{\lambda\alpha_s}{4} G^a_{\mu\rho} G^a_{\nu\sigma} \int d^4p\, Tr\left\{S(-p)\gamma^\mu \left[\frac{\partial}{\partial p^\rho}S(p)\right]\gamma^\nu \left[\frac{\partial}{\partial p^\sigma}S(p)\right]\right\} \ .$$

Introducing dimensionless variables $z = p/m_q$ we see that the integral reduces to

$$\int d^4p\ldots = \frac{1}{m_q}I^{\mu\nu\rho\sigma}$$

with $I^{\mu\nu\rho\sigma}$ independent of m_q. Thus we see that heavy quark loops lead to an effective interaction between gluons and light Higgs particles which only depends on α_s and $<H>$ and we understand the origin of the appearance of the field strength tensor.

References

[1] M.A.Shifman, Phys. Rep. **209** (1991) 341
[2] M.A.Shifman, Usp.Fiz.Nauk **157** (1989) 561
[3] N.S.Manton, Ann. Phys. **159** (1985) 220
[4] C.I.Itzykson and J.–B.Zuber, Quantum Field Theory, McGraw Hill Singapore 1985
[5] S.L.Adler, Phys. Rev. **177** (1969) 2426
[6] J.S.Bell and R.Jackiw, Nuovo Cim. **60 A** (1969) 47
[7] A.Vainshtein and V.Zakharov, Sov. Phys. Uspekhi **13** (1970) 73

QCD Sum Rules*

M.A. Shifman[1];
Notes by M. Engelhardt[2] and D. Stoll[2]

[1] Theoretical Physics Institute, University of Minnesota, Minneapolis, MN 55455, USA
[2] Institute for Theoretical Physics III, University of Erlangen–Nürnberg, Staudtstr. 7, 91058 Erlangen, Germany

1 Introduction

In this lecture we discuss QCD sum rules which is a method engineered for the sake of having an approximate calculational scheme for strong coupling QCD and in particular to account for non–perturbative effects. The basis of the method are certain ideas of the structure of the QCD vacuum and the knowledge of the short distance properties of QCD. The sum rule approach translates hadronic measurements into data characterizing vacuum properties in order to gain predictive power to be able to answer questions concerning hadronic properties. The ideas we are referring to are on the one hand the concept of the vacuum being densely filled by fluctuations of the fundamental fields which may be characterized globally by condensate values, the most important of which are

$$< 0|\frac{\alpha_s}{\pi}G^a_{\mu\nu}G^{a\,\mu\nu}|0>; \quad <0|m_q\bar{q}q|0> \ .$$

On the other hand we believe that e.g. the simplicity of the interquark interaction at short distances extends to rather large separations and eventually changes its character in a comparatively small region, similar in spirit to the bag model. For this reason only a minimal extrapolation from the well understood perturbative regime into the theoretically unknown nonperturbative one has to be done and the errors contained in the method are quite controllable. On the basis of this picture the success of the sum rule method in reproducing qualitatively the regularities in the hadronic family in turn provides new insights into the structure of the QCD vacuum. Although the method is constructed in such a way that on the one hand it is as close as possible to first principles and on the other hand allows for readjustment in order to include new experimental as well as theoretical information (e.g. from lattice QCD or heavy quark experiments) it is not a systematic perturbative expansion like the α–expansion in QED. Therefore it can not be used iteratively in order to produce results of arbitrary accuracy and rather it is intended to generate qualitative insights and results which are accurate to within 10–20%.

---

* Lectures presented at the workshop "QCD and Hadron Structure" organised by the Graduiertenkolleg Erlangen–Regensburg, held on June 9th–11th, 1992 in Kloster Banz, Germany

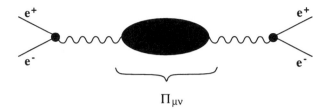

Fig. 1. *The polarization tensor in electron positron annihilation*

2 The Basic Idea (Pictorial Description)

The fundamental step in the sum rule method is a systematic separation of the short distance properties which are well known from perturbative QCD from large distance properties which can not be calculated but which may be parametrized using experimental input by means of a few average vacuum expectation values. In order to demonstrate how this separation arises (for a more formal

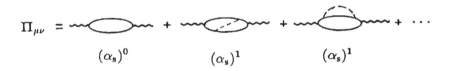

Fig. 2. *Expansion of the polarization tensor in the strong coupling constant*

derivation see appendix) we now want to consider the polarization tensor $\Pi_{\mu\nu}$ shown in Fig. 1

$$\Pi_{\mu\nu}(q) = \int <0|T\left[j^c_\mu(x)j^c_\nu(0)\right]|0> e^{iqx} dx \qquad (1)$$

obtained as vacuum expectation value of the time ordered product of charmed currents

$$j^c_\mu(x) = \; :\bar{c}(x)\gamma_\mu c(x):$$

which may be used for studying hadronic properties such as charmonium resonances. If we perform an expansion in the coupling constant α_s then we find the contributions shown in Fig. 2 where the fermion lines refer to c–quarks. In these diagrams, if the momentum q is small, we have heavy c–quarks propagating with the free fermion propagators $[p^2 + m_c^2]^{-1}$ since they are propagating over typical distances of the order of $1/m_c$ for which a perturbative treatment is applicable, since it is still far below the confinement radius. Calculating only the coupling to the electromagnetic current, but not considering further strong interactions is of course not sufficient for a prediction of properties of hadronic resonances, but a perturbative calculation of the gluonic corrections using a free gluon propagator k^{-2} is inappropriate, as well. The reason is that within the loop momentum integration long wave lengths and in particular the "on–shell" gluon momentum $k = 0$ are contained. But due to confinement the gluon can not become on–shell and therefore using the perturbative propagator, we would make an essential error in the physical description and it is clear that other contributions should also be considered. The physically correct treatment instead consists in a separation

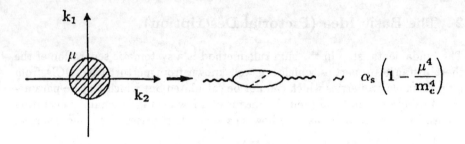

Fig. 3. *The high momentum $(k^2 > \mu^2)$ perturbative contribution to the polarization tensor*

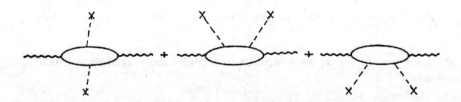

Fig. 4. *The low momentum non–perturbative contribution to the polarization tensor is dominated by coupling to gluon fluctuations in the vacuum*

of the small gluon momentum region from the large momentum region. In the k–plane this amounts to cutting out the central region and diagrammatically it means that we calculate the diagram perturbatively for high momenta down to some scale μ which gives the result shown in Fig. 3. The result for the interaction with gluons of momenta smaller than μ is determined by the diagrams shown in Fig. 4 showing that the dominant contribution arises from coupling to vacuum fluctuations and it leads to a μ–dependent correction term of the form

$$\frac{<\alpha_s G^a_{\mu\nu} G^{a\,\mu\nu}>_\mu}{m_c^4} F(\frac{q^2}{m_c^2}) \qquad (2)$$

containing the non–perturbative gluonic vacuum condensate. To see how this type of contribution can be derived formally, see appendix. Thus a physically reasonable description is one where the coupling of the quarks to low momentum gluons is not dominated by perturbative gluons but by coupling to gluonic fluctuations already present in the vacuum. As stated in the introduction these fluctuations are then parametrized by condensate values. What is important is that $< \alpha_s G^2 >$ is numerically much bigger than μ^4 in Fig. 3.

Note that the heavy quarks are for small q^2 always far off–shell and therefore no non–perturbative correction of the fermion propagators have to be taken into account. In the case of light fermions, however, the situation is different and one encounters in addition to the gluon condensate also quark condensates.

3 Pragmatic Operator Product Expansion

The theoretical tool which is almost ideally suited for the purpose of the sum rule method is the Wilson operator product expansion in which the time ordered product of currents is rewritten in the following way (using in the sequel the momentum space representation)

$$\Pi_{\mu\nu}(q) = (q_\mu q_\nu - q^2 g_{\mu\nu}) \sum_n C_n(q, m_c, \mu) \hat{O}_n(\mu) \quad (3)$$

with coefficient functions C_n containing the short distance information, in particular also the short distance singularities, and non–singular operators $\hat{O}_n$. The separation of long and short distance effects is built into this expansion, but we want more and thus come to the pragmatic version of the OPE. We want the coefficient functions to be determined perturbatively (approximatively) and we want the expectation values of the operators $< \hat{O}_n >$ to be determined entirely by non–perturbative effects. This is in fact essential for the method to work, since only if the expectation value of the operator is dominated by non–perturbative effects then the exact value of the artificial separation constant μ becomes unimportant, as will be seen below. Note that this requirement is responsible for the fact that no pragmatic OPE is possible in two–dimensional σ–models beyond the leading $1/N$ approximation. Therefore such toy models are of no help in trying to understand what happens in QCD [1]. In the previous example of the polarization operator the pragmatic OPE gives the contributions of order 1 and α_s shown in Fig. 5a and 5b which are associated with the unit operator and the contribution 5c stemming from the coupling to the vacuum fluctuations associated with the operator $G^a_{\mu\nu} G^{a\,\mu\nu}$. Since we want the pragmatic version of OPE to be approximately valid, we would like to eliminate the μ–dependence in these expressions. For this purpose we note that due to the fact that μ may be chosen large enough that all non–perturbative effects occur at $k^2 < \mu^2$ and at the same time $\mu/m_c \ll 1$ (see Fig. 6) and that we know that the artificial μ–dependence gets cancelled by perturbative contributions to $< \alpha_s G^2 >_\mu$ we can neglect the μ–dependent contribution in the perturbative part (Fig. 5b). Since

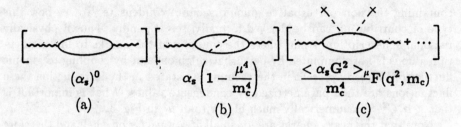

Fig. 5. Result of the pragmatic operator product expansion: (a) Polarization tensor without corrections from strong interactions; (b) Perturbative correction to the polarization tensor; (c) non–perturbative correction due to coupling to gluon fluctuations in the vacuum

on the other hand the condensate is adjusted by experimental input (it turns out that $\langle \alpha_s G^2 \rangle \gg \mu^4$ (Fig. 6) we can forget about the μ–dependence entirely, besides logarithmic corrections in the non–perturbative contribution. The coefficient function $F(q^2, m_c)$ is determined in perturbation theory (see appendix). In order for the pragmatic OPE to have a chance to be valid two important

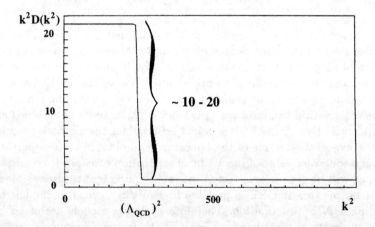

Fig. 6. Qualitative behavior of the gluon propagator in QCD

conditions have to be fulfilled[1]:

(i) The transition from the perturbative to the non–perturbative regime must be abrupt. Then, taking into account only global vacuum properties may be sufficient.

---

[1] In this example there are only two such conditions. In general they may be other conditions, as well, if other operators have to be considered in the expansion.

(ii) $\mu^4 \ll \,<\alpha_s G^2>$ in order for the exact way of separating the regions of large and small momenta to be of no importance.

In QCD both conditions are fulfilled which may be concluded from the success of the method. In addition one can easily show that the value of the gluon condensate is much bigger than what one would predict perturbatively using gaussian wave functionals. Consequently there must on the one hand exist a serious deviation from the perturbative value of the gluon 'propagator' $D(k^2)$ and on the other hand this deviation must be restricted to rather small values of k^2 extending not far beyond Λ^2_{QCD} as may be concluded from the success of perturbative calculations.

Therefore we expect $D(k^2)$

$$D_{\mu\nu}(k) = D(k^2)\left[g_{\mu\nu} - \frac{k_\mu k_\nu}{k^2}\right] \tag{4}$$

to have the qualitative behavior shown in Fig. 6 and in particular the required abrupt change from one regime to the other. Note that e.g. in two–dimensional σ–models $k^2 D(k^2)$ at $k \approx 0$ is only about two times bigger than it is at large momenta, confirming thus the impossibility of applying the pragmatic OPE [1].

4 The Fiducial Domain

Following the approximate prescription outlined above which implies that perturbation theory has to be put into the coefficient functions C_n and only non–perturbative effects into the operators $\hat{O}_n$, we find for the polarization tensor the result [2]

$$\Pi_{\mu\nu}(q) = \left(q^2 g_{\mu\nu} - q_\mu q_\nu\right) \Pi(q^2) \tag{5}$$

$$\Pi(q^2) = F_0(\frac{q^2}{m_c^2}) + \frac{<\alpha_s G^2>}{m_c^4} F_2(\frac{q^2}{m_c^2}) + \frac{<\alpha_s^{3/2} G^3>}{m_c^6} F_3(\frac{q^2}{m_c^2}) \ldots \tag{6}$$

If all terms in this series were known we possibly could perform the sum and determine the position of the poles and the residues of $\Pi(q)$ and thus find the charmonium resonances. Obviously this is not the case and on the contrary it may even be not meaningful to extend the expansion far beyond a few orders, since it is an asymptotic expansion. Therefore the idea is to keep only the first terms containing only a few vacuum condensates which characterize average vacuum properties and try to go as close to the pole as possible. The expansion is then supposed to work in the so–called fiducial domain which is characterized in the following way:

(i) One is able to stay sufficiently close to the lowest lying resonance pole, where the higher states should not contribute strongly
(ii) One can stay sufficiently far away from the lowest lying pole so that only a few terms in the expansion and the corresponding condensates are needed.

One should be aware of the fact that none of the ingredients of the sum rule method neither the pragmatic OPE nor the presence of a fiducial domain are obviously present in any theory. E.g. in two–dimensional σ–models one can explicitly show that both are absent. In QCD one is often in a better position, but not always, and there exist exceptional cases especially in connection with states carrying vacuum quantum numbers and spatially large states (high spin states and radial excitations).

In practice the lower bound in Q^2 for the fiducial domain is simply determined by the requirement that the terms in the expansion become successively smaller; the upper bound is determined by the requirement that higher excited states, the contribution of which is estimated by semiclassical methods, still do not represent a dominant effect.

The condensates which are usually used for parametrizing the non–perturbative effects are the following ones [2]

(i) $<\alpha_s G^a_{\mu\nu} G^{a\,\mu\nu}>$ determined from the mass and residue of the J/ψ-resonance
(ii) $<\alpha_s^{3/2} f^{abc} G^a_\mu{}^\nu G^a_\nu{}^\rho G^a_\rho{}^\mu>$ determined e.g. from the instanton vacuum
(iii) $<m_q \bar{q}q>$ determined from the theory of chiral symmetry breaking according to Gell–Mann–Oakes–Renner and using PCAC [2], [3], Chap. 5.5
(iv) $<\bar{q}\sigma^{\mu\nu} G_{\mu\nu} q>$ is determined from sum rules for nucleons [4]
(v) $<\bar{q}\Gamma q \bar{q} \Gamma q>$ with $\Gamma = 1, \gamma^\mu, \gamma^\mu \gamma^5, \ldots$ is calculated under the assumption of factorization [2].

Thus there are five numbers needed as input to the calculations which then allow us to determine masses, residues, magnetic moments, charge radii, etc. of all hadrons with $J \leq 2$ and without radial excitations. As already mentioned the method may fail because of the missing fiducial domain as is the case for 0^{+-}–glueballs, the η'-meson, Goldstone bosons and radial excitations.

5 A Technical Device – Borelization

In practical applications of the sum rule method there is in many cases a technical device needed in addition in order to achieve a sufficient rapid convergence of the expansion. The device we refer to is borelization (it amounts to performing a Borel transformation) and it consists in applying the following operator B_{M^2} to the function under consideration which is $\Pi(q^2)$ in the case at hand [5]

$$\hat{B}_{M^2} = \lim_{\substack{n \to \infty \\ -q^2 \to \infty}} \frac{(q^2)^n}{(n-1)!} \left(-\frac{\partial}{\partial q^2} \right)^n \tag{7}$$

$$\text{with} \quad M^2 = -\frac{q^2}{n} = fixed\,. \tag{8}$$

Application of this operator e.g. to inverse powers of q^2 gives

$$\hat{B}_{M^2}\left(\frac{1}{q^2}\right)^k = \lim_{\substack{n \to \infty \\ -q^2 \to \infty}} \frac{(q^2)^n}{(n-1)!}(q^2)^{-k-n}\frac{(k+n-1)!}{(k-1)!} \tag{9}$$

$$= \frac{1}{(k-1)!}\left(\frac{1}{M^2}\right)^k + \lim_{n \to \infty} O\left(\frac{1}{n}\right). \tag{10}$$

Thus under the Borel transformation a function $f(x)$ goes into a function $\tilde{f}(\lambda)$ with the property

$$f(x) = \sum_k \alpha_k x^k \quad \longrightarrow \quad \tilde{f}(\lambda) = \sum_k \frac{\alpha_k}{k!}\lambda^k$$

which obviously has better convergence properties. In the case at hand $-1/q^2$ plays the role of the variable x and $1/M^2$ that of λ and we want to apply the borelization to dispersion relations involving $\Pi(q^2)$ (11) which are needed in order to be able to relate the theoretical expression to experimental measurements as it is discussed in the appendix. From the dispersion relation

$$\Pi(q^2) = \text{subtr. const} + \frac{1}{\pi}\int ds \frac{Im\,\Pi(s)}{s - q^2 + i\epsilon} \tag{11}$$

$$\approx C_0 \ln(-q^2) + C_2 \frac{<O_2>}{q^4} + \ldots + C_k \frac{<O_k>}{(q^2)^k} \tag{12}$$

we obtain after borelization (a simple calculation shows that the logarithm transforms into a constant)

$$\tilde{\Pi}(M^2) = \frac{1}{\pi}\int ds \frac{1}{M^2}Im\,\Pi(s)e^{-s/M^2} \tag{13}$$

$$\approx C_0 + C_2 \frac{<O_2>}{M^4} + \ldots + C_k \frac{<O_k>}{(k-1)!(M^2)^k} \tag{14}$$

where (13) follows from borelization of the expansion

$$\hat{B}_{M^2}\frac{1}{s-q^2} = -\hat{B}_{M^2}\sum_k \frac{s^k}{(q^2)^{k+1}} \tag{15}$$

$$= -\sum_k \frac{(-1)^{k+1}s^k}{k!(M^2)^{k+1}} = \frac{1}{M^2}e^{-s/M^2}. \tag{16}$$

Thus instead of suppressing large s values in the integration by $1/s$ we have achieved after borelization an exponential suppression.

6 A Quantum Mechanical Example

In order to provide you with a concrete idea about the fiducial domain, we now want to discuss a very simple example, a single particle bound in a harmonic oscillator potential $V(r) = m\omega^2 r^2/2$. The object corresponding to the polarization tensor is in this example the recurrence amplitude for the particle being at time $t_1 = 0$ at the origin and being detected at the origin again at time $t_2 = -i\tau$. This is measured by

$$S(\tau) = \sum_{n=0,2,4,\ldots} |R_n(0)|^2 \, e^{-E_n \tau} \tag{17}$$

with: $\quad S(\tau) = 4\pi G(x_2 = 0, t_2 = -i\tau; x_1 = 0, t_1 = 0)\;.\tag{18}$

Since the exact Greens function for the harmonic oscillator is known, we have an exact result for $S(\tau)$ to compare with

$$S(\tau) = \frac{2}{\sqrt{2\pi}} \left(\frac{m\omega}{\sinh(\omega\tau)} \right)^{3/2} .$$

The ground state energy (corresponding to the pole position in the polarization tensor) can be obtained from the limit $\tau \to \infty$

$$E_0 = \lim_{\tau \to \infty} \left[-\frac{\partial}{\partial \tau} \ln(S(\tau)) \right] = \frac{3}{2}\omega\;.$$

However, we do not want to use the knowledge about the large τ-behavior but instead assume that we only know the small τ expansion from some perturbative calculations (e.g. starting from the free particle Greens function) where a large number of states contributes. We therefore have to start from the approximate result

$$-\frac{\partial}{\partial \tau} \ln(S(\tau)) = \frac{3}{2}\omega \left[\frac{1}{\omega\tau} + \frac{\omega\tau}{3} - \frac{(\omega\tau)^2}{45} + \frac{2(\omega\tau)^5}{945} \mp \ldots \right]\;.$$

It is of course not possible to extract from the finite sum the correct large τ behavior and therefore the correct ground state energy. However, there is a fiducial domain, where on the one hand not very many levels contribute and on the other hand the expansion does not yet blow up as it is shown in Fig. 7. Obviously it is still possible to find a reasonably good estimate for the ground state energy within the fiducial domain. In numbers this means that we obtain from keeping two terms in the expansion as an estimate $E_0 = \frac{2}{\sqrt{3}} E_{0\,exact}$ which deviates by 15%, from four terms we get $E_0 = 1.06 E_{0\,exact}$, i.e. a 6% error and the estimates may even be improved by using a rough model for the higher lying levels.

Keeping this simple example in mind, we come back now to the previous discussion of the polarization tensor, but discuss instead of charmed currents and the J/ψ-resonance the slightly more complicated case of the ρ-resonance. In this case we find the result shown in Fig. 8 [2]. Again there is a fiducial

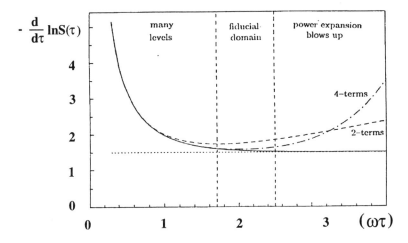

Fig. 7. Fiducial domain for the determination of the ground state energy in the harmonic oscillator

domain[2] where the sum rule approach (including fermionic condensates of light quarks) reproduces quite well the ρ–meson resonance the contribution of which is obtained by approximating the polarization tensor by the pole term only. This gives rise to the following imaginary part and transformed polarization tensor

$$Im\Pi_{\rho-meson}(s) \propto \delta(s - m_\rho^2) \Rightarrow \tilde{\Pi}(M^2)$$

and the corresponding result shown in Fig. 8. Outside of this fiducial domain the method fails for different reasons: In region I the failure is due to the incomplete summation in the polarization tensor corresponding to the failure in the harmonic oscillator example at large values of $\omega\tau$, in region II more resonances than just the ρ become important and the result may be improved by taking into account higher lying resonances in the form of a continuum contribution which sets in at $s = s_0$ and may be approximated by a perturbative calculation [2]

$$\frac{1}{\pi}\int_{s_0}^{\infty} ds \frac{1}{M^2} Im\, \Pi^{p.t.}(s) e^{-s/M^2}.$$

As shown in Fig. 8, M can be chosen as small as the ρ–meson mass, but smaller values are due not accessible to the finite expansion. At this value one is sensitive essentially only to this resonance and the nonperturbative corrections play only a miner role.

---

[2] Note that after the Borel transformation was applied the fiducial domain appears in the variable $1/M^2$ and not in $1/q^2$ anymore.

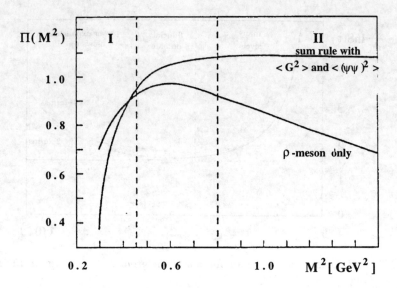

Fig. 8. Fiducial domain in the channel of the ρ-resonance. Regions I and II are explained in the text.

7 Non–factorizable Amplitudes in Weak Decays

Since the sum rule method is by now already more than ten years old and is nicely described in reviews, we now want to turn to a recent application which goes beyond a simple repetition of the more traditional ideas mentioned previously. In this example we intend to demonstrate how the old ideas can be combined with new ones, in this case concerning the weak decay of B-mesons [6].

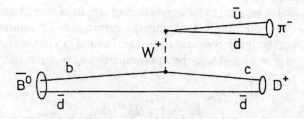

Fig. 9. B–meson decay

We want to consider the weak decay of b–quarks which is described at low energies by the effective Hamiltonian

$$H_W = \frac{G_F}{\sqrt{2}} V_{cb} V_{du} \left[c_1 \hat{O}_1 + c_2 \hat{O}_2 \right] \qquad (19)$$

with:
$$\hat{O}_1 = (\bar{c}_i \Gamma^\mu b_i)(\bar{d}_j \Gamma_\mu u_j) \quad (20)$$
$$\hat{O}_2 = (\bar{c}_j \Gamma^\mu b_i)(\bar{d}_i \Gamma_\mu u_j) \quad (21)$$

with $\Gamma_\mu = \gamma_\mu(1-\gamma^5)$ and V_{bc}, V_{du} elements of the Kobayashi–Maskawa matrix. The interaction takes place via the 'usual' operator $\hat{O}_1$ and an additional color–twisted operator $\hat{O}_2$ which has to be taken into account due to the possibility of hard gluon exchanges, which are not considered explicitly.

Now let us consider the weak decay of B–mesons

$$\bar{B}^0 \to D^+ + \pi^- \quad (b\bar{d} \to c\bar{d} + d\bar{u}) .$$

The contribution of the operator $\hat{O}_1$ in powers of $1/N_c$ (the leading one is drawn in Fig. 9) can be shown to be [7]

$$<D^+\pi^-|\hat{O}_1|\bar{B}^0> = <\pi^-|\bar{d}\Gamma^\mu u|0><D^+|\bar{c}\Gamma_\mu b|\bar{B}^0> + O(\frac{1}{N_c^2})$$

without a first order $1/N_c$ correction. Although this matrix element nicely factorizes which may also be seen from the figure, the same is not true for the matrix element of the operator $\hat{O}_2$. The reason is that this operator not only contains a singlet–singlet coupling at $1/N_c$ level but it has a contribution from an octet–octet coupling, as well

$$\hat{O}_2 = \frac{1}{N_c}(\bar{c}_i \Gamma^\mu b_i)(\bar{d}_j \Gamma_\mu u_j) + 2(\bar{c}_j \Gamma^\mu t^a_{ji} b_i)(\bar{d}_i \Gamma_\mu t^a_{ij} u_j)$$

which follows from the well–known SU(N)–identity

$$\delta_{ij}\delta_{kl} - \frac{1}{N}\delta_{il}\delta_{jk} = 2\sum_a \lambda^a_{ij}\lambda^a_{kl} .$$

Although in the literature factorization is assumed also for this operator, the non–singlet contribution does not allow such a treatment. In addition the experimental data seem to favor no $1/N_c$ corrections at all which means that in the operator $\hat{O}_2$ the factorized singlet–singlet part has to be cancelled by the non–factorizable octet–octet part.

To understand the experimental result and to demonstrate that in fact the factorizable contribution is cancelled by the non–factorizable one, we consider the heavy quarks (c,b) as very heavy $M_c \to \infty$, $M_b \to \infty$, $M_b - M_c = const$ in order to simplify the kinematics. In this limit one finds for the factorizable and non–factorizable parts in the matrix element of $\hat{O}_2$

$$<D^+\pi^-|\hat{O}_2|\bar{B}^0>_{factorizable} = -\frac{if_\pi}{N_c}2Mq_0 \quad (22)$$

$$<D^+\pi^-|\hat{O}_2|\bar{B}^0>_{non-factorizable} \approx \frac{im^2_{\sigma H}}{4\pi^2 f_\pi}2Mq_0 \quad (23)$$

where $m_{\sigma H}$ controls the B, B^* mass splitting as will be shown later. The ratio of the factorizable to the non–factorizable amplitude is then found to be

$$\frac{<D^+\pi^-|\hat{O}_2|\bar{B}^0>_{non-factorizable}}{<D^+\pi^-|\hat{O}_2|\bar{B}^0>_{factorizable}} = -\frac{N_c m_{\sigma H}^2}{4\pi^2 f_\pi^2} \approx -1 \qquad (24)$$

and therefore the two contributions to the matrix element of $\hat{O}_2$ indeed cancel almost entirely in agreement with the experimental result.

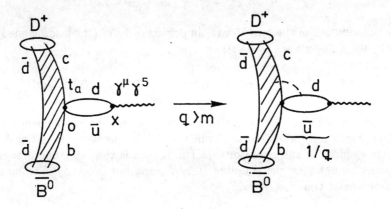

Fig. 10. *Matrix element A^β*

In order to understand this result we now consider only the octet part in $\hat{O}_2$ which is denoted $\tilde{O}_2$ and instead of concentrating on the on mass–shell amplitude we rather focus on the following matrix element

$$A^\beta = \int d^4x e^{iqx} <D^+|T\left[\tilde{O}_2(0)\bar{u}(x)\gamma^\beta\gamma^5 d(x)\right]|\bar{B}^0>$$

which is shown in Fig. 10. For large (euclidian) values of q only small values of x contribute to the integral and therefore the matrix element can be rewritten in the form

$$A^\beta = -\frac{2i}{16\pi^2}\frac{q^\alpha q^\beta}{q^2} <D^+|g\epsilon_{\alpha\mu\rho\sigma}G^{a\,\rho\sigma}\bar{c}\Gamma^\mu t^a b|\bar{B}^0> + O\left(\frac{1}{q^4}\right)$$

where the OPE has been used to replace the time ordered product of light quark fields analogously to the discussion in the appendix. In the leading order in $1/q^2$ the dual of the gluon tensor appears which is physically clear due to the fact that without coupling to an external field the matrix element of $\tilde{O}_2$ in a singlet state vanishes.

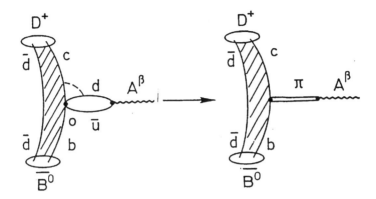

Fig. 11. *Saturation of the axial current by pions*

For the kinematics which was assumed (b,c–quarks infinitely heavy) we can simplify this expression further, since all contributions proportional to the velocity of the heavy quark v are suppressed and we obtain ($\tilde{G}^{\mu\nu}$ denotes the dual of the field strength tensor)

$$q^\alpha <D^+|g\tilde{G}^a_{\alpha\mu}\bar{c}\Gamma^\mu t^a b|\bar{B}^0> = q^0 <D^+|g\tilde{G}^a_{0i}\bar{c}\gamma^i\gamma^5 t^a b|\bar{B}^0> \qquad (25)$$
$$= -2Mq^0 <"D^+"|\bar{h}\boldsymbol{\sigma}\cdot\mathbf{B}h|"\bar{B}^0"> \ . \qquad (26)$$

We see that Γ_μ reduces to $\gamma_\mu\gamma^5$ because of parity and simplifies further to $\gamma_i\gamma^5$ due to the fact that $\gamma_0\gamma^5$ is off-diagonal and causes thus the appearance of a factor of v. Furthermore we can neglect the contributions with $G_{ji}\gamma^i$ since the corresponding matrix element must be proportional to the velocity, as well, as it is easily understood in the rest frame of the b–quark. We thus obtain (26) if we make use of the fact that the strong interaction is flavor blind such that in the limit of very heavy quarks they can be represented by a universal operator h and the corresponding states, e.g. $|"\bar{B}^0">$ without distinguishing b and c–quarks anymore. The interaction which is left is nothing but the magnetic coupling $\sigma\mathbf{B}$ in an effective heavy quark theory represented by the operator

$$H = \frac{\sigma\mathbf{B}}{2M}$$

which is known experimentally from the B, B^* mass splitting

$$m^2_{\sigma H} = -<B|\bar{h}\left(\boldsymbol{\sigma}\mathbf{B}\right)h|B> = \frac{3}{4}\left(M^2_{B^*} - M^2_B\right) \approx 0.35 GeV^2 \ .$$

Using this result we then find for A^β

$$A^\beta = 2i\frac{1}{8\pi^2}\frac{q^\beta}{q^2}m^2_{\sigma H} \qquad (27)$$

which gives the result (24) quoted previously if the pion leg is amputated from A^β. This means we assume saturation of the axial current by the pion field (Fig. 11) and obtain the decay amplitude by eliminating the vertex f_π and the pion propagator. Thus the result (27) has to be devided by i/q^2 to eliminate the propagator and by $if_\pi q^\beta$ to cancel the unwanted vertex contribution in order to obtain the decay matrix element (23).

A The Operator Product Expansion for the Polarization Tensor

This appendix, which was not part of the lectures, is intended to supplement the pictorial description of the sum rule method and in particular the pragmatic operator product expansion with the practical technical procedure [2], [5]. The aim is to show, how

 (i) fermionic condensates may arise in the pragmatic OPE
 (ii) gluon condensates arise
(iii) how the diagrams in Fig. 5 are generated.

Starting point is the time ordered product of currents in the polarization tensor which is decomposed using Wick's theorem in the form

$$T\left[j_\mu^c(x)j_\nu^c(0)\right] = -Tr\left\{<0|T\left[c(0)\bar{c}(x)\right]|0> \gamma_\mu <0|T\left[c(x)\bar{c}(0)\right]|0> \gamma_\nu\right\}$$
$$+ :\bar{c}(x)\gamma_\mu c(x)\bar{c}(0)\gamma_\nu c(0): + :\bar{c}(x)\gamma_\mu <0|T\left[c(x)\bar{c}(0)\right]|0> \gamma_\nu c(0):$$
$$+ :\bar{c}(0)\gamma_\nu <0|T\left[c(0)\bar{c}(x)\right]|0> \gamma_\mu c(x):$$

which is also used in the context of deep inelastic scattering [8]. This result is an identity independent of the nature of the state $|0>$ if the normal ordering introduced refers to this particular state, as well. Usually we are used to dealing with the vacuum state of a noninteracting theory, and calculating matrix elements of the interacting fields $\bar{c}, c$ such as $< 0|T\left[c_i(x)\bar{c}_j(0)\right]|0 >$ gives not just the free fermion propagator but all perturbative radiative corrections as well. In the sum rule approach we, however, have in mind that the vacuum is already filled with fluctuations of soft gluons and therefore we do not want to identify the vacuum with that of the free theory, but we want to have soft gluons contained in it already. Thus $|0>$ should neither be viewed as the trivial vacuum nor as the full physical vacuum since it is assumed to be trivial with respect to the heavy fermion operators and non–trivial with respect to gluonic ones. Because of normal ordering with respect to a trivial fermionic vacuum, it is already obvious from the decomposition that fermionic condensates may arise if the normal ordered operators have non–vanishing expectation values in the true physical vacuum as it is the case for light fermions. Thus the appearance of fermionic condensates is already understood.

In order to motivate the distinction further we want to start from the Lagrangian of QCD and introduce a separation of the gluon field $g_\mu = h_\mu + a_\mu$ which should be viewed as the corresponding separation into soft and hard gluons respectively which was discussed previously. Doing this we find the field strength tensor $G_{\mu\nu}$ expressed in terms of the field strength tensors $H_{\mu\nu}$ and $A_{\mu\nu}$ of the soft and hard gluons respectively and an interaction part which is not of interest for the following discussion

$$G^a_{\mu\nu}G^{a\,\mu\nu} = H^a_{\mu\nu}H^{a\,\mu\nu} + A^a_{\mu\nu}A^{a\,\mu\nu} + a_\mu - h_\mu\text{-coupling terms} \quad (28)$$
$$= H^a_{\mu\nu}H^{a\,\mu\nu} + (\partial_\mu a_\nu - \partial_\nu a_\mu)(\partial^\mu a^\nu - \partial^\nu a^\mu) + \text{interaction terms}.$$

The decomposition expresses the different treatment of the gauge field parts. While h_μ is treated as a known background field which eventually is parametrized by vacuum condensates, a_μ is treated as usual in perturbation theory. The full QCD–Lagrangian then takes the form

$$\mathcal{L} = \bar{c}(i\gamma^\mu D_\mu - m)c - \frac{1}{2}(\partial_\mu a_\nu - \partial_\nu a_\mu)(\partial^\mu a^\nu - \partial^\nu a^\mu) - \frac{1}{2}H^a_{\mu\nu}H^{a\,\mu\nu} \quad (29)$$
$$+ g\bar{c}\gamma^\mu a_\mu c + \text{interaction terms} \quad (30)$$

where $D_\mu = \partial_\mu - igh_\mu$ is the covariant derivative with respect to the background field. The difference mentioned already before expresses itself on the level of the Lagrangian in the treatment of the parts in (29) as diagonalized whereas the parts in (30) are treated as interactions being not yet taken into account. That we do not use covariant derivatives $D_\mu a_\nu$ instead of $\partial_\mu a_\nu$ is because we want to expand finally simultaneously for small α_s and for small $H_{\mu\nu}$. As a result of this difference we find, neglecting the couplings to a_μ for the moment, the following fermion propagator in the background field

$$< 0|T\left[c(y)\bar{c}(x)\right]|0> = <x|\left[\gamma^\mu(i\partial_\mu + gh_\mu - m)\right]^{-1}|y> = S_0(x,y).$$

Taking into account the coupling to the perturbative gluons, this propagator is modified in the usual way by addition of radiative corrections.

Now we want to come back to our interest of performing an operator product expansion for the polarization tensor and to this extend we have to evaluate the background field propagator. We use again the Fock–Schwinger gauge (see appendix of [9]) and obtain

$$S_0(x,0) = <x|\left[(\gamma^\mu p_\mu - m)\left(1 + (\gamma^\nu p_\nu - m)^{-1}g\gamma^\rho h_\rho\right)\right]^{-1}|0> \quad (31)$$
$$= <x|\left[1 - (\gamma^\nu p_\nu - m)^{-1}g\gamma^\rho h_\rho + (\gamma^\nu p_\nu - m)^{-1}g\gamma^\rho h_\rho(\gamma^\sigma p_\sigma - m)^{-1}\right. \quad (32)$$
$$\left. g\gamma^\tau h_\tau \mp \ldots\right](\gamma^\mu p_\mu - m)^{-1}|0>. \quad (33)$$

Expressing the gauge potential by the field strength tensor and expanding around $x = 0$ which is possible for heavy quarks, we find up to order g the result

$$S_0(x,0) = S_f(x) - \int d^4z\, S_f(x-z)g\gamma^\nu h_\nu S_f(z) \pm \ldots \quad (34)$$
$$= S_f(x) - gH_{\rho\nu}(0)\int d^4z\, S_f(x-z)\gamma^\nu x^\rho S_f(z) \pm \ldots \quad (35)$$

where S_f denotes the free fermion propagator without background field. Note also that the latter is translational invariant wheras the former is not since the Fock–Schwinger gauge breaks translation invariance.

If we finally insert the result for the background field propagator into the expression for the product of time ordered currents we obtain one contribution without gluonic field strength tensor which is just the perturbative result (Fig. 5a) and which should be corrected still by taking into account the perturbative gluons as well giving then rise to the contribution in Fig. 5b. Then there are two terms containing one gluon tensor which vanish due to the trace operation and finally there are three contributions with two gluonic tensors which exactly correspond to the diagrams in Fig. 4 and 5c. Thus we have shown how these terms arise, how they are related to the G^2 contribution in the OPE and we have demonstrated how perturbative corrections arise. In addition we pointed out where light fermion condensates may enter which however is irrelevant for the discussion of heavy quarkonia.

B The Comparison of Theoretical and Experimental Results

So far we have just shown the derivation of the polarization tensor within the sum rule approach. What is still missing is the presentation of how we usually compare to experimental results which is actually done by using dispersion relations, as we mentioned earlier. In order to get rid of the unknown subtraction constant, we write it in the form

$$\frac{d}{dq^2}\Pi(q^2) = \frac{1}{(4\pi\alpha Q_c)^2}\int ds \frac{\sigma_c(s)s}{(s-q^2)^2} \tag{36}$$

with the fine structure constant α, the c–quark charge Q_c and the cross section for charm production in e^+–e^-–annihilation $s_c(s)$. Since this cross section is measurable the right hand side of equation (36) may obviously be obtained from experiment. On the other hand it was already determined theoretically so that the comparison is straight forward. In practice, however, one usually compares moments M_n

$$M_n = \frac{1}{n!}\left(\frac{d}{dq^2}\right)^n \Pi(q^2)\bigg|_{q^2=0}$$

or performs in addition a Borel transformation [2].

References

[1] A.I.Vainshtein et al., Sov. Journ. Part. Nucl. Phys. **17** (1986) 204
[2] M.A.Shifman, A.I.Vainshtein and V.I.Zakharov, Nucl. Phys. **B 147** (1979) 385; 448
[3] T.–P.Cheng and L.–F.Li, Gauge theory of elementary particle physics, Clarendon Press Oxford 1988

[4] V.M.Belayev and B.L.Joffe, Sov. Phys. JETP **56** (1982) 493; **57** (1983) 716;
[5] V.A.Novikov et al., Fortsch. Phys. **32** (1984) 585
[6] B.Blok and M.A.Shifman, Nucl. Phys. **B 389** (1993) 534
[7] A.J.Buras, J.-M.Gérard and R.Rückl, Nucl. Phys. **B 268** (1986) 16
    M.A.Shifman, Int. Jour. Mod. Phys. **A 3** (1988) 2768
[8] T.Muta, Foundations of Quantum Chromodynamics, World Scientific Singapore 1987
[9] M.A.Shifman, Anomalies in gauge theories, this volume

The Skyrme Model*

I. Klebanov[1];
Notes by M. Engelhardt[2] and D. Stoll[2]

[1] Department of Physics, Joseph Henry Laboratories, Princeton University, Princeton, NJ 08544, USA
[2] Institute for Theoretical Physics III, University of Erlangen–Nürnberg, Staudtstr. 7, 91058 Erlangen, Germany

1 Introduction

The Skyrme model is a field theory of mesons and baryons based on a nonlinear chiral Lagrangian which only contains meson fields explicitly [1]. Baryons arise in this framework as soliton solutions. These solutions carry a topological charge, or winding number, the conservation of which explains baryon number conservation in a natural way.

At first, the Skyrme model did not attract much attention because the idea of composing baryons out of bosonic fields seemed much less intuitive than introducing elementary baryon fields. Later came the success of the concept of quarks and antiquarks as elementary particles composing all forms of hadronic matter, which was corroborated by the systematics of the hadron spectrum as well as deep inelastic scattering experiments revealing the existence of a hadron substructure.

The concept of quarks and antiquarks as fundamental fermion fields ultimately led to the formulation of Quantum Chromodynamics (QCD), thought to be the underlying theory describing strongly interacting matter. However, the connection between the quarks in QCD ("current quarks") and the original "constituent quarks" making up the hadrons in precursor models seems rather tenuous. The constituent quarks used in nonrelativistic quark models to calculate the hadron spectrum are at best rather complicated collective excitations of the underlying theory, and the successes of the nonrelativistic quark model in many respects merit the qualifier "surprising".

On the other hand, it became apparent that there also exists a rather nontrivial connection of the Skyrme model to QCD in the context of the $1/N_C$-expansion. Subsequent studies of the phenomenology of the model led to rough quantitative agreement with the properties of the low-lying baryon states observed experimentally. Even though the Skyrme model suffers from all the limitations of an effective theory and its connection with QCD is not straightforward

---

* Lectures presented at the workshop "QCD and Hadron Structure" organised by the Graduiertenkolleg Erlangen–Regensburg, held on June 9th–11th, 1992 in Kloster Banz, Germany

to establish, it is formulated as a quantum field theory and as such contains a priori more information than a more phenomenological approach like the nonrelativistic quark model. Thus, the Skyrme model represents a point of view which may be able to complement the nonrelativistic quark model in many ways.

The present account will first review large-N_C QCD and the connection of its Feynman graph expansion with the one of an interacting meson theory. Then the Skyrme model in its simplest (two flavour) form is formulated and the emergence of baryon solutions is exhibited. Baryon quantization is discussed. The model is generalized to three massless flavours and the substantial difficulties connected with the introduction of a strange quark mass are exhibited. It is shown how these difficulties can be addressed by the bound state approach to strangeness [2]. In conclusion, extensions of the model are speculated upon.

2 Large-N_C QCD

In the low-energy regime, the running coupling constant of QCD becomes large, thereby invalidating conventional perturbation theory. In search of a different expansion scheme, 't Hooft introduced a generalization of QCD to the gauge group $SU(N)$ with arbitrary N [3]. The theory simplifies greatly in the limit of

Fig. 1. *Double line notation for gluon lines.*

infinite N because only planar Feynman diagrams contribute to leading order in N, as will be seen below. The realistic case of $N = 3$ can subsequently be reached by perturbing in $1/N$. The Lagrangian is

$$\mathcal{L} = \bar{q}_\alpha^i [i\gamma^\mu(\delta^{ij}\partial_\mu - gA_\mu^{ij}) - \delta^{ij}m_\alpha]q_\alpha^j + \frac{1}{2}\text{Tr}(F_{\mu\nu}F^{\mu\nu}) \tag{1}$$

where the A_μ^{ij} are traceless, anti-hermitian $N \times N$-matrices and the field strength is

$$F_{\mu\nu} = \partial_\mu A_\nu - \partial_\nu A_\mu - g[A_\mu, A_\nu] . \tag{2}$$

The colour indices (i, j) run from 1 to N, and the flavour index α from 1 to N_f. The classification of diagrams according to order in N follows from the observation that the quark fields have $O(N)$ components, whereas the gluon field has $O(N^2)$. The gluon A^{ij} can, for the purpose of colour-index counting, be roughly thought of as a quark-antiquark pair $q^i\bar{q}^j$, which graphically amounts to introducing the "double-line notation"[1] (Fig. 1). This notation serves as a

---

[1] This identification is not meant to go any further than to give a convenient representation of the fact that the gluons are in the adjoint representation of $SU(N)$.

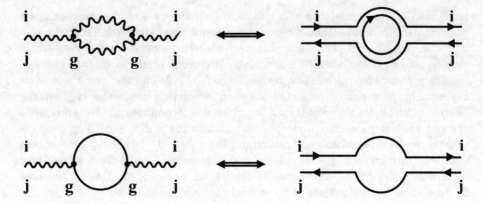

Fig. 2. Examples of Feynman diagrams in double line notation

convenient bookkeeping device when determining the order in N of a Feynman graph. E.g. in the one-loop corrections to the gluon propagator shown in Fig. 2 the gluon loop insertion contributes a factor $g^2 N$, since the index k runs over N values, whereas the contribution of the quark loop insertion merely is of order g^2. In order to achieve a smooth limit as N grows to infinity, the coupling constant has to be scaled down such that $g^2 N \to$ const. Otherwise, adding gluon loops introduces ever higher powers in N, which makes the theory diverge uncontrollably. Internal quark loops are thus suppressed by a power in N.

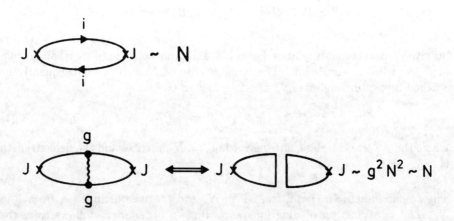

Fig. 3. N–dependencies of Feynman diagrams.

Next, one may consider the two-point function $\langle J(x) J(y) \rangle$ of quark bilinears J such as $\bar{q}q, \bar{q}\gamma_\mu q, \bar{q}\gamma_5 \gamma_\mu q$, etc. The contributions of lowest order in the interaction can be represented in the following way, where the leading dependence on N can be easily read off, taking into account that each quark loop in the double line notation contributes a factor N due to summation over the colour index (see Fig. 3). One can convince oneself that all planar graphs are of order

Fig. 4. *N–dependencies of planar diagrams.*

N, since each additional gluon line on the one hand induces a factor g^2 from the vertices, and on the other hand divides a loop into two, thus contributing a factor N (Fig. 4). Non-planar graphs, on the contrary, are suppressed by at least two powers in N, as can be seen in the following example, which in the double-line notation consists of only one quark loop, but possesses four vertices (Fig. 5). Thus, to determine the leading behaviour of the two point function in the large-

Fig. 5. *N–dependencies of non–planar diagrams.*

N limit, one needs to sum all planar graphs with only one quark-antiquark pair present at every point in time (as was mentioned above, additional quark loop insertions are also suppressed, by a factor N). This sum of planar graphs can be interpreted as creation, propagation, and annihilation of a meson. Therefore, to leading order in N, the two point function has only single-meson poles

$$\langle J(k)J(-k)\rangle = \sum_n |\langle 0|J(k)|n\rangle|^2 \frac{1}{k^2 - m_n^2} \sim N \ . \qquad (3)$$

It follows that $\langle 0|J|n\rangle \sim \sqrt{N}$, and therefore the meson decay constants are of order $O(\sqrt{N})$. Note that this simply stems from the fact that, when creating or annihilating a meson, one counts $O(N)$ possible initial and final $\bar{q}q$ pairs which differ only by the colours of their constituents; it does not mean that the coupling constant of the weak interaction becomes large. Indeed, in practice, one would rescale the effective meson field by a factor $\sqrt{N}$ to arrive at a more appropriate scaling behaviour for the decay constants.

Next, consider three-point functions of J. A typical planar contribution is shown in Fig. 6. In meson language, this graph amounts to creation of a meson, decay into two mesons, and subsequent annihilation of the two mesons. It is thus

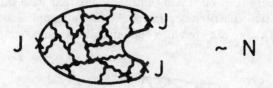

Fig. 6. *N–dependencies of three–point functions.*

proportional to $f^3 g_3$, where f is the meson decay constant and g_3 is the three-meson coupling constant (Fig. 7). Since the whole graph is $O(N)$, and $f \sim \sqrt{N}$,

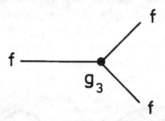

Fig. 7. *Three–meson vertex.*

the coupling constant behaves as $g_3 \sim 1/\sqrt{N}$. Similarly, one can convince oneself that the four-meson coupling behaves as $g_4 \sim 1/N$. In general,

$$g_{2+n} \sim N^{-n/2} \qquad (4)$$

Thus, mesons are free in the limit $N \to \infty$. In $N = \infty$ QCD, there is an infinite tower of stable mesons.

Furthermore, there are graphs with intermediate pure glue states (glueballs), which are suppressed (Fig. 8). This large-N argument seems to be the only simple

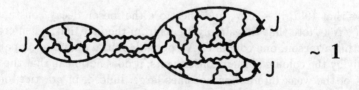

Fig. 8. *N-dependence of graphs with intermediate glueballs.*

way to explain the OZI-rule (suppression of intermediate pure glue states) which approximately holds in nature. In the meson language, this behaviour amounts to weak coupling of mesons to glueballs (Fig. 9) where the spiral line denotes a

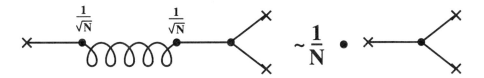

Fig. 9. *Meson–Glueball coupling.*

glueball state.

If one treats the theory beyond the leading order in N, one has to include graphs of increasing topological complexity. On the one hand, there are graphs with quark loop insertions (Fig. 10), i.e. quark-antiquark sea fluctuations, which in the meson language amount to sums over meson loops. On the other hand,

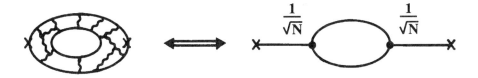

Fig. 10. *Suppressed graphs with quark loop insertions.*

there are non-planar gluon configurations which can be visualized as a planar meson with a gluon handle (Fig. 11). All new types of graphs can be translated

Fig. 11. *Suppressed non-planar configurations.*

into the language of mesons and glueballs. In this sense, QCD with a finite number of colours is a field theory of interacting mesons and glueballs.

This, however, immediately poses the question of how baryons could arise in such a picture. For the following argument, the number of flavours is assumed to be one. When $N = 3$, there does not seem to be a big difference between mesons and baryons: Meson wave functions are constructed out of $q^i \bar{q}^i$, baryon wave functions involve $\epsilon_{ijk} q^i q^j q^k$. For large N, however, the generalizations of these expressions exhibit a fundamental difference: Mesons still consist of a quark-antiquark pair, but baryons now must contain N quarks in order to be able to antisymmetrize N colour indices. Thus, meson masses are of $O(1)$, whereas the baryon mass is

$$M_B \sim N m_q + N T_q + V_{pair} \frac{N(N-1)}{2} \qquad (5)$$

where T_q is the quark kinetic energy and V_{pair} the interaction energy between a pair of quarks (12). Inspection of the graph leads to the conclusion that V_{pair} is

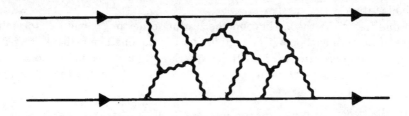

Fig. 12. *Quark–quark interaction.*

of order $O(1/N)$. As a consequence, baryons in the limit $N \to \infty$ have a mass of order N and therefore do not appear in the $1/N$-expansion. For instance, the creation of a baryon-antibaryon pair is a nonperturbative process which cannot be simply described in terms of Feynman diagrams.

Baryons are best thought of in terms of a Hartree-Fock expansion [4]. If only two-body forces between quarks are assumed, then each quark feels a force of order $O(1)$, composed of the forces due to the $N-1$ other quarks, each of order $O(1/N)$. For large N, then, a mean field approximation should be very good. In the non-relativistic Hartree-Fock ground state, each quark is in the same spatial wave function, which could be determined if the quark-quark interaction were known exactly. However, even without detailed knowledge, it is possible to derive the leading-order behaviour of many baryon properties. Whereas baryon masses are of order $O(N) \sim O(1/g^2)$, sizes and cross-sections are of order $O(1)$, and mass-splittings between low-lying baryons are of order $O(1/N) \sim O(g^2)$. This is the typical behaviour of soliton solutions (such as magnetic monopoles), which in the weak-coupling limit become heavy, rigid objects, which can be treated semi-classically. These considerations led to a revival of Skyrme's model [1] [5], who long before had already suggested the possibility of viewing baryons as soliton solutions of mesonic field theories.

3 The Two-Flavour Skyrme Model

It is extremely difficult to rewrite QCD in terms of a theory of infinitely many interacting mesons and glueballs. As a first approximation, one considers a world with two massless flavours and truncates to the lightest (massless) mesons, the pions. This, the simplest form of the Skyrme model, can be used to test the notion that the lowest-lying baryons can be interpreted as soliton solutions of a mesonic Lagrangian. The chiral pion Lagrangian is[2]

$$\mathcal{L} = \frac{f_\pi^2}{16} \text{Tr}(\partial_\mu U \partial_\mu U^\dagger) + \frac{1}{32e^2} \text{Tr}[\partial_\mu U U^\dagger, \partial_\nu U U^\dagger]^2 \tag{6}$$

where U is an element of $SU(2)$:

$$U(\mathbf{x}, t) = \exp\left(\frac{2i\tau\pi}{f_\pi}\right) \tag{7}$$

and π is the pion field. The flavour transformation properties of U can be derived via the observation that

$$U_\alpha^\beta \sim \langle \bar{q}_{Ri}^\beta q_{L\alpha}^i \rangle \tag{8}$$

where

$$q_L = \frac{1+\gamma_5}{2} q \quad, \quad q_R = \frac{1-\gamma_5}{2} q \ . \tag{9}$$

Under $SU(2)_L$ and $SU(2)_R$, respectively,

$$q_{L\alpha} \to A_\alpha^\beta q_{L\beta} \tag{10}$$

$$q_{R\alpha} \to B_\alpha^\beta q_{R\beta} \tag{11}$$

and therefore, under $SU(2) \times SU(2)$ chiral rotations,

$$U \to AUB^{-1} \ . \tag{12}$$

In the vacuum (meaning here, the classical solution of lowest energy), $U = 1$, and thus $SU(2) \times SU(2)$ is spontaneously broken down to the diagonal ($A = B$) $SU(2)_V$ subgroup, i.e., the isospin. If U is expanded in pion fluctuations near $U = 1$,

$$U = 1 + \frac{2i\tau \cdot \pi}{f_\pi} + \ldots \tag{13}$$

and the N-dependences $f_\pi \sim \sqrt{N}, 1/e \sim \sqrt{N}$ are assigned, then the pion interaction terms are consistent with the large-N rule $g_{2+n} \sim N^{-n/2}$.

---

[2] This is a similar Lagrangian as the one used in chiral perturbation theory [6]. In the simplest form of chiral perturbation theory, one starts with the first term of (6) and adds mass terms, which are then treated perturbatively. In the Skyrme model, the emphasis is different. In order to be able to study the baryon sector, it is crucial to include the second term in the chiral Lagrangian (6), without which the classical soliton solution would collapse to zero size [7]. Mass terms are introduced additionally at a later stage of the treatment.

In addition to small fluctuations around $U = 1$, which are interpreted as pions, the model has static solutions which are stable for topological reasons, with U taking values all over $SU(2)$. This can be seen as follows: One can parametrize the $SU(2)$ field as

$$U = u_0 + i\mathbf{u} \cdot \boldsymbol{\tau} . \tag{14}$$

Unitarity of U gives the constraint $u_0^2 + \mathbf{u}^2 = 1$, and therefore the $SU(2)$ group manifold possesses the topology of a 3-sphere S^3. Now, if one considers static configurations where $U \to 1$ sufficiently rapidly as $r \to \infty$, then all points with $r \to \infty$ can be identified into one point which is mapped into $U = 1$. This stands in analogy to the situation in complex function theory, where the complex plane, together with the point at infinity, can be identified with the Riemann sphere. Thus, instead of maps from $\mathbf{R}^3$ to $S^3(SU(2))$, the problem is reduced to maps from S^3(space) to $S^3(SU(2))$. These maps fall into homotopy classes labeled by an integer winding number

$$B = \frac{1}{24\pi^2} \int d^3x \epsilon^{\alpha\beta\gamma} \mathrm{Tr}(\partial_\alpha U U^\dagger \partial_\beta U U^\dagger \partial_\gamma U U^\dagger) . \tag{15}$$

The winding number is invariant under any smooth deformation of U and therefore is conserved in physical processes. This provides the motivation for Skyrme's identification of the winding number with the baryon number. One can further define the baryon number current

$$B^\mu = \frac{1}{24\pi^2} \epsilon^{\mu\alpha\beta\gamma} \mathrm{Tr}(\partial_\alpha U U^\dagger \partial_\beta U U^\dagger \partial_\gamma U U^\dagger) \tag{16}$$

and check that $\partial_\mu B^\mu = 0$ is satisfied identically, without use of the equations of motion. This is another way to see that B is conserved in smooth evolutions of U.

The simplest non-trivial configuration is a one-to-one map from space to $SU(2)$ with $B = 1$. Such a solution can be constructed via the hedgehog ansatz

$$U_0(\mathbf{x}) = \exp(iF(x)\boldsymbol{\tau} \cdot \hat{x}) = \cos F(x) + i\boldsymbol{\tau} \cdot \hat{x} \sin F(x) \tag{17}$$

with $F(0) = \pi$ and $F(x) \to 0$ as $x \to \infty$. If $F(x)$ is monotonic, this is a one-to-one map, with the origin mapped into the south pole of $SU(2)$, i.e. $U_0(0) = -1$, and the point at infinity mapped into the north pole, $U_0(\infty) = 1$. One can check by straightforward calculation that the winding number satisfies

$$B = -\frac{2}{\pi} \int_0^\infty F' \sin^2 F \, dx = 1 . \tag{18}$$

A special property of U_0 is that, for arbitrary $F(x)$, it possesses a symmetry under combined spatial and isospin rotations. If one denotes the operators of spatial and isospin rotations by $\mathbf{L}$ and $\mathbf{I}$, respectively, then U_0 satisfies

$$(\mathbf{I} + \mathbf{L})U_0 = \left[\frac{\boldsymbol{\tau}}{2}, U_0\right] - i\mathbf{x} \times \boldsymbol{\nabla} U_0 = 0 . \tag{19}$$

The classical solution within the $B = 1$ sector is found by minimizing the soliton mass with respect to the radial function F

$$M = 4\pi \frac{f_\pi}{e} \int_0^\infty r^2 dr \left[\frac{1}{8}\left(\frac{dF}{dr}\right)^2 + \frac{1}{4}\frac{\sin^2 F}{r^2}\left(1 + 2\frac{\sin^2 F}{r^2}\right) + 4\left(\frac{dF}{dr}\right)^2\right) \right] \quad (20)$$

where, for convenience, x has been replaced by $ef_\pi r$.

Next, quantum excitations of the classical soliton solution must be considered. Because of isospin invariance, if U_0 is a solution, so is $AU_0 A^{-1}$. Thus, the lowest-lying excitations are slow rigid rotations, $A \to A(t)$. A is now regarded as the quantum mechanical variable, a collective coordinate. If

$$U(\mathbf{r}, t) = A(t)U_0(\mathbf{r})A^{-1}(t) \quad (21)$$

is substituted into the action, one obtains the collective coordinate Lagrangian,

$$L = -M + \Omega \text{Tr}(\partial_0 A \partial_0 A^{-1}) \quad (22)$$

where the soliton moment of inertia Ω is

$$\Omega = \frac{2\pi}{3e^3 f_\pi} \int r^2 dr \sin^2 F \left[1 + 4\left((F')^2 + \frac{\sin^2 F}{r^2}\right)\right]. \quad (23)$$

When quantizing this Lagrangian, it is convenient to parametrize

$$A = a_0 + i\mathbf{a} \cdot \boldsymbol{\tau} \quad \text{where} \quad a_0^2 + \mathbf{a}^2 = 1. \quad (24)$$

Then the Lagrangian becomes

$$L = -M + 2\Omega \sum_{i=0}^{3} (\dot{a}_i)^2 \quad (25)$$

and the Hamiltonian takes the form

$$H = M + \frac{1}{8\Omega} \sum_i \Pi_i^2 \quad (26)$$

with the conjugate momentum $\Pi_i = 4\Omega \dot{a}_i$. Canonical quantization is achieved via $\Pi_i \to -i\partial/\partial a_i$, and thus,

$$H = M - \frac{1}{8\Omega}\nabla^2 \quad (27)$$

where ∇^2 is the Laplacian on the 3-sphere $\sum a_i^2 = 1$.

One would like to construct baryon wave functions of definite isospin and angular momentum. The generators of spatial rotations and isospin rotations are classically

$$J_k = \Omega \text{Tr}(\tau_k A^{-1} \frac{dA}{dt}) \quad (28)$$

$$I_k = \Omega \text{Tr}(\tau_k A \frac{dA^{-1}}{dt}) \quad (29)$$

which upon quantization becomes

$$J_k = \frac{i}{2}\left(a_k \frac{\partial}{\partial a_0} - a_0 \frac{\partial}{\partial a_k} - \epsilon_{klm} a_l \frac{\partial}{\partial a_m}\right) \qquad (30)$$

$$I_k = \frac{i}{2}\left(a_0 \frac{\partial}{\partial a_k} - a_k \frac{\partial}{\partial a_0} - \epsilon_{klm} a_l \frac{\partial}{\partial a_m}\right) . \qquad (31)$$

One can check that
$$J^2 = I^2 = -\frac{1}{4}\nabla^2 \qquad (32)$$

and therefore the Hamiltonian becomes
$$H = M + \frac{1}{2\Omega} J^2 . \qquad (33)$$

All wave functions satisfy the $I = J$ rule.

Since A and $-A$ give the same matrix U, as can be inferred from (21), there are two ways to quantize the soliton. One can choose between $\psi(A) = \psi(-A)$ and $\psi(A) = -\psi(-A)$. Consider as an example the wave functions[3]

$$\psi(A) = (a_0 + ia_1)^l . \qquad (34)$$

One can check that they are eigenfunctions of spin and isospin with $I = J = l/2$. If one now chooses $\psi(A) = \psi(-A)$, then l must be even, and the soliton is thus a boson. Conversely, if $\psi(A) = -\psi(-A)$, then l is odd and the soliton is a fermion.

In the two flavour case there is therefore a choice, and in the following the fermion option will be adopted. In the three flavour case, this choice will no longer exist, and the soliton will be seen to be a fermion for N odd and a boson for N even.

Once the two-flavour soliton is chosen to be a fermion, the allowed quantum numbers are $I = J = 1/2, 3/2, \ldots$. The $I = J = 1/2$ states are identified with the nucleons, and the $I = J = 3/2$ states with the deltas. The wave functions are given by

$$\langle A|p\uparrow\rangle = 1/\pi\ (a_1 + ia_2) \qquad (35)$$
$$\langle A|n\uparrow\rangle = i/\pi\ (a_0 + ia_3) \qquad (36)$$
$$\langle A|\Delta^{++}, s_z = 3/2\rangle = \sqrt{2}/\pi\ (a_1 + ia_2)^3 \qquad (37)$$
$$\langle A|\Delta^+, s_z = 1/2\rangle = -\sqrt{2}/\pi\ (a_1 + ia_2)[1 - 3(a_0^2 + a_3^2)] \qquad (38)$$

and so on. The quantum numbers of nucleons and deltas are thus successfully reproduced. Presumably the states with higher I and J are artefacts of the rigid rotator approximation, since for them the rotational energy approaches the energy of the static soliton. The masses of nucleons and deltas are

$$M_N = M + \frac{1}{2\Omega} \cdot \frac{3}{4} \ , \quad M_\Delta = M + \frac{1}{2\Omega} \cdot \frac{15}{4} \qquad (39)$$

---

[3] One can easily check that all linear functions of the a_i are automatically eigenfunctions of I^2 and J^2.

where

$$M = 36.5\frac{f_\pi}{e}, \quad \Omega = 50.9\frac{2\pi}{3e^2 f_\pi}. \quad (40)$$

Fitting the nucleon and delta masses, one arrives at $f_\pi = 129$MeV and $e = 5.45$, which can be extracted from $\pi - \pi$ scattering. This is about 30% off the experimental values. In this fit, $M(5/2) = 1730$MeV, i.e. $E_{rot} > M$, and therefore the semiclassical treatment is inapplicable because deformations of the soliton due to rotations and vibrations become dominant. In an exact treatment, these states are expected to become unstable due to their rotation. Higher N admit semiclassical treatment of more states, roughly up to $I = J \sim O(N)$.

One may now compare the phenomenology of the Skyrme model with the one obtained from a large-N nonrelativistic quark model in its simplest form, with essentially no dynamics. In Hartree-Fock approximation, low-lying baryons have all quarks in the same spatial orbital. Therefore, the spin-flavour wave function must also be totally symmetric (since the colour wave function, being a singlet, is totally antisymmetric). This constrains the spin and flavour Young tableaux, which are identical, to the form shown in Fig. 13 where the total number of

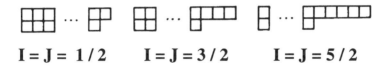

Fig. 13. *Young tableaux for the spin–flavor wave function.*

boxes has to equal N. This corresponds to the systematics found in the Skyrme model. Furthermore, if one assumes a spin-dependent pair interaction between the quarks,

$$\delta H = \lambda \sum_{i<j} \mathbf{J}_i \mathbf{J}_j = \frac{\lambda}{2}\left[\mathbf{J}^2 - \frac{3}{4}N\right] \quad (41)$$

then the energy takes the form

$$H = M + \frac{\lambda}{2}\mathbf{J}^2 \quad (42)$$

as in the Skyrme model. Thus these two very different concepts of the baryon lead to very similar predictions. Because of the dynamical nature of the Skyrme model, however, one can extract a large amount of further information from it than from the quark model. Results are shown in table 1 where μ denotes magnetic moments and g coupling constants.

Table 1. *Results from the two flavour Skyrme model*

	Prediction	Experiment
rms (I=0)	0.59 fm	0.72 fm
magnetic rms (I=0)	0.92 fm	0.81 fm
μ_p	1.87	2.79
μ_n	-1.31	-1.91
$\|\mu_p/\mu_n\|$	1.43	1.46
g_A	0.61	1.23
$g_{\pi NN}$	8.9	13.5
$g_{\pi N\Delta}$	13.2	20.3

4 The Three-Flavour Skyrme Model

In view of the encouraging success of the two-flavour Skyrme model in describing nucleons and deltas, it seems natural to extend the model to include strange particles. The standard procedure would be to start with three massless flavours and then attempt perturbation theory in the quark masses. The discussion will show that mass perturbation theory fails, and alternative treatments of finite strange quark mass will be elaborated upon in the sequel.

4.1 Massless Fermions

The Skyrme model with three massless flavours differs from the two-flavour case in some important ways. A naive generalization of the latter would be

$$\mathcal{L} = \frac{f_\pi^2}{16} \text{Tr}(\partial_\mu U \partial_\mu U^\dagger) + \frac{1}{32e^2} \text{Tr}[\partial_\mu U U^\dagger, \partial_\nu U U^\dagger]^2 \tag{43}$$

where U now is an element of $SU(3)$:

$$U = \exp\left(\frac{2i\lambda_a M^a}{f_\pi}\right) \tag{44}$$

and the components of M are the pion field, $M^a = \Pi^a$ for $a = 1, 2, 3$, the kaon field, $M^a = K^a$ for $a = 4, 5, 6, 7$, and the eta, $M^a = \eta$ for $a = 8$. Unfortunately, this Lagrangian possesses a symmetry which is not present in QCD, namely separate invariance under "naive parity" $\mathbf{x} \to -\mathbf{x}, t \to t$ and under $U \to U^\dagger$ (or equivalently, $M^a \to -M^a$), which is an operation which counts the number of Goldstone bosons modulo 2. In QCD, these symmetries are separately broken in processes like

$$K^+ K^- \to \pi^+ \pi^- \pi^0 \tag{45}$$

where an even number of Goldstone bosons decays into an odd number. QCD is only invariant under the combination of the two transformations. It is not hard to specify the correction to the equation of motion which allows processes such as (45)

$$\partial_\mu \left(\frac{1}{8}f_\pi^2 U^\dagger \partial_\mu U\right) + \ldots + \lambda \epsilon^{\mu\nu\alpha\beta} U^\dagger \partial_\mu U \ldots U^\dagger \partial_\beta U = 0 \ . \tag{46}$$

However, the term in the action, called the Wess–Zumino term, from which such a modification of the equation of motion follows, is very subtle and cannot be written in local form, as a four-dimensional integral

$$S_{WZ} = -\frac{in}{240\pi^2} \int_Q d^5x \epsilon^{\mu\nu\alpha\beta\gamma} \text{Tr}(U^\dagger \partial_\mu U \ldots U^\dagger \partial_\gamma U) \tag{47}$$

where the integral extends over a five-dimensional surface Q with space-time as its boundary and U has been smoothly continued into the interior of the surface. One feature of this term which is crucial for the description of baryons as solitons is the fact that the coefficient n must be an integer. This results from an ambiguity in the definition of the 5-surface over which the integral extends (see Fig. 14). The difference between these definitions is the integral over the

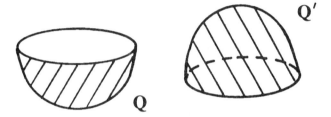

Fig. 14. *Boundaries Q, Q' for the integration in the Wess–Zumino term*

5-sphere formed by the union of Q and Q':

$$\Delta S_{WZ} = -\frac{in}{240\pi^2} \int_{Q\cup Q'} d^5x \epsilon^{\mu\nu\alpha\beta\gamma} \text{Tr}(U^\dagger \partial_\mu U \ldots U^\dagger \partial_\gamma U) \tag{48}$$

which can be shown to be a multiple of 2π if n is an integer. Only then does this difference not introduce any ambiguity in quantum mechanical amplitudes, since path integrals only involve $\exp(iS_{WZ})$.

In order to specify which value of n is compatible with QCD, one can couple the chiral theory to electromagnetism [8]. The Wess–Zumino term introduces an effective interaction describing the decay $\pi^0 \to \gamma\gamma$:

$$\mathcal{L}_{\text{int}} = \frac{ne^2}{48\pi^2 f_\pi} \pi^0 \epsilon^{\mu\nu\alpha\beta} F_{\mu\nu} F_{\alpha\beta} \tag{49}$$

which gives agreement with the amplitude calculated from the triangle diagram if one chooses $n = N$, the number of colours. Similarly, one can show that the VAAA anomaly is reproduced correctly if $n = N$. The Wess-Zumino term incorporates all the effects of QCD anomalies in low-energy processes involving photons and Goldstone bosons[4].

As mentioned above, the Wess-Zumino term is crucial for the interpretation of baryons as solitons in the $SU(3)$ Skyrme model. The soliton solution has the same form as in the two-flavour case, since it is the $SU(2)$ subgroup of $SU(3)$ which possesses the topology of a 3-sphere S^3 and which thus allows for non-trivial mappings $S^3(\text{space}) \to S^3(SU(2))$:

$$U_0(\mathbf{r}) = \exp\left(iF(r)\sum_{i=1}^{3}\lambda_i \hat{r}_i\right) . \tag{50}$$

If one considers the change of phase in the wave function of the soliton generated by S_{WZ} as it is rotated adiabatically through the angle 2π, i.e.

$$U(\mathbf{r},\tau) = \exp(i\tau\lambda_3/2)U_0(\mathbf{r})\exp(-i\tau\lambda_3/2) , \quad 0 \leq \tau < 2\pi \tag{51}$$

then one finds $S_{WZ} = \pi N$, which corresponds to a phase factor of $(-1)^N$. This means that the soliton is a fermion for odd N and a boson for even N, in agreement with QCD. To do this calculation, one has to find a smooth extension of $U(\mathbf{r},t)$ into the interior of the manifold $D_2 \times S^3$, whose boundary $S^1 \times S^3$ is space-time, in order to be able to evaluate the Wess-Zumino term [8]. Furthermore, the Wess-Zumino term imposes strong constraints on the $SU(3)_f \times SU(2)_{spin}$ quantum numbers of the baryons. This results after collective coordinate quantization

$$U(\mathbf{r},t) = A(t)U_0(\mathbf{r})A^{-1}(t) \tag{52}$$

which gives the effective Lagrangian for $A(t)$:

$$\mathcal{L} = -\frac{\Omega}{2}\sum_{j=1}^{3}(\text{Tr}\lambda_j A^{-1}\dot{A})^2 - \frac{\Phi}{2}\sum_{a=4}^{7}(\text{Tr}\lambda_a A^{-1}\dot{A})^2 - i\frac{N}{2\sqrt{3}}\text{Tr}(\lambda_8 A^{-1}\dot{A}) \tag{53}$$

where the moments of inertia Ω and Φ are functionals of the profile function $F(r)$. Numerically one has

$$\Omega = \frac{106}{f_\pi e^3} , \quad \Phi = \frac{39}{f_\pi e^3} . \tag{54}$$

Convenient coordinates are given by

$$A^{-1}\dot{A} = \frac{i}{2}\sum_{i=1}^{8}\lambda_i \dot{a}_i \tag{55}$$

---

[4] This was the original motivation for the Wess-Zumino action: to introduce the effects of the anomaly into the Lagrangian, at the classical level.

which leaves the Lagrangian

$$\mathcal{L} = \frac{\Omega}{2}\sum_{j=1}^{3}(\dot{a}_j)^2 + \frac{\Phi}{2}\sum_{a=4}^{7}(\dot{a}_a)^2 + \frac{N}{2\sqrt{3}}\dot{a}_8 \; . \tag{56}$$

If the canonical momenta to the a_i are denoted by π_i, then the Hamiltonian reads

$$H = \frac{1}{2\Omega}\sum_{j=1}^{3}\pi_j^2 + \frac{1}{2\Phi}\sum_{a=4}^{7}\pi_a^2 \tag{57}$$

and there is a constraint on the wave functions originating from the Wess-Zumino term, because $\dot{a}_8$ only appears linearly in the Lagrangian:

$$\pi_8\Psi(A) = \frac{N}{2\sqrt{3}}\Psi(A) \; . \tag{58}$$

Since the π_i generate right rotations on A, this constraint implies

$$\Psi(Ae^{i\alpha Y}) = e^{i\alpha N/3}\Psi(A) \tag{59}$$

where $Y = \lambda_8/\sqrt{3}$ is the normalized hypercharge. In order to implement the constraint on the wave function, one can write it in the general form

$$\Psi(A) = \langle I, I_3, Y | D^{(p,q)}(A) | I', I_3', Y' \rangle \; . \tag{60}$$

All the irreducible representations of $SU(3)$ can be characterized by a pair of integers p and q [9]. The elements of the (p,q) representation form a traceless tensor with p upper and q lower indices symmetric in the upper and lower indices separately. Thus every $SU(3)$ group element A can be represented by a $N_{(p,q)} \times N_{(p,q)}$ matrix $D^{(p,q)}(A)$ acting on the states of the (p,q) representation, where $N_{(p,q)}$ is the dimension of the representation. Thus the form (60) denotes a complete set of functions on the $SU(3)$ manifold, where the states are labelled by their isospin I, its third component I_3, and the hypercharge Y. Because of the group property

$$D^{(p,q)}(AB) = D^{(p,q)}(A)D^{(p,q)}(B) \tag{61}$$

the constraint (59) implies that the allowed wave functions must have the right hypercharge index $Y' = N/3$, and therefore the representation (p,q) must contain a state with hypercharge $N/3$:

$$\Psi(A) = \langle I, I_3, Y | D^{(p,q)}(A) | I', I_3', N/3 \rangle \; . \tag{62}$$

Now the $SU(3)_f \times SU(2)_{spin}$ quantum numbers of the wave functions may be derived. Under $SU(3)_f$, $U \to FUF^{-1}$, which reduces to $A \to FA$ for the rigid rotator ansatz (52), and therefore the wave functions transform as

$$\langle I, I_3, Y | D^{(p,q)}(A) | I', I_3', N/3 \rangle \to \tag{63}$$
$$\langle I, I_3, Y | D^{(p,q)}(F) | I'', I_3'', Y'' \rangle \langle I'', I_3'', Y'' | D^{(p,q)}(A) | I', I_3', N/3 \rangle$$

Thus, the left set of indices (I, I_3, Y) are the flavour indices which transform in the (p, q) representation of $SU(3)$.

Similarly, under rotations, $U_0 \to R^{-1} U_0 R$, where R is an $SU(2)_{spin}$ matrix, and therefore $A \to AR^{-1}$. Hence, the right indices (I', I_3') of the wave function transform under rotations. The angular momentum carried by the wave function is thus $(J, J_3) = (I', -I_3')$. The constraint may now be phrased in the following way: An allowed $SU(3)_f \times SU(2)_{spin}$ representation must contain a state with $Y = N/3$ and $I = J$. The smallest representations for $N = 3$ are the octet $[(1,1); J = 1/2]$, containing the nucleons, and the decuplet $[(3,0); J = 3/2]$, containing the deltas. In addition to the correct description of the quantum numbers of the lowest-lying observed baryons, the Skyrme model contains an infinite tower of exotic multiplets, which again is presumably an artefact of the rigid rotator approximation. A more complete treatment, as exhibited further below, explicitely shows that some of them are unstable.

It is useful to again compare with the large-N quark model. Just as in the two-flavour case, the spin and flavour parts of the Hartree-Fock wave function must have identical Young tableaux with N boxes each (Fig. 15) where (a) cor-

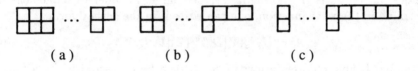

(a) (b) (c)

Fig. 15. *Young tableaux in the N–flavor model.*

responds to the large-N analogue of the octet, (b) of the decuplet, etc. Each of these representations satisfies the Skyrme model constraint. Again, the quantum numbers of the low-lying states of the Skyrme model agree with those of the quark model. However, the Skyrme model contains other representations which cannot be constructed out of quarks alone. These states have higher masses and considerations further below will show that rigid rotator methods are not trustworthy in describing these exotic baryons.

Next, the mass splittings between the multiplets must be considered. Acting on the allowed wave functions, the Hamiltonian (57) can be simplified to

$$M^{(p,q)} = \frac{1}{2\Omega} J(J+1) + \frac{1}{2\Phi} \left(C^{(p,q)} - J(J+1) - N^2/12 \right) \quad (64)$$

because for $i = 1, 2, 3$, $\pi_i = J_i$ and furthermore,

$$\sum_{i=1}^{8} \pi_i^2 \Psi^{(p,q)}(A) = C^{(p,q)} \Psi^{(p,q)}(A) \quad (65)$$

where $C^{(p,q)}$ is the value of the (quadratic) Casimir operator in the (p, q) representation. If $N = 3$, then the relevant $SU(3)_f$ multiplets are the octet, for which

$C^{(1,1)} = 3$, and the decuplet, for which $C^{(3,0)} = 6$. Therefore, for $N = 3$, the two lowest multiplets have the masses

$$M^{(1,1)} = M_{cl} + \frac{3}{8\Omega} + \frac{3}{4\Phi} \qquad (66)$$

$$M^{(3,0)} = M_{cl} + \frac{15}{8\Omega} + \frac{3}{4\Phi} . \qquad (67)$$

Just as in the two-flavour case, the mass splitting between the two lowest multiplets is $\Delta M = 3/2\Omega \sim 1/N$. More generally, one can show that the mass spectrum has the form

$$M = M_{cl} + \frac{1}{2\Omega}J(J+1) + \frac{N}{4\Phi} . \qquad (68)$$

This again agrees with the form found in the large-N quark model with $j_i j_j$ pair interactions (see the section on the two-flavour case). Thus for three massless flavours, the Skyrme model is a success.

4.2 Mass Perturbation Theory

In order to treat finite strange quark mass, one must introduce into the Lagrangian a mass term of the form [6]

$$\mathcal{L}_m = -\frac{m_K^2 f_\pi^2}{8} \operatorname{Tr} \begin{pmatrix} 0 & 0 & 0 \\ 0 & 0 & 0 \\ 0 & 0 & 1 \end{pmatrix} (U + U^\dagger - 2) \qquad (69)$$

which explicitely breaks the $SU(3)$ symmetry. In the effective Lagrangian for A, this induces a term

$$\mathcal{L}_m = -\frac{m_K^2 \alpha}{2} \operatorname{Tr} \begin{pmatrix} 0 & 0 & 0 \\ 0 & 0 & 0 \\ 0 & 0 & 1 \end{pmatrix} A \begin{pmatrix} 1 & 0 & 0 \\ 0 & 1 & 0 \\ 0 & 0 & 0 \end{pmatrix} A^{-1} \qquad (70)$$

where $\alpha \approx 107/f_\pi e^3$. In the Gell-Mann–Oakes–Renner treatment of meson masses, first order perturbation theory in the quark masses works quite well even in the three-flavour case. Using the baryon octet wave functions to attempt a similar treatment in the framework of the Skyrme model, one finds

$$m_N = m - \frac{3}{10}x \qquad m_\Lambda = m - \frac{1}{10}x \qquad (71)$$

$$m_\Sigma = m + \frac{1}{10}x \qquad m_\Xi = m + \frac{2}{10}x \qquad (72)$$

where $m = M^{(1,1)} + x$ and $x = m_K^2 \alpha/3$. Even the splittings between these masses are in bad disagreement with nature. The observed values are

$$m_\Lambda - m_N = 176\text{MeV} \qquad m_\Sigma - m_\Lambda = 78\text{MeV} \qquad 2(m_\Xi - m_\Sigma) = 250\text{MeV} \quad (73)$$

whereas the Skyrme model gives equal values for the three quantities.

Furthermore, one can extract the different flavour contents

$$\bar{s}s \sim \text{Tr} \begin{pmatrix} 0 & 0 & 0 \\ 0 & 0 & 0 \\ 0 & 0 & 1 \end{pmatrix} (U + U^\dagger - 2) \qquad (74)$$

$$\bar{u}u \sim \text{Tr} \begin{pmatrix} 1 & 0 & 0 \\ 0 & 0 & 0 \\ 0 & 0 & 0 \end{pmatrix} (U + U^\dagger - 2) \qquad (75)$$

$$\bar{d}d \sim \text{Tr} \begin{pmatrix} 0 & 0 & 0 \\ 0 & 1 & 0 \\ 0 & 0 & 0 \end{pmatrix} (U + U^\dagger - 2) \qquad (76)$$

and arrives at [10]

$$\frac{\langle \bar{s}s \rangle_N}{\langle \bar{u}u + \bar{d}d + \bar{s}s \rangle_N} = \frac{7}{30} \qquad \frac{\langle \bar{s}s \rangle_\Lambda}{\langle \bar{u}u + \bar{d}d + \bar{s}s \rangle_\Lambda} = \frac{9}{30} \ . \qquad (77)$$

Thus, a very high strangeness content of the nucleon results, which would imply that the sea is not suppressed. However, this prediction is not trustworthy, as the calculation of the mass splittings already indicates. It seems that first order perturbation theory in the masses is not sufficient and the effect of the strange quark mass on the Skyrmion wave function must be included.

In the following, N will not be fixed, but it will be treated as a variable, in the spirit of a $1/N$ expansion. Compared with the two-flavour case, the size of the $SU(3)_f$ representations grows considerably more rapidly with increasing N. Since the strange quark mass is by definition of order $O(1)$, it induces a mass difference of order $O(N)$ between the highest and the lowest strangeness members of a multiplet by simple counting of the number of strange quarks. On the other hand, in the $SU(3)$-symmetric case of zero quark masses, the splittings between the different multiplets are only of order $O(1/N)$. Thus the concept of a weakly broken $SU(3)$ symmetry fails as N grows. Instead of looking at entire multiplets, it is more meaningful to focus on states with strangeness of order $O(1)$, which are the large-N analogues of $N, \Delta, \Lambda, \Sigma, \Sigma^*$, etc. The lowest strangeness members of the representations are shown in Fig. 16. Since the multiplets

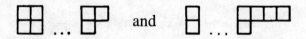

Fig. 16. *Young tableaux for states with strangeness of $O(1)$.*

contain baryons with up to $O(N)$ strange quarks, the wave functions for baryons with $O(1)$ strange quarks only extend $O(1/N)$ into the strange directions in the collective coordinate (A) space. This suggests the possibility of constructing a perturbation theory based on these deviations leading to $1/N$ expansions for various baryon observables.

4.3 Rigid Rotator and $1/N$ Expansion

Consider a soliton which rotates in the $SU(2)$ subgroup and carries out small rigid oscillations into the strange directions [11]. To describe this object, the $SU(3)$ collective coordinate A can be decomposed as

$$A(t) = B(t)S(t) \tag{78}$$

with

$$B(t) \in SU(2), \quad S(t) = \exp\left(i\sum_{a=4}^{7} d_a(t)\lambda_a\right). \tag{79}$$

Since the kinetic energy associated with B is of order $O(1/N)$, one can ignore the $SU(2)$ rotation in the $O(1)$ treatment. Expanding the collective coordinate Lagrangian, including the mass term, to second order in the strange deviations d_a, one finds

$$\mathcal{L} = \mathcal{L}_{45} + \mathcal{L}_{67} \tag{80}$$

with

$$\mathcal{L}_{ij} = 2\Phi(\dot{d}_i^2 + \dot{d}_j^2) + \frac{N}{2}(d_i\dot{d}_j - d_j\dot{d}_i) - \frac{m_K^2\alpha}{2}(d_i^2 + d_j^2). \tag{81}$$

This describes a particle of mass 4Φ on a plane (d_i, d_j) with a normal magnetic field N and a quadratic potential symmetric about the origin. The quadratic potential simply originates from the strange mass, whereas the magnetic term comes from the Wess-Zumino part. Thus, even when the strange mass is switched off, the magnetic field (which is of order $O(N)$!) prevents the particle from moving far in the (i,j) plane and perturbation theory in the deviations is appropriate. Of course, for vanishing mass, there is no reason to choose the corner of flavour space with small strange content as the one relevant for describing the lowest-lying baryons; only introducing the strange mass guarantees the corner with low strange content to be the one with the lowest energy.

The classical frequencies of the circular motion in the magnetic field are the roots of

$$4\Phi\omega^2 + N\omega = m_K^2\alpha \tag{82}$$

i.e.

$$\omega_\pm = \frac{N}{8\Phi}\left(\sqrt{1 + \left(\frac{m_K}{M_0}\right)^2} \pm 1\right), \quad M_0^2 = \frac{N^2}{16\Phi\alpha} \sim O(1) \tag{83}$$

As $m_K \to 0$, $\omega_+ \to N/4\Phi$, the cyclotron frequency, and $\omega_- \to 0$, which expresses the degeneracy of the Landau levels. If one defines a doublet

$$D = \frac{1}{\sqrt{2}}\begin{pmatrix} d_4 - id_5 \\ d_6 - id_7 \end{pmatrix} \tag{84}$$

then the Lagrangian for the strange degrees of freedom can be written as

$$\mathcal{L} = 4\Phi \dot{D}^\dagger \dot{D} + i\frac{N}{2}(D^\dagger \dot{D} - \dot{D}^\dagger D) - \alpha m_K^2 D^\dagger D \tag{85}$$

and the Hamiltonian is

$$H_0 = \frac{1}{4\Phi}\Pi^\dagger \Pi + \frac{iN}{8\Phi}(\Pi^\dagger D - D^\dagger \Pi) + \left(\frac{N^2}{16\Phi} + \alpha m_K^2\right) D^\dagger D \tag{86}$$

where $\Pi^\dagger$ is the momentum conjugate to D. To diagonalize H_0, one can transform to the basis of creation and annihilation operators

$$D = \frac{1}{\sqrt{N}}\left[1 + \left(\frac{m_K}{M_0}\right)^2\right]^{-1/4}(a + b^\dagger) \tag{87}$$

$$\Pi^\dagger = -\frac{i}{2}\sqrt{N}\left[1 + \left(\frac{m_K}{M_0}\right)^2\right]^{1/4}(a^\dagger - b) \tag{88}$$

with M_0 defined as above. In this basis,

$$H_0 = \frac{1}{2}\omega_-\{a, a^\dagger\} + \frac{1}{2}\omega_+\{b, b^\dagger\} \, . \tag{89}$$

The eigenstates form a Fock space

$$|n_s, n_{\bar{s}}\rangle = (a^\dagger)^{n_s}(b^\dagger)^{n_{\bar{s}}}|0\rangle \tag{90}$$

$$H_0|n_s, n_{\bar{s}}\rangle = \omega_- + \omega_+ + n_s\omega_- + n_{\bar{s}}\omega_+ \tag{91}$$

$|n_s, n_{\bar{s}}\rangle$ looks as if it contains n_s strange quarks and $n_{\bar{s}}$ strange antiquarks. States with $n_{\bar{s}} \neq 0$ are exotic states, which are usually presumed to be artefacts of the rigid rotator approximation. In an improved treatment, which will be exhibited below, these states disappear. Even in the rigid rotator approximation, in the limit of vanishing strange mass, it takes energy $\omega_+ \approx N/4\Phi \sim O(1)$ to create an extra strange antiquark (as opposed to replacing a u- or d-quark by an s-quark, which takes no energy).

One can now evaluate the strange content of the low-lying baryons to leading order in $1/N$

$$R = \frac{\langle \bar{s}s \rangle_B}{\langle \bar{u}u + \bar{d}d + \bar{s}s \rangle_B} \sim \langle 0|D^\dagger D|0\rangle = \frac{2}{N\sqrt{1 + (m_K/M_0)^2}} \tag{92}$$

where B denotes any finitely excited state. Thus, the mean squared deviation in the strange direction is of order $O(1/N)$, which justifies the perturbative approach presented here for large N. A standard fit gives $M_0 \approx 250$ MeV, which signifies appreciable non-linearities at $m_K = 495$ MeV, thus invalidating perturbation theory in the mass. In meson physics, all quantities are typically expanded in powers of $(m_K/4\pi f_\pi)^2 \approx 0.25$, which is why the higher powers can usually be neglected. The results presented here imply that, in contrast to meson physics, which only is concerned with small fluctuations about the vacuum, a new mass scale $M_0 \ll 4\pi f_\pi$ emerges in baryon physics. Possibly, $(m_K/M_0)^2$ should be identified with $O(m_s/\Lambda_{\rm QCD})$, which is the expected expansion parameter.

The baryon states (90) carry no definite isospin or angular momentum. In order to identify the eigenstates of I and J and to find the energy levels to order $O(1/N)$, the collective coordinate $B(t)$ must be excited. A convenient form for the $1/N$ correction to the Lagrangian can be given in terms of the angular velocity $d\alpha/dt$ defined by

$$B^{-1}\dot{B} = \frac{i}{2}\dot{\alpha}_i \tau_i \qquad (93)$$

where τ_i are the Pauli matrices:

$$\delta \mathcal{L} = \frac{1}{2}\Omega(\dot{\alpha}_i)^2 + i\dot{\alpha}_i(2\Phi - \Omega)(\dot{D}^\dagger \tau_i D - D^\dagger \tau_i \dot{D}) - \frac{1}{2}N\dot{\alpha}_i D^\dagger \tau_i D \ . \qquad (94)$$

The resulting Hamiltonian reads

$$H = H_0 + H_1 + O(1/N^2) \qquad (95)$$

with

$$H_1 = \frac{1}{2\Omega}\left[\mathbf{J}_{ud} - \frac{i}{2}(\Pi^\dagger \tau D - D^\dagger \tau \Pi)\left(1 - \frac{\Omega}{2\Phi}\right) + \frac{N\Omega}{4\Phi}D^\dagger \tau D\right]^2 \sim O(1/N) \qquad (96)$$

where $\mathbf{J}_{ud}$ is the momentum conjugate to $\boldsymbol{\alpha}$, which is the analogue of the net angular momentum of the u- and d-quarks. The full angular momentum can be obtained by considering the behaviour under spatial rotations, $U_0 \to RU_0R^{-1}$, which acts on B and D as $D \to R^{-1}D$ and $B \to BR$. Using Noether's theorem, the total angular momentum turns out to be

$$\mathbf{J} = \mathbf{J}_{ud} + \mathbf{J}_s \qquad (97)$$

where

$$\mathbf{J}_s = \frac{1}{2}(a^\dagger \tau a - b\tau b^\dagger) \ . \qquad (98)$$

Under isospin rotations, by contrast, $U \to FUF^{-1}$, and therefore $B \to FB$ and $D \to D$. Thus isospin is only carried by the $SU(2)$ rotor wave function, i.e. $\mathbf{I} = \mathbf{I}_{ud}$. Each unit of strangeness carries $1/2$ a unit of angular momentum and no isospin. Notice that these are just the quantum numbers of a strange quark.

The wave functions are sums of products of the form

$$|n_s, n_{\bar{s}}\rangle \chi_I(B) \qquad (99)$$

which carry a definite isospin I and a well-defined total angular momentum. Since the rotator wave functions χ_I satisfy the $\mathbf{J}_{ud} = \mathbf{I}$ rule, there is a further constraint on the quantum numbers: For odd N, the baryon is a fermion and must therefore carry half-integer spin. This forces states with even strangeness to carry half-integer I; states with odd strangeness must carry integer I. This leads to the actually observed baryon quantum numbers. A few of the lowest wave functions are as follows, where the labeling is $|n_s, n_{\bar{s}}\rangle|I_{ud}, J_{ud}\rangle$:

$$|N\rangle = |0,0\rangle|1/2, 1/2\rangle \qquad |\Delta\rangle = |0,0\rangle|3/2, 3/2\rangle$$
$$|\Lambda\rangle = |1,0\rangle|0, 0\rangle \qquad |\Sigma\rangle = [|1,0\rangle|1, 1\rangle]_{J=1/2} \qquad (100)$$
$$|\Sigma^*\rangle = [|1,0\rangle|1, 1\rangle]_{J=3/2}$$

Next, one can compare the predictions for the masses with the ones from the non-relativistic quark model. In the latter,

$$H_{NRQM} = m_0 + m_1|S| + m_2 \sum_{i<k} \jmath_i \jmath_k + m_2 c \sum_{i,I} \jmath_i \jmath_I + m_2 \bar{c} \sum_{I<K} \jmath_I \jmath_K \qquad (101)$$

where small indices refer to light quarks and capital ones to strange quarks. The parameter m_0 is of order $O(N)$; m_1 is the difference between the constituent masses of the strange and the light quarks, which is of order $O(1)$; m_2 measures the quark-quark interaction, which is of order $O(1/N)$. $SU(3)$ breaking means $m_1 > 0$ and, typically, $\bar{c} < c < 1$. The spectrum is of the form

$$H_{NRQM} = m_0' + m_1'Y + \frac{m_2}{2}\left(cJ(J+1) + (1-c)\left(I(I+1) - \frac{1}{4}Y^2\right)\right.$$
$$\left. + \frac{1+\bar{c}-2c}{4}Y^2\right) . \qquad (102)$$

The Gell-Mann-Okubo mass formula

$$2(M_N + M_\Xi) = 3M_\Lambda + M_\Sigma \qquad (103)$$

is fulfilled if $1+\bar{c}-2c \approx 0$. A good fit to the octet and the decuplet is achieved with $m_0' = 1062$ MeV, $m_1' = -192$ MeV, $m_2 = 2/3$ MeV, $c = 0.67$, and $\bar{c} = 0.27$. Note how far c and $\bar{c}$ deviate from the $SU(3)$-symmetric values $c = \bar{c} = 1$. Usually, the Gell-Mann-Okubo formula is justified by chiral perturbation theory. In view of the obtained values for c and $\bar{c}$, this seems to be the wrong explanation, and it may only work by accident.

In the rigid rotator model, one obtains a mass formula similar to H_{NRQM} by evaluating matrix elements of $H_1 \sim O(1/N)$. Comparing coefficients gives

$$m_2 = 1/\Omega , \qquad c = 1 - \frac{4\Omega\omega}{8\Phi\omega + N} , \qquad \bar{c} = c^2 . \qquad (104)$$

H_1, however, does not contain the full $1/N$ corrections. In order to keep track of strange-strange quark interactions, one needs to include all terms up to order D^4. Comparing with experimental data gives $\omega = 248$ MeV, $m_2 = 196$ MeV, and $c = 0.25$, which is very far from the empirical value $c = 0.67$. This induces incorrect relative splittings between Λ, Σ, and Σ^*, which is a further indication that the rigid rotator approach is insufficient.

Up to now, the rigid $SU(3)$ rotations have simply been split up into a $SU(2)$ part and one describing rigid rotation into the strange direction. However, once the kaon becomes massive, rotations into the strange direction are only approximate collective coordinates. In an exact treatment, deformations of the soliton as it rotates into the strange direction must be included. These effects become significant for sufficiently large strange mass.

4.4 The Bound State Approach to Strangeness

Due to the strong deformation effects, a treatment of general fluctuations in the strange direction about the basic $SU(2)$ Skyrmion is needed [2]. A convenient parametrization is

$$U = \sqrt{U_\pi} U_K \sqrt{U_\pi} \qquad (105)$$

where

$$U_\pi = \exp\left(\frac{2i}{f_\pi} \sum_{j=1}^{3} \lambda_j \pi^j\right) \qquad U_K = \exp\left(\frac{2i}{f_\pi} \sum_{a=4}^{7} \lambda_a K^a\right) \qquad (106)$$

where λ_i are the usual $SU(3)$ generators. Similar to the last section, one can organize the expansion in kaon fluctuations (which amounts to a $1/N$ expansion) in terms of the doublet[5]

$$K = \frac{1}{\sqrt{2}} \begin{pmatrix} K_4 - iK_5 \\ K_6 - iK_7 \end{pmatrix} = \begin{pmatrix} K^+ \\ K^0 \end{pmatrix} \qquad (107)$$

where the K^a, however, are no longer quantum mechanical variables, but fields, which allows for a description beyond the rigid rotator approximation. Furthermore, one can define a "covariant derivative"[6]

$$D_\mu K = \partial_\mu K + \frac{1}{2}[\sqrt{U_\pi^\dagger}, \partial_\mu \sqrt{U_\pi}] K \qquad (108)$$

and a quantity

$$A_\mu = \frac{1}{2}\{\sqrt{U_\pi^\dagger}, \partial_\mu \sqrt{U_\pi}\} . \qquad (109)$$

After expanding the Skyrme Lagrangian to second order in K, one arrives at the kaon-soliton interaction Lagrangian

$$\mathcal{L} = \mathcal{L}_{Sk}(U_\pi) + (D_\mu K)^\dagger D^\mu K - m_K^2 K^\dagger K$$
$$- \frac{1}{8} K^\dagger K \left\{ \mathrm{Tr}(\partial_\mu U_\pi^\dagger \partial^\mu U_\pi) + \frac{1}{e^2 f_\pi^2} \mathrm{Tr}[\partial_\mu U_\pi U_\pi^\dagger, \partial_\nu U_\pi U_\pi^\dagger]^2 \right\}$$
$$- \frac{1}{e^2 f_\pi^2} \left\{ 2(D_\mu K)^\dagger D_\nu K \mathrm{Tr}(A^\mu A^\nu) + \frac{1}{2}(D_\mu K)^\dagger D^\mu K \mathrm{Tr}(\partial_\nu U_\pi^\dagger \partial^\nu U_\pi) \right.$$
$$\left. - 6(D_\nu K)^\dagger [A^\nu, A^\mu] D_\mu K \right\}$$
$$+ \frac{iN}{f_\pi^2} B^\mu (K^\dagger D_\mu K - (D_\mu K)^\dagger K) . \qquad (110)$$

---

[5] This can be justified more generally by the theory of non-linear $SU(2) \times SU(2)$ realizations, see [12].
[6] This is motivated by the analogy to the minimal coupling in gauge theory; here one essentially treats the motion of K in the "background field" of the soliton.

The last piece originates from the Wess-Zumino term. It describes the interaction of a charged field (the charge here being the strangeness) with a "vector potential" B^μ, which is the baryon current. This is the only term which distinguishes between $s < 0$ and $s > 0$ in a fixed baryon number background, which again illustrates the crucial role of the Wess-Zumino term.

The Lagrangian (110) reduces the problem to the motion of kaons in the classical background of the $SU(2)$ soliton. For large N, the solitons rotate slowly, with velocities of order $O(1/N)$. Therefore, to find the kaon energy levels to $O(1)$, it is sufficient to consider propagation in a static background $U_\pi(\mathbf{r}) = U_0(\mathbf{r})$ which is symmetric under simultaneous spatial and isospin rotations $\mathbf{T} = \mathbf{I} + \mathbf{L}$. The kaon eigenmodes can be written as

$$K(\mathbf{r},t) = k(r,t) Y_{TLT_z} \ . \tag{111}$$

The effective Lagrangian for $k(r,t)$ is

$$L = 4\pi \int dr\, r^2 (f(r)\dot{k}^\dagger \dot{k} + i\lambda(r)(k^\dagger \dot{k} - \dot{k}^\dagger k) - h(r) k^{\dagger\prime} k' - k^\dagger k (m_K^2 + V_{eff}(r,T,L))) \tag{112}$$

where f, λ, h and V_{eff} depend on the soliton profile function $F(r)$ [2]. The corresponding variational equation reads

$$-f(r)\ddot{k} + 2i\lambda(r)\dot{k} + Ok = 0 \tag{113}$$

where

$$O = \frac{1}{r^2}\frac{d}{dr} h(r) r^2 \frac{d}{dr} - m_K^2 - V_{eff}(r) \ . \tag{114}$$

Expanding k in terms of the eigenmodes,

$$k(r,t) = \sum_{n>0} (\tilde{k}_n(r) e^{i\tilde{\omega}_n t} b_n^\dagger + k_n(r) e^{-i\omega_n t} a_n) \tag{115}$$

one obtains the eigenvalue equations

$$(f(r)\omega_n^2 + 2\lambda(r)\omega_n + O) k_n = 0 \tag{116}$$
$$(f(r)\tilde{\omega}_n^2 - 2\lambda(r)\tilde{\omega}_n + O) \tilde{k}_n = 0 \ . \tag{117}$$

Using the hermiticity of O, one finds the orthogonality relations

$$4\pi \int dr\, r^2 k_n^* k_m [f(r)(\omega_n + \omega_m) + 2\lambda(r)] = \delta_{nm} \tag{118}$$

$$4\pi \int dr\, r^2 \tilde{k}_n^* \tilde{k}_m [f(r)(\tilde{\omega}_n + \tilde{\omega}_m) - 2\lambda(r)] = \delta_{nm} \tag{119}$$

$$4\pi \int dr\, r^2 k_n^* \tilde{k}_m [f(r)(\omega_n - \tilde{\omega}_m) + 2\lambda(r)] = 0 \ . \tag{120}$$

Upon canonical quantization, these relations insure that

$$[a_n, a_m^\dagger] = \delta_{nm} \qquad [b_n, b_m^\dagger] = \delta_{nm} \ . \tag{121}$$

The Hamiltonian reads

$$H = \sum_{n>0} \left(\frac{1}{2}\omega_n \{a_n, a_n^\dagger\} + \frac{1}{2}\tilde{\omega}_n \{b_n, b_n^\dagger\} \right) \quad (122)$$

and the strangeness charge is measured by

$$S = \sum_{n>0} b_n^\dagger b_n - a_n^\dagger a_n \,. \quad (123)$$

The lowest kaon bound state is found in the $T = 1/2, L = 1$ channel ($S = -1$); numerically, its energy is $\omega_{bs} \approx 160$ MeV. This is the state on the basis of which the Λ, Σ, and Σ^* are constructed below. By contrast, its strangeness conjugate is pushed out into the continuum by the Wess-Zumino interaction. This is an example of how the exotic states appearing in the rigid rotator approximation can disappear in a more complete treatment. There is also a more weakly bound state with $T = 1/2, L = 0$ which corresponds to negative parity baryons, such as the $\Lambda(1405)$. This state constitutes a success of the bound state approach, since it generally appears at much too high energies in other models. It is also not present in the rigid rotator approximation.

To construct the Λ, Σ, and Σ^*, one must carry out the $SU(2)$ collective coordinate quantization:

$$U_0(\mathbf{r}) \to B(t) U_0(\mathbf{r}) B^{-1}(t) \qquad K(\mathbf{r}, t) \to B(t) \tilde{K}(\mathbf{r}, t) \quad (124)$$

where $\tilde{K}$ is the kaon field observed from the rotating frame. One can show that the quanta of $\tilde{K}$, which are annihilated by a_n and b_n, carry no isospin and have angular momentum given by $\mathbf{T} = \mathbf{I} + \mathbf{L}$, which is fractional. A meson bound in the $T = 1/2$, $L = 1$ orbital has $J = 1/2$, $I = 0$, $S = -1$, which are the quantum numbers of a strange quark.

The picture of the hyperfine splittings is basically the same as in the rigid rotator model and the non-relativistic quark model, except that here

$$c = 1 - \omega \frac{\int dr\, r^2 k^* k g(r)}{\int dr\, r^2 k^* k (\omega f(r) + \lambda(r))} \quad (125)$$

where

$$\lambda(r) = -\frac{Ne^2}{2\pi^2 r^2} F' \sin^2 F \quad (126)$$

$$f(r) = 1 + \frac{2\sin^2 F}{r^2} + (F')^2 \quad (127)$$

$$g(r) = \frac{4}{3} f \cos^2(F/2) - \frac{2}{r^2} \left\{ \frac{d}{dr}[r^2 F' \sin F] - \frac{4}{3} \sin^2 F \cos^2(F/2) \right\} \,. \quad (128)$$

Numerically, one now finds $c \approx 0.6$ in good agreement with experiment. The mass splittings

$$M_\Sigma - M_\Lambda = \frac{1-c}{\Omega} \qquad M_{\Sigma^*} - M_\Sigma = \frac{3c}{2\Omega} \qquad M_\Delta - M_N = \frac{3}{2\Omega} \quad (129)$$

come out with approximately the correct ratios.

Comparing with the rigid rotator treatment, one finds that a small rotation in the strange direction produces a meson with the profile function

$$k(r) \sim \sin(F(r)/2) \qquad (130)$$

in the $T = 1/2$, $L = 1$ orbital, i.e. all formulae of the bound state approach reduce to the rigid rotator approach upon substitution of (130). The advantage of the bound state approach manifests itself in the fact that it allows the kaon profile to adjust itself away from the $SU(3)$-symmetric one. This effect is clearly significant, as the shift in c from 0.25 to 0.6 shows. This is further corroborated by the significantly improved comparison with experiment. The emergence of the negative parity baryon states, as mentioned above, is especially encouraging. Besides the $I = 0$, $J^P = 1/2^-$ state which is numerically found at 1360 MeV, and which must be identified with the $\Lambda(1405)$, one finds an $I = 1$, $J^P = 3/2^-$ state at 1664 MeV, which must be identified with the $\Sigma(1670)$. A state with no confirmed empirical counterpart with $I = 1$, $J^P = 1/2^-$ is found at 1380 MeV; it may perhaps be identified with the candidate $\Sigma(1480)$, the spin of which should turn out to be 1/2 in order to agree with the present model.

Various other calculations confirm that the bound state approach may be a sensible approximation. Magnetic moment calculations by Nyman and Riska are given in table 2 (in units of μ_N) One finds many model-independent relations,

Table 2. *Results of the three flavour Skyrme model*

	Bound State Appr.	Quark Model	Experiment
Λ	-0.90	input	-0.613 ± 0.005
Σ^+	2.32	2.67	2.38 ± 0.02
			2.479 ± 0.025
Σ^0	0.72	0.79	–
Σ^-	-0.88	-1.09	-1.166 ± 0.017
Ξ^0	-1.78	-1.43	-1.250 ± 0.014
Ξ^-	-0.72	-0.49	-0.69 ± 0.04
$|\Sigma^0 \to \Lambda|$	1.60	1.63	1.82 + 0.25 − 0.18

for example $\mu(\Omega^-) = 3\mu(\Lambda)$, which also holds in the quark model. Recently, a host of baryon properties such as charge radii, magnetic moments, weak decay amplitudes, dibaryons, etc. have been calculated in the framework of the bound state approach by Rho, Scoccola, Nyman, Riska, Kunz, Mulders, Dannbom, Bjornberg, and others (for references, see [13]). There are several interesting ways to extend the treatment further:

- Some improvement is achieved by $SU(3)$ symmetry breaking in the kinetic terms in the kaon Lagrangian: $f_\pi \to f_K \approx 1.22 f_\pi$ reduces the overbinding of the kaon modes. ω_{bs} of the lowest bound state becomes about 221 MeV [14].
- Some $K-K$ interaction effects have been neglected and may alter the results for baryons with multiple strangeness.
- One of the original motivations for the bound state model was the treatment of charmed baryons. Incorporating only the pseudoscalar D and using $f_\pi \to f_D \approx 1.8 f_\pi$ in the D-Lagrangian, one finds the same quantum numbers and ordering of states as in the quark model [14] [15] [16]. However, the vector D^* is only slightly heavier than the pseudoscalar and it is necessary to include it to maintain heavy quark symmetry. This program has been started in [17], [18] and gives good results.
- Another interesting open problem is the description of dibaryon states in the bound state approach, most importantly the $S = -2$ dibaryon. In the $SU(3)$-symmetric model, it is tightly bound and absolutely stable. In the bound state approach, one would have to study kaon modes in the background of a $B = 2$ $SU(2)$ solution [13], [19]. An exact treatment should be done numerically.
- Similarly, one can study $\bar{K}$ modes in the background of $SU(2)$ nuclear matter. This may lead to $\bar{K}$ condensation [13].

Acknowledgements I am grateful to Frieder Lenz for his invitation and warm hospitality. I also thank all the participants of the workshop for pleasant and stimulating atmosphere. Special thanks to Michael Engelhardt and Dieter Stoll for preparing a great set of notes. My work is supported in part by DOE grant DE-AC02-76WRO3072, NSF Presidential Young Investigator Award PHY-9157482, James S. McDonnell Foundation grant No. 91-48, and an A.P. Sloan Foundation Research Fellowship.

References

[1] T. H. R. Skyrme, Nucl. Phys. **31** (1962) 556.
[2] C.G. Callan, I. Klebanov, Nucl. Phys. **B 262** (1985) 365;
    C.G. Callan, K. Hornbostel, I. Klebanov, Phys. Lett. **202 B** (1988) 269.
[3] G. 't Hooft, Nucl. Phys. **B 72** (1974) 461.
[4] E. Witten, Nucl. Phys. **B 160** (1979) 57.
[5] A.P. Balachandran, V. Nair, S. Rajeev, A. Stern, Phys. Rev. Lett. **49** (1982), 1124; Phys. Rev. **D 27** (1983) 1153.
[6] J. Gasser, H. Leutwyler, Ann. Phys. **158** (1984) 142.
[7] U. Meissner, I. Zahed, Adv. Nuc. Phys. **17** (1986) 143.
[8] E. Witten, Nucl. Phys. **B 223** (1983) 422 and 433.
[9] H. Georgi, Lie Algebras in Particle Physics, Benjamin/Cummings 1982.
[10] J. Donoghue, C.R. Nappi, Phys. Lett. **168 B** (1986) 105.
[11] D. Kaplan, I. Klebanov, Nucl. Phys. **B 335** (1990) 45.

[12] C. Callan, C. Coleman, J. Wess, B. Zumino, Phys. Rev. **177** (1969) 2247.
[13] I. Klebanov, Strangeness in the Skyrme Model, in: Proceedings of the NATO ASI on Hadrons and Hadronic Matter, Cargèse 1989, eds. D. Vautherin, F. Lenz, J. Negele.
[14] D.O. Riska, N.N. Scoccola, Phys. Lett. **265 B** (1991) 188.
[15] M. Rho, D.O. Riska, N.N. Scoccola, Phys. Lett. **251 B** (1990) 597;
Z. Phys. **A 341** (1992) 343.
[16] Y. Oh, D. Min, M. Rho, N.N. Scoccola, Nucl. Phys. **A 534** (1991) 493.
[17] E. Jenkins, A.V. Manohar, M.B. Wise, Nucl. Phys. **B 396** (1993) 27
[18] Z. Guralnik, M. Luke, A.V. Manohar, Nucl. Phys. **B 390** (1993) 474
E. Jenkins, A.V. Manohar, Phys. Lett. **B 294** (1992) 273
[19] J. Kunz, P.J. Mulders, Phys. Lett. **B 215** (1988) 449.

Introduction to Supersymmetry and Exact Nonperturbative Results in SUSY-QCD [*]

Notes by H.W. Grießhammer, D. Lehmann, and M. Seeger

Institut für Theoretische Physik III, Universität Erlangen-Nürnberg, Staudtstr. 7, 91058 Erlangen, Germany

1 Introduction

In the last two years, the work of Seiberg, Witten and their coworkers yielded remarkable progress in the understanding of the dynamics of supersymmetric gauge theories in four dimensions.

Supersymmetry (SUSY) is likely to play an important role in the description of the fundamental interactions beyond the Standard Model. It is a field theoretical concept which, although it can be build upon the fundamentals of quantum field theory without altering any of its basic assumptions, leads to dramatic physical and interpretational consequences. Because it is a symmetry transforming fermions into bosons and vice versa, if a theory is supersymmetric the particles in this theory always appear in pairs of one boson and one fermion.

Any supersymmetric quantum field theory which is supposed to be an expansion of or a substitute for the Standard Model must be able to account for the lack of experimental evidence for this doubling of the particle spectrum. (The explanation accepted by most physicists is the spontaneous breakdown of supersymmetry.) Furthermore, by the introduction of supersymmetry the strict separation between matter fields (fermions) and force fields (bosons) becomes obsolete; matter and force appear from a unified point of view, which is formally and esthetically very appealing.

The concept of supersymmetry entered theoretical physics in the 1970s. Soon it turned out that SUSY admitted a solution to the "hierarchy problem" of particle physics, i.e. the problem of explaining the stability of the 13 orders of magnitude between the GUT scale and the mass scale of the electro-weak bosons in perturbation theory. In supersymmetric theories, Feynman diagrams which would make such a gap unstable are cancelled by the corresponding diagrams involving the supersymmetric partners.

It is a similar mechanism of cancellation which makes supersymmetric theories interesting from a different point of view: Due to the relative minus signs

---

[*] Lectures at the workshop "Nonperturbative QCD" organised by the Graduiertenkolleg Erlangen-Regensburg, held on October 10th–12th, 1995 in Kloster Banz, Germany

between diagrams related to each other by replacing a particle in a closed loop by its super-partner, many of the notorious infinities which are usually present in quantum field theory drop out completely. The divergences of a theory become always milder by making it supersymmetric, and even more, some theories become exactly finite.

In the present notes, we will not discuss these aspects but rather treat supersymmetric gauge theories as solvable model field theories in four dimensions trying to learn the relevant lessons for the structure of QCD. In particular we will discuss the work of Seiberg and Witten on $N = 2$ SUSY gauge theories. The exact information about the infrared behavior of these theories offers important insights about the mechanisms of confinement and chiral symmetry breaking. An important methodological lesson to be learned is the necessity to use conjointly the variety of tools available in the study of quantum field theory: it seems that the algebraic structure, the renormalization group, the existence of topologically non trivial, non perturbative configurations all play an important role in the understanding of the infrared structure of the theory.

These proceedings are concerned with a thorough introduction into basic supersymmetry [1], [2], [3] and into the ideas of Seiberg and Witten. In Sect. 2, the Poincaré group and its spinorial representations are introduced, and in Sect. 3, the SUSY algebra is explained in detail. In Sect. 4, the representations of this algebra on Hilbert space are studied. Section 5 discusses the realization of supersymmetry in quantum field theories and introduces the concept of superfields. The goal of Sect. 6 is to discuss separately the "ingredients" used in the solution of the $N = 2$, SU(2) supersymmetric Yang Mills theory as found by Seiberg and Witten. Section 7 finally aims at a consistent solution of $N = 2$ SUSY-QCD.

2 Lie Algebra of Symmetries and Spinors

Supersymmetry will become most transparent when it is understood as an extension of more conventional space-time symmetries. We will start our introduction therefore with a discussion of the spacetime symmetries described by the Poincaré group. In coordinate space, a Poincaré transformation is defined by:

$$x'^\mu = \Lambda^\mu{}_\nu x^\nu + a^\mu \tag{1}$$

where Λ represents the Lorentz transformation and a the spacetime translation. To each element (Λ, a) of the Poincaré group a linear operator $\mathcal{T}(\Lambda, a)$ is associated which acts on field space as

$$\Phi'(x) = \mathcal{T}(\Lambda, a)\, \Phi(x) = \mathcal{D}(\Lambda)\, \Phi(\Lambda^{-1}(x-a))\ , \tag{2}$$

where $\mathcal{D}(\Lambda)$ is a finite dimensional representation matrix of the Lorentz group, corresponding to the spin degrees of the field Φ. The infinitesimal generators of Poincaré transformations are the four components of the four momentum $P^\mu = i\partial^\mu$ and the six independent components of the antisymmetric Lorentz generators $M^{\mu\nu}$. The latter can be decomposed into an orbital part $L^{\mu\nu}$ and a spin part $S^{\mu\nu}$:

$$M^{\mu\nu} = L^{\mu\nu} + S^{\mu\nu}, \qquad L^{\mu\nu} = x^\mu P^\nu - x^\nu P^\mu, \tag{3}$$

For example for four component Dirac fermions the spin part is given by

$$S^{\mu\nu} = \frac{1}{2}\Sigma^{\mu\nu} = \frac{i}{4}[\gamma^\mu,\gamma^\nu]. \tag{4}$$

These ten generators form a Lie algebra with the commutation relations

$$[P^\mu, P^\nu] = 0, \tag{5}$$
$$[M^{\mu\nu}, P^\rho] = i(g^{\nu\rho}P^\mu - g^{\mu\rho}P^\nu), \tag{6}$$
$$[M^{\mu\nu}, M^{\rho\sigma}] = i(g^{\nu\rho}M^{\mu\sigma} + g^{\mu\sigma}M^{\nu\rho} - g^{\mu\rho}M^{\nu\sigma} - g^{\nu\sigma}M^{\mu\rho}), \tag{7}$$

where we use $g_{\mu\nu} = g^{\mu\nu} = \text{diag}(1,-1,-1,-1)$ as Minkowski metric. With the help of the above commutation relations it is easy to see that the mass-square operator $P^2 = P_\mu P^\mu$ and the square of the Pauli-Lubanski spin vector, $W^2 = W_\mu W^\mu$, commute with all the generators, i.e. they are the Casimir operators of the Poincaré group. W^μ thereby is defined as

$$W^\mu = -\frac{1}{2}\varepsilon^{\mu\nu\rho\sigma} P_\nu M_{\rho\sigma} = -\frac{1}{2}\varepsilon^{\mu\nu\rho\sigma} P_\nu S_{\rho\sigma}, \tag{8}$$

and for massive representations its square is proportional to that of the spin operator[1]

$$W^2 = -m^2 S^2, \qquad S^i = \frac{1}{2}\varepsilon^{ijk} S^{jk}. \tag{9}$$

Besides the spacetime symmetries there are often internal symmetries like charge, flavour or colour symmetry present in field theories. An important theorem by Coleman and Mandula [7] states that in a local relativistic quantum field theory in four spacetime dimensions, the most general symmetry group of the S-matrix necessarily is a *direct* product of the Poincaré group with a compact internal Lie group, provided that [1], [3]:

− there is only a finite number of different particles associated with one-particle states of a given mass;
− there is a unique vacuum and a finite mass gap between it and the lowest one-particle state;
− we restrict ourselves to Lie groups.

Moreover, the internal symmetry group must be a direct product of a compact semisimple group with U(1) factors [3]. So, the most general Lie algebra consists of the Poincaré generators P^μ, $M^{\mu\nu}$ and a certain number of *Poincaré invariant*

---

[1] We adopt the conventions: $\varepsilon_{\mu\nu\sigma\rho} = -\varepsilon^{\mu\nu\sigma\rho}$, $\varepsilon^{0123} = 1$, $\varepsilon_{ijk} = \varepsilon^{ijk} = \varepsilon^{0ijk}$.

generators B_r of the internal symmetries, which can always be chosen Hermitean due to the compactness of the internal group:

$$[B_r, B_s] = i c^t_{rs} B_t , \qquad [B_r, P^\mu] = [B_r, M^{\mu\nu}] = 0 . \qquad (10)$$

The algebra spanned by the B_r then is the direct sum of a semisimple and an Abelian subalgebra and has the structure constants c^t_{rs}. Obviously, because of $[B_r, P^2] = [B_r, W^2] = 0$, all particles of an irreducible multiplet of the internal symmetry must have the same mass (*O'Raifeartaigh*'s theorem [6]) and the same spin.

The *Coleman-Mandula theorem* does not leave much freedom for more general symmetries, since its assumptions seem to be quite reasonable from the physical point of view. However, it only covers "bosonic" symmetries, that means symmetries whose generators do not mix fermions and bosons and therefore satisfy the *commutation* relations (10). Supersymmetry relaxes this very assumption of a Lie algebra and generalizes it to a *graded* Lie algebra, that is, allows for "fermionic" or "odd" generators which transform like spinors under Lorentz transformations and, acting on states, change their spin by a half-integer amount. Therefore, supersymmetry by definition is a symmetry between bosons and fermions; multiplets of supersymmetric theories will contain particles with different spin and statistics. The importance of supersymmetry arises from a theorem by Haag, Sohnius and Lopuszanski [8], who showed that supersymmetry is the *only* graded Lie algebra of symmetries of the S-matrix consistent with relativistic quantum field theory [1].

But before we turn to supersymmetry and explore the consequences of introducing fermionic generators, we will use the rest of this section to prepare the ground and review some properties of finite dimensional representations of the Lorentz group.

The spin generators $S^{\mu\nu}$ introduced earlier commute with the momenta P^μ and the angular part $L^{\mu\nu}$ of $M^{\mu\nu}$ and thus generate via (7) the finite dimensional Lorentz group representation $\mathcal{D}(\Lambda)$ in (2). With the aid of the spin operators S^i of (9) and the boost operators $K^i = S^{0i}$, we can rewrite the commutation relations (7) in a non-covariant form:

$$[S^i, S^j] = i\varepsilon^{ijk} S^k , \qquad (11)$$
$$[K^i, K^j] = -i\varepsilon^{ijk} S^k , \qquad (12)$$
$$[K^i, S^j] = i\varepsilon^{ijk} K^k . \qquad (13)$$

By forming the complex linear combinations

$$A = \frac{1}{2}(S + iK) , \qquad B = \frac{1}{2}(S - iK) , \qquad A^\dagger = B , \qquad (14)$$

they can be decoupled into

$$[A^i, A^j] = i\varepsilon^{ijk} A^k ,$$
$$[B^i, B^j] = i\varepsilon^{ijk} B^k , \qquad (15)$$
$$[A^i, B^j] = 0 .$$

From (15) it is clear by analogy with the rotation group SU(2) that $\boldsymbol{A}^2$ and $\boldsymbol{B}^2$ are Casimir operators with eigenvalues $a(a+1)$ and $b(b+1)$, $(a,b \in \frac{1}{2}\mathbb{N})$. We can use them to label the irreducible representations by (a,b); their dimension will then be $(2a+1)(2b+1)$. Note, however, that a *complex continuation* was necessary in (14), so that the local structure of the Lorentz group is *not* $SU(2) \otimes SU(2)$ despite of (15). In fact, because of the non-compactness of the Lorentz group, its finite dimensional representations cannot be unitary. Moreover, to be rigorous for a moment, the corresponding Lie group is not the full Lorentz group but the proper orthochronous part $\mathcal{L}_+^\uparrow$, defined by $\Lambda_0^0 > 0$, $\det(\Lambda) = +1$. The lowest dimensional non-trivial representations of $\mathcal{L}_+^\uparrow$ are obviously $(\frac{1}{2},0)$ and $(0,\frac{1}{2})$ with generators and representation matrices given below:

$$(\tfrac{1}{2},0): \quad \boldsymbol{S} = \frac{\boldsymbol{\sigma}}{2}, \quad \boldsymbol{K} = -\mathrm{i}\frac{\boldsymbol{\sigma}}{2}, \quad S^{\mu\nu} = \frac{1}{2}\sigma^{\mu\nu},$$

$$D(\Lambda) = \exp\left(\frac{\mathrm{i}}{2}\boldsymbol{\sigma}\cdot(\boldsymbol{\alpha} - \mathrm{i}\boldsymbol{\zeta})\right) = \exp\left(\frac{\mathrm{i}}{4}\omega_{\mu\nu}\sigma^{\mu\nu}\right),$$

$$(0,\tfrac{1}{2}): \quad \boldsymbol{S} = \frac{\boldsymbol{\sigma}}{2}, \quad \boldsymbol{K} = +\mathrm{i}\frac{\boldsymbol{\sigma}}{2}, \quad S^{\mu\nu} = \frac{1}{2}\bar{\sigma}^{\mu\nu},$$

$$\overline{D}(\Lambda) = \exp\left(\frac{\mathrm{i}}{2}\boldsymbol{\sigma}\cdot(\boldsymbol{\alpha} + \mathrm{i}\boldsymbol{\zeta})\right) = \exp\left(\frac{\mathrm{i}}{4}\omega_{\mu\nu}\bar{\sigma}^{\mu\nu}\right).$$

(16)

Both representations form a $SL(2,\mathbb{C})$ due to the traceless-ness of the Pauli matrices σ^i. They are related by

$$\overline{D}(\Lambda) = \varepsilon\, D(\Lambda)^* \varepsilon^{-1}, \quad \varepsilon = \mathrm{i}\sigma^2 = \begin{pmatrix} 0 & 1 \\ -1 & 0 \end{pmatrix}, \tag{17}$$

as can be seen with help of the relation $\sigma^2 \boldsymbol{\sigma}^* \sigma^2 = -\boldsymbol{\sigma}$. This means that $(0,\frac{1}{2})$ is equivalent to the complex conjugate representation of $(\frac{1}{2},0)$ and thus *inequivalent* to $(\frac{1}{2},0)$ itself. The contragredient Lorentz transformation $\Lambda^{-1\mathrm{T}}$ is represented by the contragredient $SL(2,\mathbb{C})$-matrix,

$$D(\Lambda^{-1\mathrm{T}}) = D(\Lambda)^{-1\mathrm{T}} = \varepsilon\, D(\Lambda)\, \varepsilon^{-1}. \tag{18}$$

This leads us to establish the following tensor calculus for spinors: Let $\psi = (\psi_\alpha)$ be a two component Weyl spinor transforming according to the fundamental representation $(\frac{1}{2},0)$. Then $\varepsilon\psi$ transforms contragredient and is therefore denoted by upper indices (ψ^α); due to (17), $\varepsilon\psi^*$ is a $(0,\frac{1}{2})$-spinor and denoted by "dotted" indices $(\overline{\psi}^{\dot{\alpha}})$ to distinguish it from the "undotted" $(\frac{1}{2},0)$-spinors, so that altogether we have the transformation properties

$$\psi'_\alpha = D(\Lambda)_\alpha{}^\beta \psi_\beta, \qquad \psi'^\alpha = D^{-1}(\Lambda)_\beta{}^\alpha \psi^\beta \;; (\alpha,\beta = 1,2)$$

$$\overline{\psi}'^{\dot{\alpha}} = (D^*)^{-1}(\Lambda)_{\dot{\beta}}{}^{\dot{\alpha}} \overline{\psi}^{\dot{\beta}}, \qquad \overline{\psi}'_{\dot{\alpha}} = D^*(\Lambda)_{\dot{\alpha}}{}^{\dot{\beta}} \overline{\psi}_{\dot{\beta}} \;; (\dot{\alpha},\dot{\beta} = 1,2)$$

(19)

Thus, $\varepsilon = (\varepsilon^{\alpha\beta})$ and $\varepsilon^{-1} = (\varepsilon_{\alpha\beta})$ play in spinor space an analogous role as the metric tensor $g_{\mu\nu}$ in Minkowski space and may be used to raise and lower the spinor indices

$$\psi^\alpha = \varepsilon^{\alpha\beta}\psi_\beta \, , \qquad \psi_\alpha = \varepsilon_{\alpha\beta}\psi^\beta \, ,$$
$$\overline{\psi}^{\dot\alpha} = \varepsilon^{\dot\alpha\dot\beta}\overline{\psi}_{\dot\beta} \, , \qquad \overline{\psi}_{\dot\alpha} = \varepsilon_{\dot\alpha\dot\beta}\overline{\psi}^{\dot\beta} \, . \qquad (20)$$

The raising and lowering of indices commutes with the complex conjugation, since in (17) we have defined the ε-tensor to be real and we numerically equate the dotted and undotted ε-tensors. Unlike $g_{\mu\nu}$ in Minkowski space, $\varepsilon^{\alpha\beta}$ is antisymmetric, stemming from the fact that SL(2,$\mathbb{C}$) is not only the covering group of the Lorentz group, but also equivalent to the two dimensional symplectic group Sp(2,$\mathbb{C}$). It is easily checked that $\varepsilon^{\alpha\beta}$ is invariant under Lorentz transformations. Our conventions are summarized as

$$\varepsilon^{\alpha\gamma}\varepsilon_{\gamma\beta} = \delta^\alpha{}_\beta \, , \qquad \varepsilon^{12} = \varepsilon_{21} = 1 \, ,$$
$$\varepsilon^{\dot\alpha\dot\beta} = \varepsilon^{\alpha\beta} \, , \qquad \varepsilon_{\dot\alpha\dot\beta} = \varepsilon_{\alpha\beta} \, . \qquad (21)$$

By definition, tensors of higher rank transform like products of fundamental spinors with the same index structure. Because of the contragredient transformation properties, complete contraction over upper and lower indices leads to Lorentz invariant expressions. Thus, especially

$$\psi\chi := \psi^\alpha\chi_\alpha = -\psi_\alpha\chi^\alpha = \chi\psi \, ,$$
$$\overline{\psi\chi} := \overline{\psi}_{\dot\alpha}\overline{\chi}^{\dot\alpha} = -\overline{\psi}^{\dot\alpha}\overline{\chi}_{\dot\alpha} = \overline{\chi\psi} \qquad (22)$$

are Lorentz scalars; the last equalities hold, if the spinors are treated as Grassmann (i.e. anti-commuting) variables.

Without proof we state that spinors transforming according to an irreducible representation of $\mathcal{L}_+^\uparrow$ are totally symmetric in all indices. Note also that, due to the two dimensional representation space, antisymmetry in an index pair immediately implies proportionality to an ε-tensor in these indices. Therefore, arbitrary tensors of higher rank may be decomposed by a Clebsch-Gordan expansion into a sum of irreducible spinors multiplied by some combinations of ε-tensors.

It is instructive to discuss some explicit examples which will be of importance in our derivation of the supersymmetry algebra. First, let us consider a spinor $X = (X_{\alpha\dot\beta})$ in the $(\frac{1}{2}, \frac{1}{2})$-representation, which has to transform like

$$X'_{\alpha\dot\beta} = D(\Lambda)_\alpha{}^\gamma D^*(\Lambda)_{\dot\beta}{}^{\dot\delta} X_{\gamma\dot\delta} \qquad \text{or} \qquad X' = D(\Lambda)\, X\, D^\dagger(\Lambda) \, . \qquad (23)$$

It is convenient to introduce several sets of Hermitean basis vectors in the space of complex 2×2-matrices

$$\sigma^\mu = (\sigma^\mu)_{\alpha\dot\beta} = (\mathbb{1},\, \sigma^i) \, , \qquad \sigma_\mu = g_{\mu\nu}\sigma^\nu \, ,$$
$$\bar\sigma^\mu = (\bar\sigma^\mu)^{\dot\alpha\beta} = (\mathbb{1},\, -\sigma^i) \, , \qquad \bar\sigma_\mu = g_{\mu\nu}\bar\sigma^\nu \, , \qquad (24)$$

where σ^i are the Pauli-matrices. The $\bar{\sigma}^\mu$ are dual to the σ_μ in the sense

$$(\sigma^\mu)_{\alpha\dot{\beta}}\,(\bar{\sigma}_\nu)^{\dot{\beta}\alpha} = \text{tr}\,(\sigma^\mu\,\bar{\sigma}_\nu) = 2\,\delta^\mu{}_\nu\;, \tag{25}$$

and both are related by

$$(\bar{\sigma}^\mu)^{\dot{\beta}\alpha} = \varepsilon^{\alpha\gamma}\,\varepsilon^{\dot{\beta}\dot{\delta}}\,(\sigma_\mu)_{\gamma\dot{\delta}} \quad \text{or} \quad \bar{\sigma}^\mu = -\varepsilon\,\sigma^{\mu*}\,\varepsilon\;. \tag{26}$$

We can uniquely associate an (in general complex) four component vector x^μ to every X by expanding in a basis, e.g.

$$X_{\alpha\dot{\beta}} = x^\mu\,(\sigma_\mu)_{\alpha\dot{\beta}}\;;\qquad x^\mu = \frac{1}{2}\,X_{\alpha\dot{\beta}}\,(\bar{\sigma}^\mu)^{\dot{\beta}\alpha} = \frac{1}{2}\,\text{tr}\,(X\,\bar{\sigma}^\mu)\;. \tag{27}$$

Under Lorentz transformations (23) of X it transforms like a Minkowski four vector,

$$x'^\mu = \frac{1}{2}\,\text{tr}\,(X'\,\bar{\sigma}^\mu) = \Lambda^\mu{}_\nu\,x^\nu\;,$$

$$\Lambda^\mu{}_\nu = \frac{1}{2}\,\text{tr}\,\left(D(\Lambda)\,\sigma_\nu\,D^\dagger(\Lambda)\,\bar{\sigma}^\mu\right)\;. \tag{28}$$

That the linear transformation $\Lambda^\mu{}_\nu$ is indeed a Lorentz transformation matrix, follows from the observation that the Lorentz invariant, symmetric bilinear form $X_{\alpha\dot{\beta}}\,Y^{\alpha\dot{\beta}}$ might be rewritten in terms of $g_{\mu\nu}\,x^\mu\,y^\nu$,

$$X_{\alpha\dot{\beta}}\,Y^{\alpha\dot{\beta}} = \left[x_\mu\,(\sigma^\mu)_{\alpha\dot{\beta}}\right]\,\left[y^\nu\,\varepsilon^{\alpha\gamma}\,\varepsilon^{\dot{\beta}\dot{\delta}}\,(\sigma_\nu)_{\gamma\dot{\delta}}\right] = $$
$$= x_\mu\,y^\nu\,(\sigma^\mu)_{\alpha\dot{\beta}}\,(\bar{\sigma}_\nu)^{\dot{\beta}\alpha} = \qquad\qquad a$$
$$= 2\,x_\mu\,y^\mu\;.$$

In (16) we defined the covariant $(\frac{1}{2},0)$-generators $S^{\mu\nu} = \frac{1}{2}\sigma^{\mu\nu}$. In terms of the basis vectors (24), they take the explicit form

$$\sigma^{\mu\nu} = (\sigma^{\mu\nu})_\alpha{}^\beta = \frac{i}{2}\,(\sigma^\mu\,\bar{\sigma}^\nu - \sigma^\nu\,\bar{\sigma}^\mu)\;, \tag{29}$$

which is the Weyl spinor analogue of (4). The $(0,\frac{1}{2})$-generators $\frac{1}{2}\bar{\sigma}^{\mu\nu}$ are related to them via (17),

$$\bar{\sigma}^{\mu\nu} = (\bar{\sigma}^{\mu\nu})^{\dot{\alpha}}{}_{\dot{\beta}} = -\varepsilon\,\sigma^{\mu\nu*}\,\varepsilon^{-1} = $$
$$= \frac{i}{2}\,(\bar{\sigma}^\mu\,\sigma^\nu - \bar{\sigma}^\nu\,\sigma^\mu) = \sigma^{\mu\nu\dagger}\;, \tag{30}$$

and it is not difficult to check by explicit calculation that the following relations hold:

$$\sigma^\mu\,\bar{\sigma}^\nu = g^{\mu\nu} - i\sigma^{\mu\nu}\;, \tag{31}$$
$$\bar{\sigma}^\mu\,\sigma^\nu = g^{\mu\nu} - i\bar{\sigma}^{\mu\nu}\;. \tag{32}$$

Next, let us consider the $(\frac{1}{2},0)\otimes(\frac{1}{2},0)$-representation with spinors $\tilde{X}_{\alpha\beta}$. They might be decomposed by a Clebsch-Gordan expansion into an irreducible (and thus symmetric) $(1,0)$-contribution $X_{(\alpha\beta)}$ and an antisymmetric $(0,0)$-contribution $X_{[\alpha\beta]}$,

$$\tilde{X}_{\alpha\beta} = X_{(\alpha\beta)} + X_{[\alpha\beta]} , \qquad X_{[\alpha\beta]} = \varepsilon_{\alpha\beta}\left(\frac{1}{2}\varepsilon^{\gamma\delta}\tilde{X}_{\delta\gamma}\right) . \tag{33}$$

We are now going to show that the $(1,0)$-spinor representation is equivalent to an antisymmetric rank 2 Minkowski tensor representation of the Lorentz group. To this end we map the symmetric spinors $X_{(\alpha\beta)}$ one to one onto traceless spinors by raising the second index,

$$X_\alpha{}^\beta = X_{(\alpha\gamma)}\,\varepsilon^{\gamma\beta} , \qquad X_\alpha{}^\alpha = 0 . \tag{34}$$

Traceless 2×2-matrices $X = (X_\alpha{}^\beta)$, however, are nothing but the Lie algebra of $SL(2,\mathbb{C})$ and so may be expanded uniquely in terms of the generators $\sigma_{\mu\nu}$,

$$X_\alpha{}^\beta = x^{\mu\nu}\,(\sigma_{\mu\nu})_\alpha{}^\beta , \tag{35}$$

thereby defining the real antisymmetric coefficients $x^{\mu\nu}$. We are left to derive their Lorentz transformation properties from the transformation property

$$X'_\alpha{}^\beta = D(\Lambda)_\alpha{}^\gamma\,D^{-1}(\Lambda)_\delta{}^\beta\,X_\gamma{}^\delta \quad \text{or} \quad X' = D(\Lambda)\,X\,D^{-1}(\Lambda) \tag{36}$$

of the $(1,0)$-spinors. From (28), we have

$$\begin{aligned} D(\Lambda)\,\sigma_\mu\,D^\dagger(\Lambda) &= \Lambda^\rho{}_\mu\,\sigma_\rho , \\ \overline{D}(\Lambda)\,\bar{\sigma}_\mu\,\overline{D}^\dagger(\Lambda) &= \Lambda^\rho{}_\mu\,\bar{\sigma}_\rho , \end{aligned} \tag{37}$$

and using $D^{-1} = \overline{D}^\dagger$ we obtain the Lorentz transformation behavior of the $\sigma_{\mu\nu}$

$$D(\Lambda)\,\sigma_{\mu\nu}\,D^{-1}(\Lambda) = \Lambda^\rho{}_\mu\,\Lambda^\sigma{}_\nu\,\sigma_{\rho\sigma} , \tag{38}$$

which in fact proves that the coefficients $x^{\mu\nu}$ transform as Minkowski tensors,

$$x'^{\rho\sigma} = \Lambda^\rho{}_\mu\,\Lambda^\sigma{}_\nu\,x^{\mu\nu} . \tag{39}$$

3 Supersymmetry Algebra

The key idea underlying supersymmetry is to extend the Lie algebra formed by the bosonic generators $\mathcal{B}_i = P^\mu, M^{\mu\nu}, B_l$ to a graded Lie algebra by supplementing it with a set of fermionic generators $\mathcal{F}_\alpha$. A graded Lie algebra is defined by the (anti-)commutation relations:

$$[\,\mathcal{B}_i\,,\mathcal{B}_j\,] = \mathrm{i}\,c_{ij}^k\,\mathcal{B}_k , \qquad c_{ij}^k = -c_{ji}^k ; \tag{40}$$

$$\{\mathcal{F}_\alpha, \mathcal{F}_\beta\} = d_{\alpha\beta}^i\,\mathcal{B}_i , \qquad d_{\alpha\beta}^i = d_{\beta\alpha}^i \tag{41}$$

$$[\,\mathcal{F}_\alpha, \mathcal{B}_i\,] = s_{\alpha i}^\beta \mathcal{F}_\beta . \tag{42}$$

The Jacobi identities generalize to

$$[[\mathcal{B}_1, \mathcal{B}_2], \mathcal{B}_3] + [[\mathcal{B}_3, \mathcal{B}_1], \mathcal{B}_2] + [[\mathcal{B}_2, \mathcal{B}_3], \mathcal{B}_1] = 0 \, , \tag{43}$$

$$[[\mathcal{F}_1, \mathcal{B}_1], \mathcal{B}_2] + [[\mathcal{B}_2, \mathcal{F}_1], \mathcal{B}_1] + [[\mathcal{B}_1, \mathcal{B}_2], \mathcal{F}_1] = 0 \, , \tag{44}$$

$$[\{\mathcal{F}_1, \mathcal{F}_2\}, \mathcal{B}_1] + \{[\mathcal{B}_1, \mathcal{F}_1], \mathcal{F}_2\} - \{[\mathcal{F}_2, \mathcal{B}_1], \mathcal{F}_1\} = 0 \, , \tag{45}$$

$$[\{\mathcal{F}_1, \mathcal{F}_2\}, \mathcal{F}_3] + [\{\mathcal{F}_3, \mathcal{F}_1\}, \mathcal{F}_2] + [\{\mathcal{F}_2, \mathcal{F}_3\}, \mathcal{F}_1] = 0 \, . \tag{46}$$

Such a graded Lie algebra structure occurs almost naturally, if the generators – collectively called $\mathcal{G}$ for the moment – are of a bilinear form

$$\mathcal{G} = \sum_{ij} \int dp\, dq\, a_i^\dagger(p)\, K_{ij}(p,q)\, a_j(q) \, , \tag{47}$$

which is reasonable at least in non-interacting theories. We use a slightly symbolic notation here, and $a_i(p)$ may denote either bosonic $b_i(p)$ or fermionic $f_i(p)$ annihilation operators. By a bosonic (or even) generator we mean an operator that consists of a pure bosonic and a pure fermionic contribution but does not mix fermions and bosons:

$$\mathcal{B} = \sum_{ij} \int dp\, dq\, b_i^\dagger(p)\, K_{ij}^1(p,q)\, b_j(q) + \sum_{ij} \int dp\, dq\, f_i^\dagger(p)\, K_{ij}^2(p,q)\, f_j(q) \; ; \tag{48}$$

acting on states, it will not change their statistics. Correspondingly a fermionic (or odd) generator has the form

$$\mathcal{F} = \sum_{ij} \int dp\, dq\, b_i^\dagger(p)\, K_{ij}^3(p,q)\, f_j(q) + \sum_{ij} \int dp\, dq\, f_i^\dagger(p)\, K_{ij}^4(p,q)\, b_j(q) \tag{49}$$

and *does* change the spin of a one-particle state by a half integer amount and thus its statistics. If we assume standard (anti-)commutation relations for the creation and annihilation operators,

$$\begin{aligned} [b_i(p), b_j^\dagger(q)] &= i\,\delta_{ij}\,\delta(p-q) \, , \\ \{f_i(p), f_j^\dagger(q)\} &= \delta_{ij}\,\delta(p-q) \, , \end{aligned} \tag{50}$$

the left hand sides of (40)-(42) are the only combinations that lead again to bilinear generators (47) of the appropriate statistics.

Once having accepted this graded Lie algebra structure, which relates bosonic and fermionic generators intimately, we can exploit the knowledge of the bosonic subsector provided by the Coleman-Mandula theorem to determine the fermionic sector. We will denote the fermionic generators by Q from now on and assume that with Q also $Q^\dagger$ is a generator of the graded algebra.

First, we will show that the generators Q must necessarily be spin 1/2 Weyl spinors, if the underlying Hilbert space has a positive definite metric [1]. To this end we note that every Q may be decomposed into a sum of $\mathcal{L}_+^\uparrow$-irreducible (a,b)-representations,

$$Q = \sum Q_{(\alpha_1...\alpha_a),(\dot\beta_1...\dot\beta_b)} \, , \tag{51}$$

where the $Q_{(\alpha_1...\alpha_a),(\dot\beta_1...\dot\beta_b)}$ are totally symmetric both in the dotted and undotted indices separately as indicated by the brackets. Now consider the anti-commutator

$$\left\{ Q_{(\alpha_1...\alpha_a),(\dot\beta_1...\dot\beta_b)}, Q^\dagger_{(\dot\gamma_1...\dot\gamma_a),(\delta_1...\delta_b)} \right\} \tag{52}$$

and equate all occurring indices. Since the resulting object will then be totally symmetric, it must be a component of the irreducible $[\frac{1}{2}(a+b), \frac{1}{2}(a+b)]$-representation. However, for $a+b > 1$ there is no generator in the bosonic sector which transforms accordingly and might be used to close the algebra. So, (52) has to vanish, whenever all indices are equal. But since it is a positive operator in Hilbert space, the $Q_{(\alpha...\alpha),(\dot\alpha...\dot\alpha)}$ themselves have to vanish, by the positivity of the metric. What is more, they have to vanish in *every frame*, and so because of the irreducibility all the $Q_{(\alpha_1...\alpha_a),(\dot\beta_1...\dot\beta_b)}$ have to vanish. This completes our proof that no representations with spin higher than 1/2 are possible for the fermionic generators Q. We denote the $(\frac{1}{2}, 0)$-spinors by Q^i_α and the $(0, \frac{1}{2})$ ones by $\overline{Q}^j_{\dot\alpha}$ where $\alpha, \dot\alpha = 1, 2$ are Weyl spinor indices. The indices $i, j = 1, \ldots, N$ label different kinds of generators and usually refer to some representation of an internal symmetry group. The case $N = 1$ is often called *simple supersymmetry*, whereas the $N > 1$ situation is usually referred to as *extended supersymmetry*. The Hermitean conjugate of each $(\frac{1}{2}, 0)$-spinor Q^i_α must be one of the $(0, \frac{1}{2})$-generators $\overline{Q}^j_{\dot\beta}$ and vice versa, so that by appropriate labelling

$$\overline{Q}^i_{\dot\alpha} = Q^i_\alpha{}^\dagger . \tag{53}$$

As a next step, we will fix the structure constants $d^i_{\alpha\beta}$ and $s^\beta_{\alpha i}$ in (41), (42) as far as possible. The line of argumentation will be, first to determine the most general expression compatible with the Lorentz structure of the (anti-)commutators. The generators appearing at the right hand side of (41)-(42) are then identified by their Lorentz transformation properties. If there is no generator sitting in the corresponding representation, the contribution has to vanish. Additionally, we may then exploit Jacobi identities to further restrict the structure constants.

The anti-commutator $\{Q^i_\alpha, \overline{Q}^j_{\dot\beta}\}$ sits in the $(\frac{1}{2}, 0) \otimes (0, \frac{1}{2}) = (\frac{1}{2}, \frac{1}{2})$-representation of the Lorentz group, which is equivalent to Minkowski space as we have seen in (27), (28), so that P^μ, the only spin 1 generator in the game, has to close the algebra:

$$\{Q^i_\alpha, \overline{Q}^j_{\dot\beta}\} = 2 c^{ij} P_{\alpha\dot\beta} = 2 c^{ij} (\sigma_\mu)_{\alpha\dot\beta} P^\mu .$$

Evidently, the matrix (c^{ij}) is Hermitean and may be diagonalized by a unitary redefinition of the Q's. If all eigenvalues of (c^{ij}) are positive, we can renormalize the odd generators such that

$$\{Q^i_\alpha, \overline{Q}^j_{\dot\beta}\} = 2\delta^{ij} (\sigma_\mu)_{\alpha\dot\beta} P^\mu . \tag{54}$$

Negative eigenvalues would give rise to a minus sign at the right hand side of (54). Taking the trace over the spinor indices with arbitrary but fixed $i = j$ gives

$$\sum_\alpha \{Q^i_\alpha, \overline{Q}^i_{\dot\alpha}\} = \sum_\alpha \{Q^i_\alpha, Q^{i\,\dagger}_\alpha\} = 2\,\mathrm{tr}\,(\sigma_\mu P^\mu) = 4\,P^0 \, , \tag{55}$$

which immediately implies that with the l.h.s. of (55) also the Hamiltonian $H = P^0$ has to be positive definite. Note that an indefinite matrix (c^{ij}) would only be compatible with $H = 0$ and thus has to be discarded on physical grounds, leaving (54) as the only sensible possibility.

The anti-commutator of two undotted odd generators lies in the $(\frac{1}{2}, 0) \otimes (\frac{1}{2}, 0)$-representation and can be decomposed into an antisymmetric $(0, 0)$- and a symmetric $(1, 0)$-part, cf. (33)

$$\{Q^i_\alpha, Q^j_\beta\} = 2\,\varepsilon_{\alpha\beta} Z^{ij} + M^{\mu\nu}\,(\sigma_{\mu\nu})_{\alpha\beta}\,Z^{(ij)} \, ,$$

with Z^{ij} ($Z^{(ij)}$) being antisymmetric (symmetric) Lorentz scalars and thus linear combinations of the generators of the internal symmetries,

$$Z^{ij} = (a^r)^{ij}\,B_r \, . \tag{56}$$

However, we will show below that Q^i_α has to commute with P^μ so that the $M^{\mu\nu}$-part must vanish in fact, $Z^{(ij)} = 0$, yielding

$$\{Q^i_\alpha, Q^j_\beta\} = 2\,\varepsilon_{\alpha\beta} Z^{ij} \, , \tag{57}$$

$$\{\overline{Q}^i_{\dot\alpha}, \overline{Q}^j_{\dot\beta}\} = 2\,\varepsilon_{\dot\alpha\dot\beta} Z^{ij\,*} \, . \tag{58}$$

The vanishing of the commutator of Q^i_α with the momentum P^μ is more involved to derive by formal arguments, but rather obvious from the fact that spacetime translations act only on the spacetime arguments and not on the spinorial degrees of freedom. As we already know, P^μ sits in the $(\frac{1}{2}, \frac{1}{2})$-representation and we can associate a spinor $P_{\beta\dot\gamma} = P^\mu\,(\sigma_\mu)_{\beta\dot\gamma}$ with it. The commutator $[Q^i_\alpha, P_{\beta\dot\gamma}]$ then is of the type $(\frac{1}{2}, 0) \otimes (\frac{1}{2}, \frac{1}{2}) = (0, \frac{1}{2}) \oplus (1, \frac{1}{2})$. Since there are no spin $3/2$ generators present and the spin $1/2$ component must be antisymmetric in the undotted indices, the only possibility is

$$[Q^i_\alpha, P_{\beta\dot\gamma}] = 2\,c^{ij}\,\varepsilon_{\alpha\beta}\,\overline{Q}^j_{\dot\gamma} \, , \tag{59}$$

Projecting on P^μ by contracting both sides with $(\bar\sigma^\mu)^{\dot\gamma\beta}$ (see (25)) gives

$$[Q^i_\alpha, P^\mu] = -c^{ij}\,(\sigma^\mu)_{\alpha\dot\beta}\,\overline{Q}^{\dot\beta j} \, , \tag{60}$$

$$[\overline{Q}^{\dot\alpha i}, P^\mu] = -c^{ij\,*}\,(\bar\sigma^\mu)^{\dot\alpha\beta}\,Q^j_\beta \, . \tag{61}$$

We can now use the Jacobi identity (44) to show that

$$[[Q^i_\alpha, P^\mu], P^\nu] = [[Q^i_\alpha, P^\nu], P^\mu] \, . \tag{62}$$

Explicit evaluation of the double commutator, however, yields

$$[[Q^i_\alpha, P^\mu], P^\nu] = c^{ik} c^{kj*} (\sigma^\mu \bar\sigma^\nu)_\alpha{}^\beta Q^j_\beta , \qquad (63)$$

which is not symmetric in the Minkowski indices – cf.(31) – unless

$$CC^* = 0 .$$

The matrix $C = (c^{ij})$ is symmetric, as can be seen by calculating

$$0 = [Z^{ij}, P^\mu] = \frac{1}{2}\varepsilon^{\beta\alpha} [\{Q^i_\alpha, Q^j_\beta\}, P^\mu] \sim (c^{ij} - c^{ji}) ,$$

and we have $CC^\dagger = 0$ or $c^{ij} = 0$. So, finally we have proven:

$$[Q^i_\alpha, P^\mu] = 0 , \qquad (64)$$

$$[\overline{Q}^i_{\dot\alpha}, P^\mu] = 0 . \qquad (65)$$

The commutators of the odd generators with the Lorentz generators follow immediately from the transformation property of Weyl spinors,

$$Q'^i_\alpha = \exp\left(-\frac{i}{2}\omega_{\mu\nu}M^{\mu\nu}\right) Q^i_\alpha \exp\left(+\frac{i}{2}\omega_{\mu\nu}M^{\mu\nu}\right) = \exp\left(\frac{i}{4}\omega_{\mu\nu}\sigma^{\mu\nu}\right)_\alpha{}^\beta Q^i_\beta ,$$

which implies for infinitesimal transformations

$$[Q^i_\alpha, M^{\mu\nu}] = -\frac{1}{2}(\sigma^{\mu\nu})_\alpha{}^\beta Q^i_\beta , \qquad (66)$$

$$[\overline{Q}^i_{\dot\alpha}, M^{\mu\nu}] = \frac{1}{2}\overline{Q}^i_{\dot\alpha} (\bar\sigma^{\mu\nu})^{\dot\beta}{}_{\dot\alpha} . \qquad (67)$$

We are left with the commutation relations between Q^i_α and the Hermitean generators B_r of the internal symmetry. They are easily determined to be

$$[Q^i_\alpha, B_r] = b_r^{ij} Q^j_\alpha \qquad (68)$$

$$[\overline{Q}^i_{\dot\alpha}, B_r] = -\overline{Q}^j_{\dot\alpha} b_r^{ji} , \qquad (69)$$

saying that the Q^i_α carry indeed a representation of the internal symmetry group. In (69) we have made use of the Hermiticity of the matrix (b_r^{ij}), which can be shown by applying the Jacobi identity (45) to $[\{Q^i_\alpha, \overline{Q}^j_{\dot\beta}\}, B_r] = 0$ and inserting (68) and (54).

Let us now consider the Poincaré invariant operators Z^{ij} introduced in (56). If we express them via (57) in terms of odd generators, $Z^{ij} = \frac{1}{2}\varepsilon^{\beta\alpha}\{Q^i_\alpha, Q^j_\beta\}$, and apply the Jacobi identity (45) to $[\{Q^i_\alpha, Q^j_\beta\}, B_r]$, we obtain with the help of (68), (69) and (56)

$$[Z^{ij}, B_r] = b_r^{ik} Z^{kj} + b_r^{jk} Z^{ik} , \qquad (70)$$

$$[Z^{ij}, Z^{kl}] = (a^r)^{kl} b_r^{im} Z^{mj} + (a^r)^{kl} b_r^{jm} Z^{im} . \qquad (71)$$

This means that the Z^{ij} form an invariant subalgebra of the internal symmetry algebra. Starting with $[\{Q_\alpha^i, Q_\beta^j\}, \overline{Q}_{\dot\gamma}^k]$ we obtain by a similar reasoning

$$[\overline{Q}_{\dot\alpha}^i, Z^{jk}] = 0 , \qquad (72)$$

hence

$$(a^r)^{ij} b_r^{kl} = 0 , \qquad (73)$$

and thus

$$[Q_\alpha^i, Z^{ij}] = 0 , \qquad (74)$$
$$[Z^{ij}, Z^{kl}] = 0 , \qquad (75)$$

which implies that the invariant subalgebra spanned by the Z^{ij} is Abelian. Since the internal symmetry algebra is a direct sum of a semisimple and an Abelian part, the Z^{ij} must be in the Abelian sector and thus commute with all the generators B_r,

$$[Z^{ij}, B_r] = 0 . \qquad (76)$$

So, finally we have found that the Z^{ij} commute with *all* the generators of the SUSY-algebra. For this reason they are called *central charges*. Note that due to the antisymmetry in the indices, central charges can only exist in extended supersymmetries, $N > 1$. In the absence of central charges, the internal symmetry group is the largest possible, namely U(N), since the algebra is obviously invariant under a unitary substitution

$$Q'^i_\alpha = U^{ij} Q^i_\alpha , \qquad \overline{Q}'^i_{\dot\alpha} = U^{ij*} \overline{Q}^i_{\dot\alpha} .$$

The effect of central charges is to reduce this symmetry to a smaller group. In the case of simple supersymmetry ($N = 1$) the internal symmetry group is U(1) and commonly referred to as $\mathcal{R}$-*symmetry*,

$$[Q_\alpha, \mathcal{R}] = Q_\alpha ,$$
$$[\overline{Q}_{\dot\alpha}, \mathcal{R}] = -\overline{Q}_{\dot\alpha} . \qquad (77)$$

Since $Q \leftrightarrow \overline{Q}$ under parity, it is a chiral symmetry.

Let us conclude this section with a summary of the most general supersymmetry algebra compatible with relativistic quantum field theory:

$$[P^\mu, P^\nu] = 0 \tag{5}$$
$$[M^{\mu\nu}, P^\rho] = i(g^{\nu\rho}P^\mu - g^{\mu\rho}P^\nu) \tag{6}$$
$$[M^{\mu\nu}, M^{\rho\sigma}] = i(g^{\nu\rho}M^{\mu\sigma} + g^{\mu\sigma}M^{\nu\rho} - g^{\mu\rho}M^{\nu\sigma} - g^{\nu\sigma}M^{\mu\rho}) \tag{7}$$

$$[B_r, P^\mu] = 0$$
$$[B_r, M^{\mu\nu}] = 0 \tag{10}$$
$$[B_r, B_s] = i c_{rs}^t B_t$$

$$\{Q_\alpha^i, \overline{Q}_{\dot\beta}^j\} = 2\delta^{ij}(\sigma_\mu)_{\alpha\dot\beta}P^\mu \tag{54}$$

$$\{Q_\alpha^i, Q_\beta^j\} = 2\varepsilon_{\alpha\beta}Z^{ij}, \quad \{\overline{Q}_{\dot\alpha}^i, \overline{Q}_{\dot\beta}^j\} = 2\varepsilon_{\dot\alpha\dot\beta}Z^{ij*} \tag{57, 58}$$

$$[Q_\alpha^i, P^\mu] = 0, \quad [\overline{Q}_{\dot\alpha}^i, P^\mu] = 0 \tag{64, 65}$$

$$[Q_\alpha^i, M^{\mu\nu}] = \tfrac{1}{2}(\sigma^{\mu\nu})_\alpha{}^\beta Q_\beta^i, \quad [\overline{Q}_{\dot\alpha}^i, M^{\mu\nu}] = -\tfrac{1}{2}\overline{Q}_{\dot\alpha}^i(\bar\sigma^{\mu\nu})^{\dot\beta}{}_{\dot\alpha} \tag{66, 67}$$

$$[Q_\alpha^i, B_r] = b_r^{ij} Q_\alpha^j, \quad [\overline{Q}_{\dot\alpha}^i, B_r] = -\overline{Q}_{\dot\alpha}^j (b_r)^{ij*} \tag{68, 69}$$

$$[Z^{ij}, \text{any generator}] = 0$$

No further restrictions on the structure constants follow from the Jacobi identities, as can be seen by checking them all explicitly.

4 Representations of the Supersymmetry Algebra

The unitary irreducible representations of the SUSY algebra are spanned by sets of one-particle states; this is completely analogous to the representation theory of the Poincaré group.

Because every representation of the SUSY algebra is trivially a representation of the Poincaré algebra (the latter is a subalgebra of the former), its representation vectors can be labelled by eigenvalues of the mass-square operator $P^\mu P_\mu = M^2$ and the square of the Pauli-Lubanski spin vector $W^\mu W_\mu$. The latter ones eigenvalues, $-M^2 s(s+1)$, with s an integer or half-integer can, in the massive case be used to further specify the representation while in the massless case the spin is replaced by the helicity which is also an integer or half-integer.

The SUSY generators commute with $P^\mu P_\mu$ but not with $W^\mu W_\mu$; an irreducible representation of the SUSY algebra will therefore contain different spins or felicities but only one mass.

The easiest case is of course the "simple" supersymmetry, i.e. SUSY with one fermionic generator. To construct the representations of this algebra let us make use of the Wigner trick: Go to a fixed Lorentz frame characterized by a certain value of the four momentum P_μ, and study the invariance group — the "little group" — of this vector.

Let us first investigate the massive case in which we choose the rest frame of the particle, $P_\mu = (M, \mathbf{0})$, $M \neq 0$. In this frame, the SUSY algebra (54) reduces to

$$\{Q_\alpha, \overline{Q}_{\dot\beta}\} = 2M\delta_{\alpha\beta}. \tag{78}$$

If we now define $a_\alpha = \frac{Q_\alpha}{\sqrt{2M}}$, we have two ordinary fermionic creation and annihilation operators obeying the Clifford algebra

$$\{a_\alpha, a_\beta^\dagger\} = \delta_{\alpha\beta}; \tag{79}$$

all the other anti-commutators vanish.

As is known from many-body theory, the irreducible representations of a Clifford algebra contain one vector $|0\rangle$ (the "Clifford vacuum") which is annihilated by all the a_α. (Because $|0\rangle$ is a one-particle state with mass M and a given spin it is of course not the vacuum state of a quantum field theory.)

Acting with our creation operators on this "vacuum" we obtain four independent states,

$$|0\rangle, \quad a_1^\dagger|0\rangle, \quad a_2^\dagger|0\rangle, \quad a_1^\dagger a_2^\dagger|0\rangle. \tag{80}$$

Let us now choose the "Clifford vacuum" to have spin 0. Because the SUSY generators have spin $\frac{1}{2}$ (see (66)), the above states can be interpreted as containing one scalar, two fermionic and one pseudoscalar degree of freedom, respectively. This irreducible representation is also known as the "chiral massive multiplet"; in the simplest form of SUSY-QCD, e.g., it is used to describe a (left handed) quark and its superpartner, the squark.

Table 1. The $N = 1$ massive multiplet

spin	0	$\frac{1}{2}$
nr. of states	2	2

We also observe a very general result, valid for *every* SUSY representation of non-vanishing four momentum: The number of fermionic states (with odd fermion number N_F) has to be equal to the number of bosonic states (with even fermion number). The proof of this theorem is based on the operator $(-1)^{N_F}$, whose trace measures the difference between the number of bosonic states and the number of fermionic states. It is clear that it anti-commutes with the SUSY generators,

$$(-1)^{N_F} Q_\alpha = -Q_\alpha (-1)^{N_F}, \tag{81}$$

and therefore, using the cyclic property of the trace,

$$\text{tr}[(-1)^{N_F}\{Q_\alpha, \overline{Q}_{\dot\beta}\}] = 0. \tag{82}$$

If we now insert the fundamental SUSY anti-commutator (54) we obtain

$$\text{tr}[(-1)^{N_F} P_\mu] = 0, \tag{83}$$

and, because the four momentum is fixed and non-vanishing,

$$\text{tr}[(-1)^{N_F}] = 0, \tag{84}$$

which proves the theorem.

In the massless case we choose the frame with $P^\mu = (E, 0, 0, E)$ as our reference frame. Because of

$$\{Q, \overline{Q}\} = 2P_\mu \sigma^\mu = 2\begin{pmatrix} P_0 + P_3 & P_1 - iP_2 \\ P_1 + iP_2 & P_0 - P_3 \end{pmatrix} = \begin{pmatrix} 0 & 0 \\ 0 & 4E \end{pmatrix} \tag{85}$$

only the component with $\alpha = 2$ can be used as a generator; if we define $a_2 = \frac{Q_2}{\sqrt{2E}}$ we obtain only two independent states,

$$|0\rangle, \ a_2^\dagger |0\rangle, \tag{86}$$

with helicities h and $h + \frac{1}{2}$, respectively.

As any Lorentz covariant, local field theory will exhibit PCT symmetry, one has to include another doublet of states with opposite helicities.

Choosing e.g. $h = \frac{1}{2}$, one obtains the one-particle states of a massless vector particle together with a massless Majorana fermion. In QCD, this multiplet is called the "gauge supermultiplet" and contains the gluon and its superpartner, the gluino.

Table 2. *The $N = 1$ gauge multiplet*

helicity	-1	$-\frac{1}{2}$	$\frac{1}{2}$	1
nr. of states	1	1	1	1

We can now proceed to "extended" supersymmetry; we will only be concerned with the case where the number N of SUSY generators is equal to 2. Because it is antisymmetric, the central charge matrix has to be proportional to the ϵ-tensor,

$$Z^{ij} = \epsilon^{ij} Z. \tag{87}$$

We may also observe that our extended SUSY algebra possesses a new symmetry, $U(2) = SU(2) \times U(1)$, acting in internal space. This symmetry is called "$\mathcal{R}$-symmetry"; in non-extended supersymmetry it reduces to $U(1)$.

In the construction of the representations of the algebra, let us first consider the case of vanishing central charge, $Z = 0$.

In the massive case, one has $2N = 4$ fermionic generators $a_\alpha^{i\dagger}$, which can be used to create $2^{2N} = 16$ states,

$$|0\rangle, \ a_1^{1\dagger}|0\rangle, \ldots, \ a_1^{1\dagger}a_2^{1\dagger}a_1^{2\dagger}a_2^{2\dagger}|0\rangle. \tag{88}$$

If the Clifford ground state is again chosen to have spin 0, these states transform as five scalars, two Dirac fermions and one massive vector particle under the Poincaré group, altogether yielding 16 physical degrees of freedom.

Exactly as above, in the massless case the number of independent generators is reduced by a factor of two, i.e. we have two operators $a_2^{i\dagger}$ and four states

$$|0\rangle, \ a_2^{1\dagger}|0\rangle, \ a_2^{2\dagger}|0\rangle, \ a_2^{1\dagger}a_2^{2\dagger}|0\rangle. \tag{89}$$

With a spin 0 ground state these states have the respective helicities 0, $\frac{1}{2}$ and 1. Again the inclusion of the PCT-conjugate multiplet will be necessary to ensure Lorentz covariance. In this way we end up with the (in QCD terminology) so-called "$N = 2$ gauge multiplet" which consists of the gluon and its superpartners, two Weyl spinors and two scalars. The $\mathcal{R}$- symmetry acts on the two spinors which transform as a doublet while the other fields are invariant.

Switching back to $N = 1$ language, we can decompose this multiplet into two $N = 1$ sub-multiplets: a chiral one containing the scalars and one of the spinors (which one doesn't matter because of $\mathcal{R}$-symmetry) and a gauge multiplet with the vector particle and the remaining spinor.

Table 3. *The $N = 2$ gauge multiplet*

helicity	-1	$-\frac{1}{2}$	0	$\frac{1}{2}$	1
nr. of states	1	2	2	2	1
$N = 1$ gauge multiplet	1	1		1	1
$N = 1$ chiral multiplet		1	$1+1$	1	

We now turn to the case of a non-vanishing central charge Z which by a redefinition of the phase of the SUSY generators we choose to be real. Considering the massive case first, we change over to the linear combinations

$$a_\alpha = \frac{1}{\sqrt{2}}(Q_\alpha^{1\dagger} + \epsilon_{\alpha\beta}Q_\beta^{2\dagger}), \tag{90}$$

$$b_\alpha = \frac{1}{\sqrt{2}}(Q_\alpha^{1\dagger} - \epsilon_{\alpha\beta}Q_\beta^{2\dagger}), \tag{91}$$

and obtain the algebraic relations

$$\{a_\alpha, a_\beta\} = \{b_\alpha, b_\beta\} = \{a_\alpha, b_\beta\} = 0 \tag{92}$$

and

$$\{a_\alpha, a_\beta^\dagger\} = 2\delta_{\alpha\beta}(M + Z), \tag{93}$$

$$\{b_\alpha, b_\beta^\dagger\} = 2\delta_{\alpha\beta}(M - Z). \tag{94}$$

From the positive definiteness of the l.h.s. we conclude that

$$|Z| \leq M. \tag{95}$$

If $|Z| \neq M$, the algebra is isomorphic to the $Z = 0$, $M \neq 0$ case; the particle content is the same as constructed above.

However, if $|Z| = M$, either the a's or the b's cease to be independent operators and the one-particle subspace becomes $2^N = 4$-dimensional. In this way, one can construct a "hyper-"multiplet containing four scalar and four spinorial degrees of freedom. The doubling of the degrees of freedom as compared to the $N = 1$ massive multiplet can be shown to be a consequence of the $N = 2$ supersymmetry [3].

Table 4. *The Hypermultiplet*

spin	0	$\frac{1}{2}$
nr. of states	$2 + 2$	$2 + 2$

In the massless situation one can show that the central charge necessarily has to vanish [3], so there are no additional multiplets in this case.

5 Field Theories Built on SUSY Representations

In this section we are going to tackle the primary goal of our introduction, namely the formulation of supersymmetric field theory. The task consists of determining a multiplet of fields forming an (off shell) representation of the supersymmetry algebra for each of the SUSY multiplets relevant to us. Additionally, we have to construct supersymmetric Lagrangeans for them. One way to achieve this would be to start with an arbitrary field contained in the multiplet of interest, apply subsequent SUSY transformations to it and identify the other fields belonging to the multiplet in what comes out. If necessary, one has to introduce auxiliary fields to enforce the transformations to close (this reflects the fact, that we are dealing with *off-shell* representations; once having constructed the Lagrangean one knows the equations of motion and can eliminate the auxiliary fields with their help to gain an on-shell representation). For details and examples of this procedure we refer to the literature (e.g. [1], [2], [3]) and leave it as an exercise to the reader. We will present a more elegant and powerful method of facilitating this task, namely the use of *superfields*. To keep things as transparent as possible, we restrict here to simple ($N = 1$) supersymmetry.

In both approaches it is advantageous to turn over from the algebra to a group by exponentiating the generators. However, in a graded Lie algebra one is faced with a problem: the appearance of anti-commutators seems to rule out the application of the Baker-Campbell-Hausdorff formula which in turn is necessary to prove that subsequent transformations do not leave the group manifold. The problem is solved by the introducing *Grassmann* valued, spacetime independent Weyl spinor parameters θ^α and $\bar\theta_{\dot\alpha} = \theta^*_\alpha$ which multiply the fermionic generators,

$$\theta Q = \theta^\alpha Q_\alpha , \qquad \overline{\theta Q} = \bar\theta_{\dot\alpha} \bar Q^{\dot\alpha} . \tag{96}$$

By definition, they satisfy

$$\{\theta_\alpha, \theta_\beta\} = \{\bar\theta_{\dot\alpha}, \bar\theta_{\dot\beta}\} = \{\theta_\alpha, \bar\theta_{\dot\beta}\} = 0 . \tag{97}$$

Additionally, they commute with all bosonic SUSY generators $\mathcal{B}$ and anticommute with all fermionic ones $\mathcal{F}$,

$$[\theta_\alpha, \mathcal{B}] = 0 , \qquad \{\theta_\alpha, \mathcal{F}\} = 0 . \tag{98}$$

With their help, the supersymmetry algebra can be rewritten entirely in terms of commutators, e.g.[2]

$$[\theta Q, \xi Q] = [\overline{\theta Q}, \overline{\xi Q}] = 0 ,$$
$$[\theta Q, \overline{\xi Q}] = 2(\theta \sigma_\mu \bar\xi) P^\mu , \tag{99}$$

defines a closed sub-algebra, where the ξ are another set of Grassmann parameters. Group elements of this "Lie group with anti-commuting parameters" are gained just as in ordinary Lie groups by exponentiating the algebra

$$G(x, \theta, \bar\theta) = \exp\left(i\theta Q + i\overline{\theta Q} + ix \cdot P\right) , \tag{100}$$

and the Baker-Campbell-Hausdorff formula,

$$\exp(A) \exp(B) = \exp(A + B + \frac{1}{2}[A, B] + \ldots) \tag{101}$$

may now be invoked to determine the motion in parameter space under a (left) multiplication with a group element $G(y, \xi, \bar\xi)$:

$$G(y, \xi, \bar\xi) G(x, \theta, \bar\theta) = G(x', \theta', \bar\theta')$$

$$\begin{aligned}
x'^\mu &= x^\mu + y^\mu + i\xi\sigma^\mu\bar\theta - i\theta\sigma^\mu\bar\xi , \\
\theta' &= \theta + \xi , \\
\bar\theta' &= \bar\theta + \bar\xi .
\end{aligned} \tag{102}$$

---

[2] We use here and in the following the short hand notation $\theta\sigma_\mu\bar\xi = \theta^\alpha (\sigma_\mu)_{\alpha\dot\beta} \bar\xi^{\dot\beta}$, which transforms under Lorentz transformations like a Minkowski four vector.

Note that the higher order contributions in (101) (indicated by the dots) vanish identically in this case. The parameter space, given as the direct sum of the four dimensional Minkowski space and the four dimensional manifold spanned by the Grassmann parameters θ^α, $\overline{\theta}_{\dot\beta}$, is referred to as *superspace*. A function $S(x,\theta,\overline{\theta})$ defined on superspace is called a *superfield*. It can be expanded into a power series with respect to θ and $\overline{\theta}$ which truncates at finite order due to $\theta_\alpha^2 = 0$, $\overline{\theta}_{\dot\beta}^2 = 0$. The coefficients of this expansion, referred to as component fields, are ordinary fields depending on spacetime. Thus, a superfield might be understood as a collection of certain component fields. It is not too difficult to convince oneself that the most general Lorentz scalar superfield is of the form

$$S(x,\theta,\overline{\theta}) = c(x) + \theta\phi(x) + \overline{\theta\chi}(x) + \theta\theta\, m(x) + \overline{\theta\theta}\, n(x)$$
$$+\theta\sigma^\mu\overline{\theta}\, A_\mu(x) + \theta\theta\,\overline{\theta\omega}(x) + \overline{\theta\theta}\,\theta\psi(x) + \theta\theta\,\overline{\theta\theta}\, d(x) \ , \quad (103)$$

containing the complex Lorentz scalar component fields $c(x)$, $m(x)$, $n(x)$ and $d(x)$, the $(\frac{1}{2},0)$-spinor fields $\phi(x)$ and $\psi(x)$, the $(0,\frac{1}{2})$-spinor fields $\overline{\chi}(x)$ and $\overline{\omega}(x)$ and the complex Lorentz vector field $A_\mu(x)$, so altogether 16 bosonic (off-shell) degrees of freedom and an equal number of fermionic fields illustrating the "fermions = bosons" rule. The mass dimension of Q and $\overline{Q}$ is fixed by the SUSY algebra to the value one half, see (54), which in turn implies the mass dimensions

$$[\theta] = [\overline{\theta}] = -\frac{1}{2} \quad (104)$$

for the Grassmann parameters since θQ and $\overline{\theta Q}$ must be dimensionless quantities in (100). So, in general component fields of higher powers in θ and $\overline{\theta}$ will either have higher mass dimensions or be spacetime derivatives of fields with lower mass dimension.

According to (102), a SUSY transformation acts on a general superfield like

$$G(y^\mu, \xi, \overline{\xi})\, S(x^\mu, \theta, \overline{\theta}) = S(x^\mu + y^\mu + i\xi\sigma^\mu\overline{\theta} - i\theta\sigma^\mu\overline{\xi}, \theta + \xi, \overline{\theta} + \overline{\xi}) \ . \quad (105)$$

Considering an infinitesimal transformation by Taylor expanding the r.h.s. yields a linear representation of the generators Q_α, $\overline{Q}_{\dot\beta}$ and P^μ as differential operators on superspace:

$$P_\mu = i\,\partial_\mu$$
$$i\,Q_\alpha = \frac{\partial}{\partial\theta^\alpha} + i\,(\sigma^\mu)_{\alpha\dot\beta}\,\overline{\theta}^{\dot\beta}\,\partial_\mu \ , \quad (106)$$
$$i\,\overline{Q}_{\dot\beta} = -\frac{\partial}{\partial\overline{\theta}^{\dot\beta}} - i\,\theta^\alpha\,(\sigma^\mu)_{\alpha\dot\beta}\,\partial_\mu \ .$$

It is easily checked that they indeed fulfill the SUSY algebra (54), (57), (58).

We define the fermionic derivatives

$$D_\alpha = \frac{\partial}{\partial \theta^\alpha} - i(\sigma^\mu)_{\alpha\dot\beta}\overline{\theta}^{\dot\beta}\partial_\mu ,$$

$$\overline{D}_{\dot\beta} = -\frac{\partial}{\partial \overline{\theta}^{\dot\beta}} + i\theta^\alpha(\sigma^\mu)_{\alpha\dot\beta}\partial_\mu ,$$
(107)

which also satisfy the algebra

$$\{D_\alpha, D_\beta\} = \{\overline{D}_{\dot\alpha}, \overline{D}_{\dot\beta}\} = 0 ,$$
(108)

$$\{D_\alpha, \overline{D}_{\dot\beta}\} = 2i(\sigma^\mu)_{\alpha\dot\beta}\partial_\mu ;$$
(109)

they are (supersymmetry-)covariant, since they anticomm-ute with all SUSY generators:

$$\{D_\alpha, Q_\beta\} = \{D_\alpha, \overline{Q}_{\dot\beta}\} = \{\overline{D}_{\dot\alpha}, Q_\beta\} = \{\overline{D}_{\dot\alpha}, \overline{Q}_{\dot\beta}\} = 0 .$$
(110)

Superfields are a linear representation of the supersymmetry algebra in the sense that products and linear combinations of superfields are again superfields. However, a general superfield will be a reducible representation. To gain a specific irreducible representation, i.e. a specific supermultiplet, one has to impose appropriate, SUSY covariant constraints on the superfield which eliminate certain component fields.

5.1 Chiral Superfields

The simplest irreducible $N = 1$ SUSY representation is the chiral supermultiplet consisting of (left handed) spin 1/2 Weyl spinors and their supersymmetric spin 0 partners. It is described by a *chiral superfield* $\Phi(x,\theta,\overline{\theta})$, which is defined by the covariant constraint

$$\overline{D}_{\dot\alpha}\Phi(x,\theta,\overline{\theta}) = 0 .$$
(111)

This constraint can be simplified by transforming to the new coordinates $y^\mu = x^\mu - i\theta\sigma^\mu\overline{\theta}$, θ and $\overline{\theta}$, where it takes on the form

$$-\frac{\partial}{\partial \overline{\theta}^{\dot\alpha}}\Phi(y,\theta,\overline{\theta}) = 0 ;$$
(112)

obviously, the most general solution to (112) is an arbitrary function of y^μ and θ alone and given by the θ-expansion

$$\Phi(y,\theta) = \phi(y) + \sqrt{2}\,\theta\psi(y) + \theta\theta\, F(y) ;$$
(113)

we assumed here that the chiral superfield is a Lorentz scalar quantity since we are interested in a chiral multiplet containing a Lorentz scalar (spin 0) field as the component field with the lowest mass dimension. In principle one could build up the chiral multiplet on top of a higher spin state, leading to a superfield transforming according to a higher spin representation of the Lorentz group.

An example for this case is the field strength superfield discussed later in this introduction, which then carries a Weyl spinor index.

Transforming back to the original coordinates $(x, \theta, \bar{\theta})$ in (113) yields

$$\Phi(x,\theta,\bar{\theta}) = \phi(x) + \sqrt{2}\,\theta\psi(x) + \theta\theta\, F(x) - \mathrm{i}\,\theta\sigma^\mu\bar{\theta}\,\partial_\mu\phi(x) \\ + \frac{\mathrm{i}}{\sqrt{2}}\,\theta\theta\,\partial_\mu\psi(x)\sigma^\mu\bar{\theta} - \frac{1}{4}\,\theta\theta\,\overline{\theta\theta}\,\partial_\mu\partial^\mu\,\phi(x)\ . \tag{114}$$

Note that the constraint (111) did only restrict the *number* of component fields compared to the most general Lorentz scalar superfield (103) but not their space-time dependence.

The Hermitean conjugate $\Phi^\dagger(x,\theta,\bar{\theta})$ of a chiral superfield is called antichiral superfield and satisfies the constraint

$$D_\alpha\,\Phi^\dagger(x,\theta,\bar{\theta}) = 0\ . \tag{115}$$

Under an infinitesimal SUSY transformation, the chiral superfield is changed by an amount $\delta\Phi$,

$$\Phi(x,\theta,\bar{\theta}) \to \Phi(x,\theta,\bar{\theta}) + \delta\Phi(x,\theta,\bar{\theta})\ , \\ \delta\Phi(x,\theta,\bar{\theta}) = \mathrm{i}\left(\xi Q + \bar{\xi}\bar{Q}\right)\Phi(x,\theta,\bar{\theta})\ . \tag{116}$$

Inserting (106) and the expansion (114) allows to identify the variations of the component fields:

$$\delta\phi(x) = \sqrt{2}\,\xi\psi(x)\ , \\ \delta\psi_\alpha(x) = \sqrt{2}\,\xi_\alpha F(x) - \mathrm{i}\sqrt{2}\,(\sigma^\mu)_{\alpha\dot{\beta}}\,\bar{\xi}^{\dot{\beta}}\,\partial_\mu\,\phi(x)\ , \tag{117} \\ \delta F(x) = \mathrm{i}\sqrt{2}\,\partial_\mu\,\psi(x)\,\sigma^\mu\,\bar{\xi}\ .$$

An important observation, true for any superfield, can be made in (117): The component field with the highest mass dimension, in this case the $\theta\theta$- or F-term $F(x)$, transforms into a total derivative so that its spacetime volume integral is SUSY invariant. This will be of relevance in the search for possible candidates for Lagrange densities.

5.2 Vector Superfields

A *vector superfield* is defined through the constraint

$$V(x,\theta,\bar{\theta}) = V^\dagger(x,\theta,\bar{\theta})\ , \tag{118}$$

which is covariant due to the unitarity of SUSY transformations (100). A glance at the general Lorentz scalar superfield (103) shows that there is exactly one *complex* Lorentz vector component field present which is rendered to be *real* by the constraint. Other consequences of (118) are

$$c(x), d(x) \in \mathbb{R} , \quad m^*(x) = n(x) \in \mathbb{C} ,$$
$$\phi(x) = \chi(x) , \quad \omega(x) = \psi(x) , \tag{119}$$

in the notation of (103), so that the number of off-shell degrees of freedom is reduced by a factor 2 compared to the general scalar superfield. Trivial rewriting and implementing the constraint yields the standard form of a vector superfield:

$$\begin{aligned}V(x,\theta,\bar\theta) &= C(x) + i\,\theta\chi(x) - i\,\overline{\theta\chi}(x) + \theta\sigma^\mu\bar\theta\, A_\mu(x)\\ &+ \frac{i}{2}\theta\theta\,[\,M(x) + iN(x)\,] - \frac{i}{2}\overline{\theta\theta}\,[\,M(x) - iN(x)\,]\\ &+ i\,\theta\theta\,\bar\theta[\,\bar\lambda(x) - \frac{i}{2}\bar\sigma^\mu\partial_\mu\chi(x)\,] - i\,\overline{\theta\theta}\,\theta[\,\lambda(x) - \frac{i}{2}\sigma^\mu\partial_\mu\bar\chi(x)\,]\\ &+ \frac{1}{2}\theta\theta\,\overline{\theta\theta}\,[\,D(x) - \frac{1}{2}\partial_\mu\partial^\mu C(x)\,] \,.\end{aligned} \tag{120}$$

With a bit of work one can determine the transformation properties of the components under infinitesimal SUSY transformations starting from $\delta V = i\,(\,\xi Q + \bar\xi\bar Q\,)\,V$, e.g.

$$\begin{aligned}\delta C(x) &= i\,(\,\xi\chi(x) - \overline{\xi\chi}(x)\,) ,\\ \delta\lambda_\alpha(x) &= i\,D(x)\xi_\alpha + \frac{1}{2}(\sigma^\mu\bar\sigma^\nu)_\alpha{}^\beta\xi_\beta\, F_{\mu\nu}(x) ,\\ \delta A_\mu(x) &= i\,(\,\xi\sigma_\mu\bar\lambda(x) - \lambda(x)\sigma_\mu\bar\xi\,) + \partial_\mu\,(\,\xi\chi(x) + \overline{\xi\chi}(x)\,) ,\\ \delta D(x) &= \partial_\mu\,(\,\xi\sigma^\mu\bar\lambda(x) + \lambda(x)\sigma^\mu\bar\xi\,) ,\end{aligned} \tag{121}$$

where $F_{\mu\nu}(x)$ is the U(1) field strength tensor, $F_{\mu\nu} = \partial_\mu A_\nu - \partial_\nu A_\mu$, transforming like

$$\delta F_{\mu\nu}(x) = i\,\partial_\mu\,(\,\xi\sigma_\nu\bar\lambda(x) - \lambda(x)\sigma_\nu\bar\xi\,) - i\,\partial_\nu\,(\,\xi\sigma_\mu\bar\lambda(x) - \lambda(x)\sigma_\mu\bar\xi\,) \,. \tag{122}$$

Thus, it follows that λ_α, $\bar\lambda_{\dot\alpha}$, $F_{\mu\nu}$ and D form the desired irreducible representation of the supersymmetry algebra, namely the *vector* or *gauge supermultiplet*. Moreover, the component field with the highest mass dimension, the $\theta\theta\,\overline{\theta\theta}$- or D-term $D(x)$, transforms into a total spacetime derivative.

Starting from the observation that given a chiral superfield Λ, $\bar D_{\dot\alpha}\Lambda = 0$, one can a construct a particular vector superfield by forming the combination $i\,(\,\Lambda - \Lambda^\dagger\,)$ we define a U(1) *gauge transformation* of a vector superfield V by

$$V(x,\theta,\bar\theta) \to V(x,\theta,\bar\theta) + i\,(\,\Lambda(x,\theta,\bar\theta) - \Lambda^\dagger(x,\theta,\bar\theta)\,) \,. \tag{123}$$

By inserting the θ-expansions (120) and (114) we deduce the behavior of the component fields under this gauge transformation (123):

$$A_\mu \to A_\mu + \partial_\mu (\phi + \phi^\dagger) \;, \tag{124}$$
$$\lambda_\alpha \to \lambda_\alpha \;, \tag{125}$$
$$D \to D \;, \tag{126}$$
$$C \to C + \mathrm{i}(\phi - \phi^\dagger) \;, \tag{127}$$
$$\chi \to \chi + \sqrt{2}\,\psi \;, \tag{128}$$
$$M + \mathrm{i}N \to M + \mathrm{i}N + 2F \;. \tag{129}$$

The vector field A_μ has indeed the desired transformation property (124). The fields belonging to the vector supermultiplet, λ_α, $\bar{\lambda}_{\dot{\alpha}}$, D and the field strength $F_{\mu\nu}$ are gauge invariant, whereas the component fields C, χ, M and N are gauge variant and may be eliminated by an appropriate gauge choice. The *Wess-Zumino gauge* is defined by

$$C = 0 \;, \quad \chi = 0 \;, \quad M = 0 \;, \quad N = 0 \tag{130}$$

and still allows for arbitrary gauge transformations of the vector field A_μ. In this gauge, the vector superfield takes on the simple form

$$V_{\mathrm{WZ}}(x, \theta, \bar{\theta}) = \theta \sigma^\mu \bar{\theta}\, A_\mu(x) + \mathrm{i}\,\theta\theta\,\bar{\theta}\bar{\lambda}(x) - \mathrm{i}\,\bar{\theta}\bar{\theta}\,\theta\lambda(x) + \frac{1}{2}\theta\theta\,\bar{\theta}\bar{\theta}\, D(x) \;. \tag{131}$$

with the powers

$$V_{\mathrm{WZ}}^2(x,\theta,\bar{\theta}) = \frac{1}{2}\theta\theta\,\bar{\theta}\bar{\theta}\,A_\mu(x)\,A^\mu(x) \;, \qquad V_{\mathrm{WZ}}^n(x,\theta,\bar{\theta}) = 0 \quad (n > 2) \;. \tag{132}$$

Note, however, that fixing the Wess-Zumino gauge breaks SUSY invariance.

5.3 Lagrangeans for Chiral Superfields

We are now going to construct renormalisable supersymmetric Lagrangeans involving chiral superfields $\Phi_i(x,\theta,\bar{\theta})$, where the index i labels the various chiral superfields in the theory. As we have seen in the previous sections, the $\theta\theta$- or F-term of an arbitrary chiral superfield Φ, in the following denoted by $\Phi|_{\theta\theta}$, and the $\theta\theta\,\bar{\theta}\bar{\theta}$- or D-term of an arbitrary vector superfield V, denoted by $V|_{\theta\theta\,\bar{\theta}\bar{\theta}}$, are SUSY invariant up to total spacetime derivatives and thus good candidates for supersymmetric Lagrange densities. Formally, these components may be projected out by appropriate Grassmann integrals:

$$\Phi(x,\theta,\bar{\theta})\big|_{\theta\theta} = \int \mathrm{d}^2\theta\, \Phi(x,\theta,\bar{\theta}) \;, \tag{133}$$

$$V(x,\theta,\bar{\theta})\big|_{\theta\theta\,\bar{\theta}\bar{\theta}} = \int \mathrm{d}^2\theta\,\mathrm{d}^2\bar{\theta}\, V(x,\theta,\bar{\theta}) \;, \tag{134}$$

where we defined the measures[3]

$$\mathrm{d}^2\theta = -\frac{1}{4}\mathrm{d}\theta^\alpha\,\mathrm{d}\theta^\beta\,\varepsilon_{\alpha\beta}\,, \qquad \mathrm{d}^2\overline{\theta} = -\frac{1}{4}\mathrm{d}\overline{\theta}^{\dot\alpha}\,\mathrm{d}\overline{\theta}^{\dot\beta}\,\varepsilon_{\dot\alpha\dot\beta}\,. \qquad (135)$$

Given chiral superfields $\Phi_i(x,\theta,\overline{\theta})$ any product of them will again be chiral due to the linearity of the operators $\overline{D}_{\dot\alpha}$. Their $\theta\theta$-components are given by

$$\int \mathrm{d}^2\theta\,\Phi_i = F_i\,,$$

$$\int \mathrm{d}^2\theta\,\Phi_i\Phi_j = \phi_i F_j + \phi_j F_i - \psi_i\psi_j\,, \qquad (136)$$

$$\int \mathrm{d}^2\theta\,\Phi_i\Phi_j\Phi_k = \phi_i\phi_j F_k + \phi_i F_j\phi_k + F_i\phi_j\phi_k$$
$$-\psi_i\psi_j\phi_k - \psi_i\psi_k\phi_j - \psi_j\psi_k\phi_i\,.$$

Higher powers of Φ_i need not be taken into account, since their $\theta\theta$-components would have a mass dimension greater than four and therefore would lead to non-renormalisable expressions (note that $\theta\theta$ itself has mass dimension -1). The most general *superpotential* $W(\Phi)$ then reads

$$W(\Phi) = \sum_i \lambda_i\,\Phi_i + \sum_{ij}\frac{1}{2}m_{ij}\,\Phi_i\Phi_j + \sum_{ijk}\frac{1}{3}g_{ijk}\,\Phi_i\Phi_j\Phi_k\,, \qquad (137)$$

with symmetric coupling constants m_{ij} and g_{ijk}. The combination $\Phi_i^\dagger\Phi_i$ is Hermitean and thus a vector superfield. Its $\theta\theta\,\overline{\theta}\overline{\theta}$-component,

$$\int \mathrm{d}^2\theta\,\mathrm{d}^2\overline{\theta}\,\Phi_i^\dagger\Phi_i = F_i^\dagger F_i + \frac{1}{2}\partial_\mu\phi_i^\dagger\,\partial^\mu\phi_i - \frac{1}{4}\phi_i^\dagger\partial_\mu\partial^\mu\phi_j$$
$$\qquad (138)$$
$$-\frac{1}{4}\partial_\mu\partial^\mu\phi_i^\dagger\phi_i + \frac{\mathrm{i}}{2}\overline{\psi}_i\overline{\sigma}^\mu\partial_\mu\psi_i - \frac{\mathrm{i}}{2}\partial_\mu\overline{\psi}_i\overline{\sigma}^\mu\psi_i\,,$$

contains spacetime derivatives of the relevant component fields and therefore serves as kinetic part of the Lagrangean. So, altogether, the most general renormalisable supersymmetric Lagrangean that can be constructed from chiral superfields only is given by

$$\mathcal{L} = \int \mathrm{d}^2\theta\,\mathrm{d}^2\overline{\theta}\,\sum_i \Phi_i^\dagger\Phi_i + \left(\int \mathrm{d}^2\theta\,W(\Phi) + \text{h.c.}\right)\,, \qquad (139)$$

or in terms of component fields and neglecting surface terms

$$\mathcal{L} = \partial_\mu\phi_i^\dagger\partial^\mu\phi_i + \mathrm{i}\,\overline{\psi}_i\overline{\sigma}^\mu\partial_\mu\psi_i + F_i^\dagger F_i + \Big(\lambda_i F_i + m_{ij}\,\phi_i F_j$$
$$\qquad (140)$$
$$-\tfrac{1}{2}m_{ij}\,\psi_i\psi_j + g_{ijk}\,\phi_i\phi_j F_k - g_{ijk}\psi_i\psi_j\phi_k + \text{h.c.}\Big)\,,$$

---

[3] Remember that a Grassmann integral is defined by $\int \mathrm{d}\theta^\alpha\,\theta^\beta = \delta^{\alpha\beta}$ and $\mathbb{C}$-linearity; it is equivalent to the derivative: $\int \mathrm{d}\theta^\alpha\,f(\theta) = \frac{\partial}{\partial\theta^\alpha}f(\theta)$.

where a summation convention is used. We now see that there is no kinetic term for the F-field present, indicating that this field is an auxiliary field and may now be eliminated with help of the equations of motion following from (140).

5.4 Supersymmetric Gauge Invariant Interactions

The simplest supersymmetric gauge theory is the SUSY-Maxwell theory; the supersymmetric extension of the vector field A_μ is a vector superfield V.

The field strength belonging to this vector field should, of course, be invariant under gauge transformations

$$V \longrightarrow V + i(\Lambda - \Lambda^\dagger). \tag{141}$$

One may easily check that

$$W_\alpha = -\frac{1}{4}\overline{DD}D_\alpha V \tag{142}$$

fulfills this condition. It is also chiral,

$$\overline{D}_{\dot\beta} W_\alpha = 0, \tag{143}$$

because the product of three $\overline{D}$'s vanishes identically.

The component representation of this field is given by

$$W_\alpha = -i\lambda_\alpha(y) + \theta_\beta \left[\delta^\beta_\alpha D(y) - \frac{i}{2}(\sigma^\mu\overline\sigma^\nu)^\beta_\alpha F_{\mu\nu}(y)\right] - \\ - \theta\theta\sigma^\mu_{\alpha\dot\alpha}\partial_\mu\overline\lambda^{\dot\alpha}(y) \tag{144}$$

where

$$F_{\mu\nu} = \partial_\mu A_\nu - \partial_\nu A_\mu \tag{145}$$

is the well known Maxwell field strength. To obtain the Lagrangean density, one has to take the square of this object and to pick out the component quadratic in θ. This can be achieved by a simple Grassmannian integration:

$$\mathcal{L}^{N=1}_{\text{Maxwell}} = \frac{1}{4}\int d^2\theta\, W^\alpha W_\alpha + \frac{1}{4}\int d^2\overline\theta\, \overline{W}_{\dot\alpha}\overline{W}^{\dot\alpha} = \\ = \frac{1}{2}D^2 - \frac{1}{4}F_{\mu\nu}F^{\mu\nu} + \frac{i}{2}\lambda\sigma^\mu\overleftrightarrow{\partial}_\mu\overline\lambda, \tag{146}$$

which transforms as a density under SUSY transformations, is therefore the supersymmetric generalization of the free Maxwell Lagrangean. In this equation we have used the symbol

$$A\overleftrightarrow{\partial}_\mu B = A\partial_\mu B - (\partial_\mu A)B. \tag{147}$$

There is a second Hermitean supersymmetric combination of the field strengths,

$$i\left(\int d^2\theta\, W^\alpha W_\alpha - \int d^2\bar{\theta}\, \overline{W}_{\dot{\alpha}}\overline{W}^{\dot{\alpha}}\right) = \frac{1}{2}\epsilon_{\kappa\lambda\mu\nu}F^{\kappa\lambda}F^{\mu\nu} - 2\partial_\mu(\lambda\sigma^\mu\overline{\lambda}), \quad (148)$$

which breaks parity. This combination can be written as a total divergence and therefore does not appear in the usual SUSY-QED Lagrangean because it does not contribute to the equations of motion. However, it plays an important rôle in the effective action of SUSY-QCD in the context of θ-terms.

The generalization containing the θ-term can easily be written down if we define the complex coupling constant

$$\tau = \frac{4\pi i}{g^2} + \frac{\theta}{2\pi}. \quad (149)$$

We obtain

$$\mathcal{L}_\theta = \frac{g^2}{8\pi}\mathrm{Im}\left[\tau \int d^2\theta\, W^\alpha W_\alpha\right] = \frac{1}{2}D^2 - \frac{1}{4}F_{\mu\nu}F^{\mu\nu} + \frac{i}{2}\lambda\sigma^\mu\overset{\leftrightarrow}{\partial}_\mu\overline{\lambda} -$$
$$- \frac{\theta g^2}{16\pi^2}\left(\frac{1}{4}\epsilon_{\kappa\lambda\mu\nu}F^{\kappa\lambda}F^{\mu\nu} - \partial_\mu(\lambda\sigma^\mu\overline{\lambda})\right), \quad (150)$$

where we have used the imaginary part notation for later convenience.

The coupling of the supersymmetric Maxwell theory to matter, e.g. to a chiral superfield Φ, can be introduced in a way which is very similar to ordinary QED: Under the gauge transformations

$$\Phi \longrightarrow e^{-2iq\Lambda}\Phi, \quad (151)$$

where Λ has to be chiral to ensure that the r.h.s. is also a chiral superfield, the kinetic term in the Lagrangean behaves as

$$\Phi^\dagger\Phi \longrightarrow \Phi^\dagger\Phi e^{2iq(\Lambda^\dagger - \Lambda)}, \quad (152)$$

and is therefore not gauge invariant.

The minimal supersymmetric gauge invariant extension of this Lagrangean may be obtained by simply replacing

$$\Phi^\dagger\Phi \longrightarrow \Phi^\dagger e^{2qV}\Phi. \quad (153)$$

As before, the correct SUSY transformation law is only obeyed by the part which is quartic in the Grassmann variables. Picking out this one, we obtain the minimally coupled kinetic part of the matter Lagrangean

$$\int d^2\theta \int d^2\bar{\theta}\, \Phi^\dagger e^{2qV}\Phi = (\mathcal{D}_\mu\phi)^\dagger \mathcal{D}^\mu\phi + \frac{i}{2}\psi\sigma^\mu\overset{\leftrightarrow}{\mathcal{D}}_\mu\overline{\psi} + F^*F +$$
$$+ i\sqrt{2}q(\phi^\dagger\psi\lambda - \phi\overline{\psi}\overline{\lambda}) + qD\phi^\dagger\phi \quad (154)$$

with the covariant derivative

$$\mathcal{D}_\mu = \partial_\mu + iqA_\mu. \tag{155}$$

Our next task is to generalize this construction to non-Abelian gauge groups. Consider the behavior of a matter field in the fundamental representation under a gauge transformation,

$$\Phi \longrightarrow e^{-2ig\Lambda}\Phi, \tag{156}$$

where Λ has to belong to the adjoint representation,

$$\Lambda = \Lambda^a T^a. \tag{157}$$

Here the T^a are the generators of the gauge group, obeying the Lie algebra relations

$$[T^a, T^b] = it^{abc}T^c \tag{158}$$

and the orthogonality condition

$$\text{tr}[T^a T^b] = \frac{1}{2}\delta^{ab}. \tag{159}$$

For the term $\Phi^\dagger e^{2gV}\Phi$ to be gauge invariant, the exponential of the vector field must transform as

$$e^{2gV} \longrightarrow e^{-2ig\Lambda^\dagger} e^{2gV} e^{2ig\Lambda}. \tag{160}$$

Here we have defined the super Lie algebra valued vector field

$$V = V^a T^a. \tag{161}$$

Taking the logarithm of this equation in a closed expression is not possible, but expanding in powers of Λ one obtains for the Lorentz vector component of V

$$A_\mu^a \longrightarrow A_\mu^a + \partial_\mu(\beta^a + \beta^{a\dagger}) + gt^{abc}(\beta^b + \beta^{b\dagger})A_\mu^c + O(\Lambda^2) \tag{162}$$

if β denotes the scalar component of Λ. This is of course the well known expression for the gauge transformation of a Yang-Mills field.

Exactly as in SUSY electrodynamics, the minimally coupled part of the matter Lagrangean is obtained by picking out the term which is quartic in the Grassmann variables,

$$\int d^2\theta \int d^2\bar\theta\, \Phi^\dagger e^{2gV}\Phi = (\mathcal{D}_\mu \phi)^\dagger \mathcal{D}^\mu \phi + \frac{i}{2}\psi\sigma^\mu \overset{\leftrightarrow}{\mathcal{D}}_\mu \overline{\psi} + F^*F +$$
$$+i\sqrt{2}g(\phi^\dagger T^a \psi \lambda^a - \overline{\lambda}^a \phi T^a \overline{\psi}) + gD^a \phi^\dagger T^a \phi \tag{163}$$

with the covariant derivative in the fundamental representation

$$\mathcal{D}_\mu = \partial_\mu + igA_\mu^a T^a. \tag{164}$$

Because the derivatives of D do not appear in the Lagrangean, it is an auxiliary field which can be eliminated by a Gaussian integration. (In the canonical formalism one would simply insert the equation of motion for D.) The result is a potential for the scalar field ϕ,

$$U = \frac{1}{2}g^2(\phi^\dagger T^a \phi)^2, \qquad (165)$$

the so-called superpotential.

Note that this potential is uniquely fixed by the requirement of supersymmetry.

If supersymmetry is unbroken, the vacuum has zero energy. Therefore also the potential energy of the scalar has to vanish.

$$\phi^\dagger T^a \phi = 0. \qquad (166)$$

One can always find the trivial solution

$$\phi = 0. \qquad (167)$$

The non-zero solutions of vanishing potential energy are called "flat directions".

If, e.g., ϕ transforms in the adjoint representation of SU(2),

$$\phi^\dagger T^a \phi = \epsilon^{abc} \phi^{b\dagger} \phi^c, \qquad (168)$$

one such flat direction would be given by

$$\phi^b = \delta^{b3} \tilde{\phi}. \qquad (169)$$

We can conclude that, like e.g. in the Goldstone model, we have a range of (classical) vacua, each one characterized by a certain value of $\tilde{\phi}$. But unlike the Goldstone model, the different solutions are not related by symmetries; they lead to different physics, observables like the spectrum or correlation functions will depend on $\tilde{\phi}$.

So far we have not yet constructed a kinetic term for the super-Yang-Mills field V. A natural requirement for its field strength is that it should transform covariantly under gauge transformations. To find such an object, consider the transformation property of the following expression:

$$e^{-2gV} D_\alpha e^{2gV} \longrightarrow e^{-2ig\Lambda}(e^{-2gV} D_\alpha e^{2gV})e^{2ig\Lambda} + e^{-2ig\Lambda} D_\alpha e^{2ig\Lambda}. \qquad (170)$$

The first term on the r.h.s. looks covariant; the second term contains only chiral superfields and is therefore annihilated by the operator $\overline{D}$. This leads us to the ansatz

$$W_\alpha = -\frac{1}{8g}\overline{DD} e^{-2gV} D_\alpha e^{2gV}, \qquad (171)$$

which reduces to the SUSY-Maxwell expression in the case where everything (anti-)commutes.

A SUSY transformation on W_α results in the transformation laws

$$\delta_\zeta F_{\mu\nu}^a = i[\zeta\sigma_\nu \mathcal{D}_\mu \overline{\lambda}^a + \overline{\zeta}\overline{\sigma}_\nu \mathcal{D}_\nu \lambda^a] - (\mu \leftrightarrow \nu)$$
$$\delta_\zeta \lambda^a = i\zeta D^a + \sigma^{\mu\nu}\zeta F_{\mu\nu}^a$$
$$\delta_\zeta D^a = \overline{\zeta}\overline{\sigma}_\mu \partial_\mu \lambda^a - \zeta\sigma_\mu \partial_\mu \overline{\lambda}^a \tag{172}$$

for the component fields. The supersymmetric Yang-Mills theory is therefore based on the Lagrangean density

$$\mathcal{L}_{\text{YM}}^{N=1} = \frac{1}{2}\int d^2\theta \, \text{tr}[W_\alpha W^\alpha] + \text{h.c.} =$$
$$= \frac{1}{2}D^{a2} - \frac{1}{4}F_{\mu\nu}^a F^{\mu\nu a} + \frac{i}{2}\lambda^a \sigma^\mu \overset{\leftrightarrow}{\mathcal{D}}_\mu \overline{\lambda}^a \tag{173}$$

where we have defined the covariant derivative in the adjoint representation

$$\mathcal{D}_\mu^{ab} = \partial_\mu \delta^{ab} - g t^{acb} A_\mu^c \tag{174}$$

as well as the non-Abelian field strength

$$F_{\mu\nu}^a = \partial_\mu A_\nu^a - \partial_\nu A_\mu^a - g t^{abc} A_\mu^b A_\nu^c. \tag{175}$$

This Lagrangean contains the usual three and four gluon couplings as well as a gluon-gluino-vertex.

We now turn to the $N = 2$ SU(2) gauge theory. As we have seen in Sect. 4, the $N = 2$ gauge multiplet can be decomposed into an $N = 1$ gauge multiplet $(A_\mu^a, \lambda_\alpha^a)$ and an $N = 1$ chiral multiplet (ψ_α^a, ϕ^a) ($a = 1, 2, 3$).

The $N = 2$ super-Yang-Mills Lagrangean is therefore given by an $N = 1$ super-Yang-Mills Lagrangean minimally coupled to a $N = 1$ chiral multiplet in the adjoint representation. It reads:

$$\mathcal{L}_{\text{YM}}^{N=2} = -\frac{1}{4}F_{\mu\nu}^a F^{\mu\nu a} + \frac{i}{2}\lambda^a \sigma^\mu \overset{\leftrightarrow}{\mathcal{D}}_\mu \overline{\lambda}^a + \frac{i}{2}\psi^a \sigma^\mu \overset{\leftrightarrow}{\mathcal{D}}_\mu \overline{\psi}^a +$$
$$(\mathcal{D}_\mu \phi)^{a\dagger}\mathcal{D}^\mu \phi^a + +\sqrt{2}g\epsilon^{abc}(\phi^{a\dagger}\psi^b \lambda^c + \text{h.c.}) +$$
$$+ g^2(\epsilon^{abc}\phi^{b\dagger}\phi^c)^\dagger \epsilon^{ade}\phi^{d\dagger}\phi^e. \tag{176}$$

The two supersymmetries can be obtained by using the SU(2) $\mathcal{R}$-symmetry acting on (λ, ψ) together with the $N = 1$ transformation laws (172). One deduces, e.g.,

$$\delta_\zeta F_{\mu\nu}^a = i[\zeta_1 \sigma_\nu \mathcal{D}_\mu \overline{\lambda}^a + \overline{\zeta}_1 \overline{\sigma}_\nu \mathcal{D}_\mu \lambda^a + \zeta_2 \sigma_\nu \mathcal{D}_\mu \overline{\psi}^a + \overline{\zeta}_2 \overline{\sigma}_\nu \mathcal{D}_\nu \psi^a] - (\mu \leftrightarrow \nu). \tag{177}$$

In addition to supersymmetry, this Lagrangean has even more symmetries, the so-called "$\mathcal{R}$-symmetries", mentioned already above. It is invariant under an SU(2)$_\mathcal{R}$ acting on (λ, ψ) as well as under the replacements

$$\text{U}(1)_J: \quad \Phi(\theta) \longrightarrow \Phi(e^{-i\alpha}\theta) \tag{178}$$

and

$$\text{U}(1)_\mathcal{R}: \quad \begin{cases} \Phi(\theta) \longrightarrow e^{2i\beta}\Phi(e^{-i\beta}\theta) \\ W_\alpha(\theta) \longrightarrow e^{i\beta}W_\alpha(e^{-i\beta}\theta) \end{cases}, \tag{179}$$

if one denotes by Φ the $N = 1$ chiral superfield containing ψ_α and ϕ and by W_α the vector superfield strength, i.e. the chiral superfield containing A_μ and λ_α.

6 The Structure of $N=2$ Supersymmetric Yang Mills Theories

Our main purpose is to describe the exact quantum behavior in the infrared of the $N = 2$, SU(2) supersymmetric Yang Mills theory found by Seiberg and Witten. We will first discuss separately the "ingredients" used in the solution and then proceed to the actual construction.

6.1 Lagrangean and Superalgebra

It is useful to define $\lambda =: \Psi^1$, $\psi =: \Psi^2$ and rewrite the Lagrangean (176) as

$$\mathcal{L} = -\frac{1}{2}\mathrm{tr}\,[F_{\mu\nu}]^2 - 2\mathrm{i}\,\mathrm{tr}\,[\Psi^{i\dagger}\bar{\sigma}^\mu \mathcal{D}_\mu \Psi^i] + 2\,\mathrm{tr}\,[(\mathcal{D}_\mu\phi)^\dagger (\mathcal{D}^\mu\phi)] - \\ -\sqrt{2}\,g\,\mathrm{tr}\,[\mathrm{i}\phi^\dagger \epsilon^{ij}[\Psi^i, \Psi^j] + \mathrm{h.\ c.}\,] - 2g^2\,\mathrm{tr}\,[\phi,\phi^\dagger]^2 \,, \tag{180}$$

where the trace is taken in the adjoint representation of SU(2).

A nice exercise is to calculate the conserved supersymmetric Noether current from the transformation properties (177) of the fields [9]

$$S^i_\mu(x) = \mathrm{tr}\,[\sqrt{2}F_{\rho\sigma}\sigma^{\rho\sigma}\sigma_\mu \Psi^i + 2\epsilon^{ij}(\mathcal{D}_\nu\phi)\sigma^\nu\sigma_\mu\Psi^j + 2\sqrt{2}\,g[\phi,\phi^\dagger]\sigma_\mu\Psi^i] \,. \tag{181}$$

Using the canonical (anti-)commutation relations, one may test that the supersymmetry algebra is indeed fulfilled and show that the central charge (57) is given by

$$Z = \sqrt{2}\int \mathrm{d}^3x\, \partial_i\left[\phi^a(\mathbf{x})\Big(E^a_i(\mathbf{x}) + \mathrm{i}B^a_i(\mathbf{x})\Big)\right]\,, \tag{182}$$

where $E^a_i = F^{0i\,a}$ is the colour electric and $B^a_i = \frac{1}{2}\varepsilon^{ijk}F^{jk\,a}$ the colour magnetic field. In order the total surface integral Z to be nonzero and the algebra to be extended, the theory has to allow for finite energy configurations which fall off slowly at infinity. We will investigate more closely their existence in the next sections.

6.2 Supersymmetric QCD as Higgs Model

As seen in (55), the vacuum has zero energy in a theory in which supersymmetry is not broken. Hence, ϕ has to be a constant in spacetime, so that the kinetic terms in the Lagrangean (180) are zero, and the superpotential has to vanish in the classical vacuum:

$$\mathrm{tr}\,[\phi,\phi^\dagger]^2 = 0 \tag{183}$$

Whenever ϕ and $\phi^\dagger$ commute with each other, i. e. are aligned, and hence ϕ and $\phi^\dagger$ lie in the maximal Abelian subalgebra of SU(2), this equation holds with nonzero vacuum expectation value of the scalar field. These directions in the Lie algebra space are are called *flat directions*:

$$\langle \text{vac}|\phi^b(\mathbf{x})|\text{vac}\rangle = a\hat{\phi}^b(\mathbf{x}) \ , \quad \hat{\phi}^b \| \hat{\phi}^{b\,\dagger} \tag{184}$$

$$a^2 = \phi^2 \ , \quad a \in \mathbb{C} \ , \quad \partial_\mu a = 0 \ ; \quad \hat{\phi}^b = \frac{\phi^b}{a}$$

Note that the choice of $\hat{\phi}(\mathbf{x})$ at every spacetime point corresponds to a choice of gauge. The (nearly) gauge invariant eigenvalue a of the vacuum field configuration is an arbitrary, x-independent complex constant, and the vacuum breaks a symmetry of the Lagrangean spontaneously by developing a nonzero vacuum expectation value a. This hint for a Higgs mechanism parallel to the one in the standard model suggests to investigate the residual symmetry group the broken vacuum state allows for.

For given $\hat{\phi}$, (184) fixes the gauge partially for nonzero a, since it permits only two kinds of residual gauge transformations: The first kind is a U(1) group $\exp ig\alpha^a(\mathbf{x})\hat{\phi}^a(\mathbf{x})$ with arbitrary gauge function $\alpha^a(\mathbf{x})$ and corresponds to rotations about the $\hat{\phi}$ axis in colour space at each spacetime point. The second kind is given by the Weyl reflection group $\mathcal{W}$ of SU(2) and describes rotations in colour space such that the direction $\hat{\phi}$ points to in the Lie algebra space is reversed.

The analysis becomes more familiar in the U gauge [5], [10], in which the field $\hat{\phi}$ is diagonal,

$$\hat{\phi}(\mathbf{x}) = \frac{\sigma^3}{2} \ , \tag{185}$$

and which may be reached by diagonalizing $\hat{\phi}$ at every spacetime point. Then the U(1) part of the residual symmetry group is given by all functions $\exp ig\alpha(\mathbf{x})\frac{\sigma^3}{2}$ with arbitrary $\alpha(\mathbf{x})$ and by the Weyl reflection group $\{\mathbb{1}, R := i\sigma^2\}$. R describes rotations around the σ^2 axis about the angle π and reflects the arbitrariness one has in the ordering of the eigenvalues, $R\sigma^3 R^\dagger = -\sigma^3$. Under this transformation, a reverses sign, and therefore the expectation values a and $-a$ have to be identified. The correct gauge invariant parameter differentiating between different flat directions is hence (at least locally)

$$u := \text{tr}\,\phi^2 = \frac{1}{2}a^2 \ , \tag{186}$$

and w. r. t. this variable the residual gauge group is U(1). Nonetheless, it will be more convenient in what follows to work with a as variable labelling the different vacua.

In the standard language, the original SU(2) gauge symmetry "breaks down" to U(1) because of the nonzero vacuum expectation value a, and as in the standard model one expects the Higgs mechanism to give mass to two charged vector mesons, leaving the gauge boson which lives in the σ^3 (respectively $\hat{\phi}$) direction massless. Indeed, expanding in the U gauge about the classical vacuum,

$$\phi = a\frac{\sigma^3}{2} + \tilde{\phi} \ , \tag{187}$$

the kinetic energy term of the scalar fields,

$$2\,\mathrm{tr}\,[\left(\mathcal{D}_\mu \phi\right)^\dagger \left(\mathcal{D}^\mu \phi\right)] = g^2 |a|^2 \left[\left(A_\mu^1\right)^2 + \left(A_\mu^2\right)^2\right] + ... \tag{188}$$

shows that the offdiagonal fields $W_\mu^\pm := \frac{1}{\sqrt{2}}(A_\mu^1 \pm i A_\mu^2)$ of charge one w. r. t. the residual U(1) symmetry acquire semiclassically a mass $m_W^2 = 2g^2|a|^2$. At least different values of $|a|^2$ will hence label different vacuum sectors of the theory.

One obtains a Lagrangean similar to the supersymmetric version of the Georgi Glashow model [5], [10], SU(2) Yang Mills theory with a real Higgs field ϕ in the adjoint representation of the gauge group. From the SUSY mass relation between particles in the same multiplet (O'Raifeartaigh's theorem) (57), one concludes that all the offdiagonal fields $A_\mu^{1,2}, \lambda^{1,2}, \psi^{1,2}$ acquire the same nonvanishing mass for nonzero a, and only the "photon" A_μ^3 together with its supersymmetric partners stays massless. In general, $\hat{\phi}$ prescribes in a gauge invariant manner the direction in the Lie algebra space at each spacetime point in which the electromagnetic field lives,

$$F_{\mu\nu}^{\mathrm{em}} := \hat{\phi}^b F_{\mu\nu}^b \; . \tag{189}$$

We will not bother to write down the (lengthy) interactions between massive and massless as well as amongst the massive fields induced by the spontaneous symmetry breakdown, since they will not play a role in what follows.

Classically, the $U(1)_\mathcal{R}$ symmetry (179) changes the phase of the Higgs field, and hence shows that only the length $|a|$ of the Higgs field is relevant. The classical moduli space of infinitely many physically inequivalent ground states (flat directions) is therefore the half line $|a| \in [0; \infty[$ of real dimension one. Different values of $|a|$ yield indeed different masses for the charged gauge bosons. For $a = 0$, additional particles become massless and one expects therefore that the classical moduli space has a singularity.

One can show that the flat directions remain flat when one takes into account quantum corrections. The *quantum moduli space* of infinitely many, physically inequivalent ground states will nonetheless change drastically, because the classical $U(1)_\mathcal{R}$ symmetry (179) is broken in the quantum theory due to nonperturbative, instanton, effects to a $\mathbb{Z}_4$ symmetry which maps a to $\pm a, \pm ia$. For u, which is the better variable to take to describe the quantum moduli space, the symmetry is broken to $\mathbb{Z}_2$, identifying u and $-u$. Therefore, the quantum moduli space is labelled by the complex value u and will now be of complex dimension one. Different $u, u' \neq \pm u$ yield different masses and correlation functions, and hence different physics. The occurence and position of singularities in the quantum moduli space can differ drastically from the classical one, as will be demonstrated below, but the system will still be in the Higgs phase for nonzero a. The base manifold of u is the right half of the complex plane, and this will prove very powerful in the analysis of the theory. In contradistinction to the Georgi Glashow model, in which the Higgs potential is $\frac{1}{2}\mu^2 \phi^a \phi^a - \frac{1}{4}\lambda(\phi^a \phi^a)^2$ and the vacuum manifold of flat potentials is given by all fields with length $a = \sqrt{2}\mu/\sqrt{\lambda}$, i.e. by the 2-sphere S^2, in the SUSY model it is much larger: Any value of a yields a flat direction.

Therefore the effective low energy theory of $N = 2$ supersymmetric pure SU(2) gauge theory, which is given by integrating out the heavy modes propagating only over short distances, is $N = 2$ supersymmetric QED. The Georgi Glashow model is embedded in pure $N = 2$ Yang Mills theory and hence can give valuable hints for SUSY QCD. The light fields dominate the long range behavior and therefore are also the relevant degrees of freedom w. r. t. the central charge (182). Z is nonzero in the Higgs phase because fields approach their vacuum value at infinity, $\lim_{r \to \infty} \phi^a(\mathbf{x}) = a\hat{\phi}^a(\mathbf{x})$.

6.3 Central Charge and Topology

If there exist physical particles which are not the fundamental fields, $W^\pm$ and the photon(with their SUSY partners) or their bound states, the Hilbert space is larger than the one that can be reached by perturbation theory about the trivial vacuum. Indeed, topology shows that the Hilbert space of the quantum theory decomposes into disjoint sectors not only of different electric, but also of different magnetic total charge, and of different values for an effective ϑ vacuum angle.

Magnetic Monopoles The second term in (182) measures the total magnetic flux through the surface at infinity and hence will be nonzero only in the presence of monopoles. At first sight, this seems to be ruled out because the naive QED magnetic field is divergence free thanks to the Bianchi identity, $\partial_i B_i^{\text{em}} = 0$. Because of the nontrivial dependence of the magnetic field on $\hat{\phi}$ as it is defined in the Higgs phase (189), it will indeed *not* obey the Bianchi identity of QED and magnetic monopoles may occur, although the non-Abelian Bianchi identity $\mathcal{D}_i^{ab} B_i^b = 0$ is valid. This should not come as a surprise, since our guide, the Georgi Glashow model, is known to allow for finite energy solutions of nonzero total magnetic flux, namely the 't Hooft Polyakov monopoles [11].

Their strength can be derived from topological arguments: ϕ can approach any vacuum expectation value $a\hat{\phi}$ at infinity, and $\hat{\phi}^\dagger$ must be aligned to $\hat{\phi}$. In order to obtain all inequivalent configurations $\hat{\phi}$, the electromagnetic gauge group has to be factored out since – as already mentioned – rotations about the $\hat{\phi}$ axis leave it invariant. Therefore, $\hat{\phi}$ at infinity can have any value in SU(2)/U(1). Gauge inequivalent configurations exist, if arbitrary $\hat{\phi}(\mathbf{x})$ cannot be rotated to point into the same direction in colour space on the whole sphere S^2 at spatial infinity. Such gauge transformations which continuously diagonalize every $\hat{\phi}(\mathbf{x})$ are not present in general since the mapping which defines the diagonalization, i.e. the alignment of the vacuum expectation value of the Higgs field at infinity, $D : S^2 \to \text{SU}(2)/\text{U}(1)$, decomposes into infinitely many distinct equivalence classes[4], each of which is labelled by an arbitrary integer winding number

$$n_\mathrm{m} = \frac{g}{4\pi} \oint_{S^2 \text{at} \infty} d^2\Sigma^i \, B^{ia}\hat{\phi}^a \in \mathbb{Z} \,. \tag{190}$$

The central charge (182) of monopoles is therefore given by

$$Z_\text{monopole} = \sqrt{2}a\frac{4\pi\mathrm{i}}{g}n_\mathrm{m} \,, \tag{191}$$

and the resulting quantization condition for electric and magnetic charges, $g_\mathrm{m} \in \frac{4\pi}{g}\mathbb{Z}$, is twice the quantization condition of the Dirac monopole. This can be traced back to the fact that there are no fields present which transform under the fundamental representation of SU(2) [5], [10].

As we will discuss in the following, due to the asymptotic freedom of the theory, for large a we are in the semiclassical regime. In this region monopoles can be reliably constructed as field configurations following the solution of Bogomolnyi, Prasad and Sommerfield for the Georgi Glashow model [14]. One starts in a gauge in which all fields are time independent, $\partial_0 A^a_\mu = 0$, $\partial_0 \phi = 0$. When ϕ is real, the Higgs potential vanishes ($[\phi,\phi^\dagger] = 0$) and therefore ϕ lies in the moduli space automatically. From the bosonic part of the total energy of a static field configuration without colourelectric fields ($E^a_i = 0$), one can show by the Bogomolnyi trick

$$\begin{aligned}E &:= \int d^3x \left[\frac{1}{2}\left(B^a_i\right)^2 + \left(\mathcal{D}_i\phi\right)^a\left(\mathcal{D}_i\phi\right)^a\right] = \\ &= \int d^3x \left[\frac{1}{\sqrt{2}}B^a_i \pm \left(\mathcal{D}_i\phi\right)^a\right]^2 \mp \int d^3x \sqrt{2}\, \partial_i\left(B^a_i\phi^a\right)\end{aligned} \tag{192}$$

(using the non-Abelian Bianchi identity) that the energy, and hence the mass of the static monopole has a lower bound (called Bogomolnyi bound)

$$M_\text{monopole} = E \geq \sqrt{2}|a\frac{4\pi\mathrm{i}}{g}n_\mathrm{m}| = |Z_\text{monopole}| \,. \tag{193}$$

The monopole will be stable when E is minimal. For monopole charge one, the Higgs field should be mapped with winding number one to the sphere at infinity, which suggests the hedgehog ansatz $\phi^a \| x^a$ for it, and the magnetic field should fall like $1/r^2$ at infinity. The Wu-Yang-'t Hooft-Polyakov-Julia-Zee ansatz alignes also B^a_i and $\hat{\phi}^a$:

$$\begin{aligned}A^a_i &= \varepsilon^a{}_{ij}\frac{x_j}{gr^2}\left(1 - K(r)\right) \,, \quad \lim_{r\to\infty} K(r) = 0 \\ \phi &= \frac{x^a\sigma^a}{2gr^2}H(r) \,, \quad \lim_{r\to\infty} \phi = a\hat{\phi}^a\frac{x^a}{r} \,.\end{aligned} \tag{194}$$

---
[4] More on topological considerations can be found e. g. in [5], [10], [12], [13].

Minimizing (192) to obtain the stable solution, the following equations of motion result

$$r^2 K'' = K(K^2 - 1) + KH^2$$
$$r^2 H'' = 2HK^2 \qquad (195)$$

(primes denote differentiation w. r. t. r) which are solved by

$$K(r) = \frac{gar}{\sinh gar} \to \begin{cases} 1 + \mathcal{O}((gar)^2) \text{ for } r \to 0 \\ gar \, e^{-gar} \text{ for } r \to \infty \end{cases}$$
$$H(r) = gar \coth(gar) - 1 \to \begin{cases} \mathcal{O}((gar)^2) \text{ for } r \to 0 \\ gar \text{ for } r \to \infty \end{cases} \qquad (196)$$

(assuming $a \geq 0$) and constitute an exact solution for the 't Hooft Polyakov monopole in the limit of vanishing Higgs potential with $\sqrt{\mu}/\lambda = 1$ fixed. The solution is regular everywhere and has size $1/ga$. The gauge invariant QED-magnetic field at infinity is indeed

$$B_i \to \frac{x_i}{gr^3} \text{ for } r \to \infty , \qquad (197)$$

and the BPS monopole saturates the Bogomolnyi bound

$$M_{\text{BPS}} = \sqrt{2} \left| a \frac{4\pi}{g} n_{\text{m}} \right| . \qquad (198)$$

We remark that this relation coincides with the condition (95) between masses and central charges for small representations of the SUSY algebra. Therefore, the monopole in the large a limit belongs to a small representation of $N = 2$. Since the number of states in a representation cannot change abruptly, the monopole will stay in a small representation for any a.

If one wants to rotate the solution (194) to the U gauge (185), the gauge transformation diagonalising the Higgs field can only be continuous everywhere except on a line from the origin, the position of the monopole, to infinity [5], [10]. Otherwise the topological winding number (190), which is just constructed to be *gauge invariant* would be changed. So, the 't Hooft Polyakov monopole acquires a Dirac string and a singularity at the origin which give rise to a violation of the Bianchi identity of QED in the U gauge. Nonetheless, and in contradistinction to the Dirac monopole case, the BPS monopole has finite energy, and the magnetic field strength remains finite even at its centre. Hence, the solution cannot be rotated to the U gauge by a nonsingular gauge transformation.

Dyons Monopoles carrying electric charge as well are *dyons* and can be constructed as above, when one includes the electric field term $\int d^3x \frac{1}{2}(E_i^a)^2$ in (192) and supplements the ansatz (194) with one for the zero component of the gauge field which yields the electric field, $E_i^a = -\partial_i A_0^a$,

$$A_0^a = \frac{x^a}{gr^2} J(r) \ . \tag{199}$$

In this case the field equations

$$\begin{aligned} r^2 K'' &= K(K^2 - 1) + K(H^2 - J^2) \\ r^2 J'' &= 2JK^2 \ , \quad r^2 H'' = 2HK^2 \end{aligned} \tag{200}$$

are solved by

$$K(r) = \frac{gar}{\sinh gar}$$
$$J(r) = \sinh\gamma \Big[gar \coth(gar) - 1\Big] \tag{201}$$
$$H(r) = \cosh\gamma \Big[gar \coth(gar) - 1\Big]$$

(again $a \geq 0$; γ is an arbitrary constant) and yield as mass for the stable dyon of electric and magnetic charge one

$$M_{\text{dyon}} = \sqrt{2}\, \frac{4\pi}{g} |a| \cosh^2 \gamma. \tag{202}$$

As will be seen from the next paragraph, in the small coupling limit, which by asymptotic freedom (cf. a following section on the perturbative regime) is also the classical limit, one should set

$$\cosh^2 \gamma \to 1 + \frac{g^2}{4\pi} \quad \text{for } g \to 0 \tag{203}$$

to obtain

$$M_{\text{dyon}} = \sqrt{2}\, g|a| \sqrt{1 + \left(\frac{4\pi}{g^2}\right)^2} \ . \tag{204}$$

Again, the dyon saturates the Bogomolnyi bound and relation (95) for particle masses in short SUSY representations.

Electric Charges and Dyons The first term in the expression for the central charge (182) counts the total number of electric charges minus anticharges n_e in classical QED, as can be seen from Gauß's law

$$\partial_i E_i^{\text{em}} = g\rho_e \, , \tag{205}$$

ρ_e being the electric charge density, so that classically

$$Z = \sqrt{2}\, ga(n_e + \tau_{\text{class}} n_m) \, , \quad \tau_{\text{class}} := \frac{4\pi i}{g^2} \, . \tag{206}$$

Witten [15] found that the electric charge of a monopole doesn't have to be integer in a theory which allows for a nonzero CP violating ϑ vacuum angle. The ϑ angle of the original non-Abelian theory is irrelevant, because of the presence of massless fermionic fields Ψ^i and the U(1) anomaly [5], [10].

Nonetheless, the effective low energy theory, which must be given by the most general $N = 2$ supersymmetric QED Lagrangean due to above considerations, does have an additional term

$$S_{\vartheta^{\text{em}}} = \frac{\vartheta^{\text{em}} g^2}{64\pi^2} \int d^4x \, \varepsilon^{\mu\nu\rho\sigma} F_{\mu\nu}^{\text{em}} F_{\rho\sigma}^{\text{em}} = -\frac{\vartheta^{\text{em}} g^2}{8\pi^2} \int d^4x \, E_i^{\text{em}} B_i^{\text{em}} \tag{207}$$

incorporating an effective ϑ^{em} angle. The effective action represents the exact low energy theory. Therefore the effective ϑ cannot be rotated away. Because the electromagnetic Gauß's law is given as the functional derivative of the full action (146) w. r. t. A_0^{em}, it now has an additional term

$$\partial_i E_i^{\text{em}} = g\left(\rho_e + \frac{\vartheta^{\text{em}} g}{8\pi^2} \partial_i B_i^{\text{em}}\right) \, , \tag{208}$$

which is nonzero in the presence of monopoles (190). Integration therefore shows that the electric flux at infinity is not necessarily an integer because of ϑ^{em}.

$$\int d^3x \, \partial_i \left[E_i^a \phi^a\right] = ag\left(n_e + \frac{\vartheta^{\text{em}}}{2\pi}\right) \tag{209}$$

This is the effect reflected by the arbitrary parameter γ in the dyon solution (201), as can be tested by calculating its electric charge from the electric field

$$E_{\text{dyon}}^i \to \frac{x^i}{gr^3} \sinh\gamma \quad \text{for } r \to \infty$$
$$\Rightarrow n_e = \frac{4\pi \sinh\gamma}{g^2} \, . \tag{210}$$

In the classical, small coupling limit ($\vartheta^{\text{em}} = 0$), (203) indeed shows that $\sinh\gamma \to g^2/4\pi$ and hence the electric charge of the dyon (201) is one.

Saturation of the Central Charge As a result of these considerations, the central charge (182) is semiclassically given by (redefining $ga \to a$ as [16] does)

$$Z = \sqrt{2}\,(an_e + a_D n_m) \,, \quad n_e, n_m \in \mathbb{Z}$$
$$\text{with } a_D := a\tau_{\text{scl}} \,, \quad \tau_{\text{scl}} := \left(\frac{\vartheta^{\text{em}}}{2\pi} + \frac{4\pi i}{g^2}\right) \tag{211}$$

Note that the classically purely imaginary parameter τ_{class} (206) becomes complex in the quantum theory.

An at first sight surprising, but very important result is that in the semiclassical regime *all* particles, charged and uncharged, BPS monopoles (198) and dyons (202), obey the saturation equation for the Bogomolnyi bound and the relation (95) for particle masses of short SUSY representations, i. e. SUSY multiplets in which the central charge is saturated:

$$M = |Z| \,, \quad Z = \sqrt{2}\,(an_e + a_D n_m) \,. \tag{212}$$

The parameters a, a_D of the quantum moduli space and the electric and magnetic charges of the particles in the physical spectrum determine their masses uniquely, and because τ is complex, different a yield now indeed different physics, as has been expected from the breakdown of the classical $U(1)_\mathcal{R}$ symmetry.

To shed more light on (212), consider the number of states before and after spontaneous symmetry breaking. First, all fields are massless, and every SUSY multiplet contains $2^N = 4$ helicity states, while for massive fields, after the breakdown, one would usually expect $2^{2N} = 16$ states, cf. the section on representations of the SUSY algebra. On the other hand, one knows [5], [10] that the number of degrees of freedom is left unchanged by the Higgs mechanism: The electrically charged gauge fields $W_\mu^\pm$ "eat up" the would-be Goldstone bosons of the Higgs field and incorporate them as their longitudinal modes, leaving four states in the Higgs phase as well. How to resolve this apparent contradiction on the number of degrees of freedom?

The way out is to observe that when the central charge is saturated, i. e. the Bogomolnyi bound is reached, one remains with 4 physical states in each multiplet even after the spontaneous breakdown of the symmetry, as shown above. This is the deeper reason why (212) must be true in the semiclassical approximation for *all* physical particles [9].

6.4 The Full Quantum Theory

It is conceivable that relation (212) will not even be changed by nonperturbative quantum corrections, when one substitutes renormalized quatities for the bare ones. The above semiclassical reasoning may be wrong concerning the exact spectrum of particle masses in QCD, but one would trust its answers to qualitative questions like the number of particle states (four), the representations of the symmetry algebrae involved (SUSY) and the structure of the SUSY algebra itself in a theory in which SUSY is not broken. This fixes us to the Bogomolnyi bound by the above considerations. Therefore, in SUSY Yang Mills theories in the Higgs phase, the classical mass spectrum does not feel any quantum corrections and (212) is an exact result of the quantum theory. Nonetheless, one expects that relation (211) between a and a_D, which has been obtained semiclassically, has to be given up due to quantum corrections.

So, (212) remains valid, showing that particles without electric and magnetic charge (the photon and its $N = 2$ superpartners) are massless even beyond perturbation theory.

We start our discussion of the full quantum theory by enumerating the physical parameters which distinguish the theories. Classically, we have a dimensionless coupling constant, a microscopic ϑ parameter and the the modulus a. The theory is asymptotically free, the coefficient of g^3 in the β-function being -4 (this will be derived later). Therefore, through dimensional transmutation, the coupling constant is traded for a scale (e. g. a or u) in which all the dimensional quantities are measured but otherwise does not have any dynamical significance. Due to the presence of massless fermions the ϑ parameter can be rotated away. By doing a chiral change of variables of the fermionic field, through the axial anomaly a term with the same structure as the ϑ-term is produced.

As a consequence, physical quantities depend on one complex parameter, corresponding to the classical a. The exact definition of this renormalized, dimensional (measured in the unit specified above) complex parameter is convention dependent. We will discuss it in the following, but first turn again to (212) determining the mass of the particles.

6.5 The Duality Group

Two more properties carry over from semiclassics to the full quantum theory. The first one is that the ϑ^{em} vacuum angle of the effective theory can only be defined modulo 2π and so a shift

$$\vartheta^{\text{em}} \to \vartheta^{\text{em}} + 2\pi m \, , \quad m \in \mathbb{Z}$$
$$\text{i. e. } \tau \to \tau + m \, , \quad a_D \to a_D + ma \tag{213}$$

must be a symmetry of the spectrum as well,

$$M_{\vartheta^{\text{em}}} = M_{\vartheta^{\text{em}}+2\pi m} \, , \tag{214}$$

which implies that one has to redefine what one considers to be the electric and magnetic charge of a particle.

$$n_e \to n_e - m n_m \, , \quad n_m \to n_m \tag{215}$$

In the absence of magnetic monopoles, this symmetry is void.

Introducing the two vectors

$$\mathbf{n} := \begin{pmatrix} n_m \\ n_e \end{pmatrix} \, , \quad \mathbf{v} := \begin{pmatrix} a_D \\ a \end{pmatrix} \, : \, M = \sqrt{2} \, |\mathbf{n} \cdot \mathbf{v}| \, , \tag{216}$$

the mass formula (212) as well as the formula for the central charge (211) is invariant under any transformation

$$\mathbf{n}^T \to \mathbf{n}^T A^{-1} \ , \quad \mathbf{v} \to A\mathbf{v} \tag{217}$$

where the invertible 2×2 matrix A has to have integer entries in order to map arbitrary $\mathbf{n}$ into vectors with integer entries. Therefore $\det A, \det A^{-1} \in \mathbb{Z}\setminus\{0\}$, and with $\det A^{-1} = 1/\det A$, A has determinant ± 1 and hence must be a member of the product of the group of linear matrices with integer entries and unit determinant, SL(2,$\mathbb{Z}$), generated by

$$S := \begin{pmatrix} 0 & -1 \\ 1 & 0 \end{pmatrix} \text{ and } T := \begin{pmatrix} 1 & 1 \\ 0 & 1 \end{pmatrix}$$
$$\text{with } T^m = \begin{pmatrix} 1 & m \\ 0 & 1 \end{pmatrix}, \quad T^m T^{-m} = \mathbb{1}, \ m \in \mathbb{Z} \tag{218}$$

and the two element group

$$\{\mathbb{1}, \begin{pmatrix} 0 & 1 \\ 1 & 0 \end{pmatrix}\} \ . \tag{219}$$

T^m implements the transformation (213/215) of the ϑ^{em} angle, showing again that its entries must be integers.

The second element of the two element group can be ruled out to be a physically relevant transformation of the theory: It maps $\tau_{\text{scl}} \to \frac{1}{\tau_{\text{scl}}}$ as can be seen from (211) and hence violates the constraint that physical coupling constants must be real because of unitarity,

$$g^2 \geq 0 \ , \ \text{Im}\, \tau \geq 0 \ . \tag{220}$$

The matrix S interchanges electric and magnetic charges,

$$\begin{aligned} n_m \to -n_e \ , & \quad n_e \to n_m \\ a_D \to -a \ , & \quad a \to a_D \\ \text{i. e. } \tau \to & -\frac{1}{\tau} \ , \end{aligned} \tag{221}$$

and is compatible with the positivity constraint (220). It implements the other property of the quantum sector which is extremely important: *Duality*.

Consider first the classical theory: Since one knows that the effective low energy theory allows for monopoles and dyons, Maxwell's equations read

$$\partial^\mu F^{\text{em}}_{\mu\nu} = -g j^e_\nu \ , \quad \partial^\mu F^{\text{em}\,*}_{\mu\nu} = -g^{\text{class}}_m j^m_\nu \tag{222}$$

where $g^{\text{class}}_m = \frac{4\pi}{g}$. $j^{e/m}_\mu$ is the electric/magnetic current four vector, and $F^{\text{em}\,*}_{\mu\nu} = \frac{1}{2}\varepsilon_{\mu\nu}{}^{\rho\sigma} F^{\text{em}}_{\rho\sigma}$ is the dual field strength tensor. The second equation replaces the naive Bianchi identity of QED. Dirac was the first to notice that these equations remain unchanged under the replacement

$$F^{\text{em}}_{\mu\nu} \to F^{\text{em}*}_{\mu\nu} \ , \quad F^{\text{em}*}_{\mu\nu} \to -F^{\text{em}}_{\mu\nu}$$
$$gj^{\text{e}}_\mu \to g_m j^{\text{m}}_\mu \ , \quad g_m j^{\text{m}}_\mu \to -gj^{\text{e}}_\mu \tag{223}$$

where the minus sign is necessary because $(F^{\text{em}*}_{\mu\nu})^* = -F^{\text{em}}_{\mu\nu}$. This interchange of what one calls electric and magnetic fields ($E^{\text{em}}_i \to B^{\text{em}}_i$, $B^{\text{em}}_i \to -E^{\text{em}}_i$) and currents is called *duality transformation*. It interchanges also the role of electric and magnetic coupling constants by reversing as in (221). Therefore, S maps the sector of small electric coupling g to one of large magnetic coupling constant $\frac{4\pi}{g}$ and vice versa, and interchanges the role of fundamental particles and solitons. Hence it is *not* a symmetry of the theory, it merely shows the *same* theory in a *different* picture: For example, strongly interacting electric particles will better be described by weakly interacting monopoles and dyons.

In order to deepen the understanding of duality in a quantum theory, to stress once more that it is not a symmetry and to prepare calculations in the following sections, we discuss now the basic features of eletcromagnetic duality for the path integral in the simplest context, i.e. the free Abelian theory. Consider the Maxwell Lagrangean

$$\mathcal{L} = -\frac{1}{4g^2} F_{\mu\nu} F^{\mu\nu} \ , \tag{224}$$

where $F_{\mu\nu} = \partial_\mu A_\nu - \partial_\nu A_\mu$ and $g^2(x)$ is an x-dependent coupling constant. The generating functional $\mathcal{Z}$ is given by

$$\mathcal{Z} = \int \mathcal{D}A \exp\left(i \int d^4 x \mathcal{L}\right) \ . \tag{225}$$

Using the properties of Gaussian integration, one can alternatively represent $\mathcal{Z}$ as

$$\mathcal{Z} = \int \mathcal{D}A_\mu \int \mathcal{D}G_{\mu\nu} \exp\left(i \int d^4 x \left[\frac{1}{2} G_{\mu\nu} F^{\mu\nu} + \frac{g^2(x)}{4} G_{\mu\nu} G^{\mu\nu}\right]\right) \ . \tag{226}$$

Integrating first over A_μ we obtain a δ-distribution

$$\delta\left(\partial^\mu G_{\mu\nu}\right) \ , \tag{227}$$

whose solution is

$$G_{\mu\nu} = \varepsilon_{\mu\nu\rho\sigma} \partial^\rho B^\sigma \ , \tag{228}$$

where B_ρ is a "magnetic" gauge potential. Therefore, the generating functional has the alternative expression

$$\mathcal{Z} = \int \mathcal{D}B_\rho \exp\left(-i \int d^4 x \frac{g^2(x)}{4} \left(\partial_\rho B_\sigma - \partial_\sigma B_\rho\right)\left(\partial^\rho B^\sigma - \partial^\sigma B^\rho\right)\right) \ . \tag{229}$$

Introducing sources, we can easily prove that in gauge invariant correlators the electric and magnetic fields are interchanged in the two pictures, i.e.

$$F_{\mu\nu} = \varepsilon_{\mu\nu\rho\sigma}\partial^\rho B^\sigma \ . \tag{230}$$

The coupling constants have also a simple transformation rule interchanging weak and strong coupling as above, however the relation between the potential A_μ and B_ρ is extremely nonlocal.

The simple calculations above show again the special features of the duality transformation distinguishing it from a physical symmetry. The "electric" and "magnetic" photons do not form a multiplet but give two equivalent pictures of the same physical content. The same features can be seen also in the operatorial formalism. If one works in the $A_0 = 0$ gauge, the potentials A_i and the electric field E_i are conjugate variables and the physical Hilbert space is constrained by the Gauß law $\nabla \cdot \boldsymbol{E} = 0$. The physical, transverse photons are created by the magnetic field $\boldsymbol{B}$. Under a duality transformation, $\boldsymbol{E}$ and $\boldsymbol{B}$ are rotated into each other, (223), and the Gauß law and the $\nabla \cdot \boldsymbol{B} = 0$ operatorial equation are interchanged. The physical Hilbert space of transverse photons remains the same and both dual pictures are realized on it.

In the presence of matter, the generic situation remains analogous to the one described above. The duality transformation gives equivalent descriptions in electric or magnetic languages of the same physical content. Electric and magnetic charges of matter should be interchanged simultaneously with the change from electric to magnetic photons in the quantum theory. Electric charges and monopoles cannot be the fundamental particles simultaneously, i.e. in the same picture.

In very special situations (like e.g. $N = 4$ SUSY gauge theories) it could happen that in a given picture one has both electrically charged particles and monopoles present. Then the duality transformation could relate their properties, acting in this sense as an *ordinary* symmetry.

Duality is therefore an extremely powerful tool to connect weak and strong coupling sectors by redefining what one considers to be the fundamental field: Electrically charged particles which couple locally to photons or magnetically charged particles which couple locally to "magnetic" photons.

To conclude, the physically relevant group under which the central charge and mass formula (212) remain invariant, called the *(full) duality group*, is expected to be SL(2,$\mathbb{Z}$), inducing

$$\mathbf{v} \to \begin{pmatrix} a & b \\ c & d \end{pmatrix} \mathbf{v} \quad \text{and} \quad \tau \to \frac{a\tau + b}{c\tau + d} \tag{231}$$
$$\text{with} \quad ad - bc = 1 \ , \quad a, b, c, d \in \mathbb{Z} \ .$$

7 Towards a Solution of Pure $N=2$ SUSY QCD

This chapter gives an introduction to the line of arguments Seiberg and Witten [16] proposed for a construction of the complete mass spectrum of the quantum theory.

7.1 The Effective Theory

In the Seiberg-Witten approach, exact statements are made about the quantum effective action of the massless degrees of freedom.

As we discussed above, for large moduli the semiclassical approximation is reliable. We know, therefore, that the model is in the Higgs phase, i.e. the massless degrees of freedom form the $N = 2$, U(1) gauge supermultiplet. The multiplet contains, in a $N = 1$ notation, the gauge superfield W_α, and a chiral superfield Φ.

The effective Lagrangean represents the exact interaction of these massless degrees of freedom, and it is formally defined as the generating functional of the one particle irreducible Green's function involving these fields. The effective action respects all the non-anomalous symmetries of the theory, and knowing it, one has solved the theory. Clearly, if at some value of the modulus additional massless fields appear in the theory, the action should be changed, i.e. such points represent singularities in the modulus-space. The understanding of these points and therefore of the phase structure of the theory is one of the main results of the Seiberg-Witten approach.

The effective action is an infinite expansion in the derivatives of the fields. Seiberg and Witten limit themselves to terms containing at most two derivatives, motivated as follows.

The demand for perturbative renormalisability allows only terms in the effective action with not more than two derivatives and at most a four fermion interaction. The high symmetry constrains the vector multiplet part of the $N = 2$ SUSY QED effective Lagrangean to depend only on one single holomorphic functionof the $N = 2$ vectormultiplet Ψ^α [17]

$$\mathcal{L}_{\text{eff}} = \frac{1}{4\pi} \operatorname{Im} \int d^2\theta \int d^2\bar\theta \, \mathcal{F}(\Psi) \;, \tag{232}$$

or in terms of the $N=1$ chiral multiplet Φ and vector multiplet W^α which make up Ψ^α

$$\mathcal{L}_{\text{eff}} = \frac{1}{4\pi} \operatorname{Im} \left[\int d^2\theta \, \frac{1}{2} \frac{\partial^2 \mathcal{F}}{\partial \Phi^2} W^\alpha W_\alpha + \int d^2\theta \int d^2\bar\theta \, \frac{\partial \mathcal{F}}{\partial \Phi} \bar\Phi \right] \;. \tag{233}$$

Here, in line with redefining the Higgs field ϕ to include the coupling, g was absorbed into the definition of V, Φ and W^α. The *prepotential* $\mathcal{F}(\Phi)$ is a general holomorphic function. The relation between the first ("gauge") and second

("Kähler") term in (233) is a consequence of $N = 2$ SUSY. It can be easily proven by requiring that the interaction between the scalar field and its fermionic partner (coming from the first term) will be related by the $SU(2)_\mathcal{R}$ symmetry to the interaction between the scalar and the photino coming from the second term.

The effective action (233) has a flat direction: $\mathcal{F}$ can be expanded around an arbitrary complex value of Φ, a. This represents the exact, quantum definition of the modulus, i.e. (233) represents the effective action for the full moduli space of the theory. The knowledge of $\mathcal{F}(\Phi)$ gives simultaneously the information about the moduli dependence of the effective action and for a fixed modulus the dependence on the chiral superfield expanded around the modulus.

To calculate the masses (212), it is enough to construct a and a_D, which Seiberg and Witten show to be possible from the knowledge of the global properties of $\mathcal{F}$. The astonishing fact that these can be determined from the low energy effective theory at points at which certain fields become massless, as will be seen in the next section, is intimately connected to the fact that $\mathcal{F}$ is holomorphic.

Using the semiclassical Lagrangean (173), we read up that

$$\mathcal{F}_{\text{scl}}(\Psi) = \frac{1}{2}\tau_{\text{scl}}\Psi^2$$

$$\frac{\partial \mathcal{F}_{\text{scl}}}{\partial \Phi} = \tau_{\text{scl}}\Phi e^V \, , \quad \left.\frac{\partial \mathcal{F}_{\text{scl}}}{\partial \Phi}\right|_{\Phi=a} = a\tau_{\text{scl}} = a_D \tag{234}$$

$$\left.\frac{\partial^2 \mathcal{F}_{\text{scl}}}{\partial \Phi^2}\right|_{\Phi=a} = \tau_{\text{scl}} \, .$$

Together with the mass formula (212)

$$M = |Z| \, , \quad Z = \sqrt{2}\left(an_e + \left.\frac{\partial \mathcal{F}}{\partial \Psi}\right|_{\Psi=a} n_m\right) \, , \tag{235}$$

it is evident that in the full quantum theory

$$\tau(a) := \left.\frac{\partial^2 \mathcal{F}}{\partial \Psi^2}\right|_{\Psi=a}$$

$$\text{Im}\,\tau(a) =: \frac{4\pi}{g_{\text{eff}}^2(a)} \, , \quad \text{Re}\,\tau(a) =: \frac{\vartheta_{\text{eff}}^{\text{em}}(a)}{2\pi} \tag{236}$$

$$a_D(a) := \left.\frac{\partial \mathcal{F}_{\text{class}}}{\partial \Psi}\right|_{\Psi=a} \quad \text{and hence} \quad \frac{\partial a_D}{\partial a} = \tau$$

plays the rôle of the effective coupling constant in the Abelian theory, i.e. determines the (effective) $\vartheta_{\text{eff}}^{\text{em}}$ angle and coupling g_{eff} in the quantum theory.

We stress that all physical quantities will depend only on the complex parameter a.

7.2 Duality in the Effective Quantum Theory

Next, we discuss the duality properties of (233), i.e. the various transformations which relate effective actions with different $\mathcal{F}(\Phi)$ without changing the physical content of the theory. We start with the transformation which generalizes the electro-magnetic duality to the $N = 2$ SUSY situation. Defining

$$\Phi_D \equiv h(\Phi) := \frac{\partial \mathcal{F}}{\partial \Phi} , \qquad (237)$$

we see from (229) that the second term in (233) transforms as

$$\frac{d\Phi_D}{d\Phi} W_\alpha W^\alpha \to -\frac{1}{\frac{d\Phi_D}{d\Phi}} W_{D,\alpha} W_D^\alpha . \qquad (238)$$

In the same notation the first term in (233) can be written as

$$\operatorname{Im}\left(\Phi_D \overline{\Phi}\right) , \qquad (239)$$

which is the same as

$$-\operatorname{Im}\left(\Phi \overline{\Phi}_D\right) . \qquad (240)$$

Therefore we showed that there is a transformation of variables

$$W_\alpha \to W_{D,\alpha}$$
$$\Phi_D \to -\Phi \equiv -h^{-1}(\Phi_D) \qquad (241)$$
$$\Phi \to \Phi_D \equiv h(\Phi) ,$$

which gives a new effective action equivalent to (233).

The physical content is not changed if the function h is replaced by the function $-h^{-1}$. Even though the independent information is expressed in terms of the change of one function, it is more convenient to treat Φ and $\Phi_D \equiv h(\Phi)$ in a symmetric fashion as a doublet, as has been done for a and a_D in (216). In this way the transformation is expressed as

$$\begin{pmatrix} \Phi_D \\ \Phi \end{pmatrix} \to \begin{pmatrix} 0 & -1 \\ 1 & 0 \end{pmatrix} \begin{pmatrix} \Phi_D \\ \Phi \end{pmatrix} . \qquad (242)$$

Another transformation which leaves the physics unchanged is changing $\Phi_D \equiv h(\Phi)$ by a linear term in Φ with real coefficient, i.e.

$$\Phi_D \to \Phi_D + b\Phi$$
$$\Phi \to \Phi . \qquad (243)$$

Then the first term in (233) is changed by

$$\operatorname{Im}\left(b\Phi\overline{\Phi}\right) ,$$

while the second is changed by

$$b \, \text{Im} \left(\int d^2\theta \, W_\alpha W^\alpha \right) \sim b \, F_{\mu\nu} F^{\mu\nu} \ . \tag{244}$$

From (233) we see that b changes ϑ_{eff}. In an Abelian theory, (233) is a total derivative which cannot get contributions at infinity. However, when monopoles are present b has a physical influence since it shifts the charge of the monopoles. Therefore the transformation (243) leaves the physics invariant only if b is an integer and the transformation is accopanied by relabelling of the monopole electric charge. These legal transformations are in accordance with (218) generated by

$$\begin{pmatrix} \Phi_D \\ \Phi \end{pmatrix} \to \begin{pmatrix} 1 & 1 \\ 0 & 1 \end{pmatrix} \begin{pmatrix} \Phi_D \\ \Phi \end{pmatrix} \ . \tag{245}$$

The 2×2 matrices appearing in (242) and (245) generate the full SL(2,$\mathbb{Z}$) group of 2×2 matrices with integer entries and unit determinant, which is therefore the duality group of (233), and so one has shown that the duality group of Sect. (6.5) survives also in the quantum theory.

An important point related to duality is the exact definition of the central extensions in the $N = 2$ SUSY algebra. Using (233), one can calulate the generators of the algebra and their commutation relations. One expands around a value a of the modulus and defines $a_D := h(a)$. Then, repeating the calculation of Sect. (6.3), the electric and magnetic charge contributions to the central extension are normalized by a and a_D respectively. In particular if the massive fields are added to the effective action, for a short multiplet the mass will be given by (the coupling is absorbed in a and a_D)

$$M = \sqrt{2} \, | \, n_e a + n_m a_D \, | \tag{246}$$

where n_e, n_m are the electric and magnetic charges carried by the multiplet. This result, being algebraic, is exact. Equation (246) gives a definition of the quantum modulus a and of a_D through an actual physical measurement.

As seen in Sect. (6.5), in the presence of charged matter a duality transformation requires also a relabelling of the charges of the matter fields in such a way that the physics remains unchanged. From (246) it follows that if (Φ, Φ_D) transform according to a SL(2,$\mathbb{Z}$) transformation M,

$$\begin{pmatrix} \Phi_D \\ \Phi \end{pmatrix} \to M \begin{pmatrix} \Phi_D \\ \Phi \end{pmatrix} \ , \tag{247}$$

the electric and magnetic charges should be relabelled according to

$$\begin{pmatrix} n_m \\ n_e \end{pmatrix} \to M^{-1\,T} \begin{pmatrix} n_m \\ n_e \end{pmatrix} \ , \tag{248}$$

where M^T is the transpose of M. Therefore, the same theory characterized by a point in the moduli space can be described by different looking effective Lagrangeans accompanied by matter fields having different charge assignments.

The deep and surprising observation by Seiberg and Witten is that the aforementioned ambiguity in the description is a necessary feature of the $N = 2$ SUSY gauge theories. This class of theories have monodromies, i.e. there are closed paths in moduli space for which – when one returns to the initial point – one finds the effective action written in a different picture. The existence of monodromies is a mathematical consequence of the existence of singular points in moduli space in the functions $\mathcal{F}(\Phi)$ or $h(a)$. A general argument for the existence of singularities follows from (220),

$$\frac{4\pi}{g_{\text{eff}}^2} = \text{Im}\frac{dh(a)}{da} \geq 0 . \tag{249}$$

The imaginary part of an analytic function cannot have a well defined sign unless it is a constant (which would correspond to the classical theory). Since there are quantum corrections, $h(a)$ must have singularities in order to fulfill the inequality.

7.3 A Monodromy at Infinity

An explicit example of the appearence of monodromies is obtained if we take a closed path which stays all the time in the region where the modulus has a large absolute value. In such a domain, due to asymptotic freedom, perturbation theory is reliable and – the theory being $N = 2$ SUSY – there is an important *non-renormalization theorem* [17]: In SUSY $N = 2$ QED and QCD, the one loop calculation yields the exact result to every order in perturbation theory. Because of the high symmetry, fermionic and bosonic divergences cancel exactly beyond one loop. Furthermore, the result is uniquely determined by the transformation properties the effective action has to have in order the $U(1)_\mathcal{R}$ to be broken with the right remaining $\mathbb{Z}_4$ symmetry:

$$\mathcal{F}(\Psi) = \frac{1}{2}\tau_{\text{class}}\Psi^2\left[1 + \frac{g^2}{4\pi^2}\ln\frac{\Psi^2}{\Lambda^2}\right] + \sum_{k=1}^{\infty}\mathcal{F}_k\left(\frac{\Lambda}{\Psi}\right)^{4k} \tag{250}$$

The last term comes from possible nonperturbative k instanton contribution which drop out in the classical limit, as does $\vartheta^{\text{em}}(u \to \infty) \to 0$.

To get a feeling for this result, calculate the effective coupling τ from (236) for $|u|, |a| \to \infty$

$$a_D(a) = \left.\frac{\partial\mathcal{F}}{\partial\Phi}\right|_{\Phi=a} \to a\tau_{\text{class}}\left[\left(1 + \frac{g^2}{4\pi^2}\right) + \frac{g^2}{4\pi^2}\ln\frac{a^2}{\Lambda^2}\right] \tag{251}$$

$$\tau(a) = \left.\frac{\partial^2\mathcal{F}}{\partial\Phi^2}\right|_{\Phi=a} \to \tau_{\text{class}}\left[1 + \frac{g^2}{4\pi^2}\left(3 + \ln\frac{a^2}{\Lambda^2}\right)\right] , \tag{252}$$

whose imaginary part gives the running coupling

$$\frac{1}{g_{\text{eff}}^2(a)} \to \frac{1}{g^2}\left[1 + \frac{g^2}{4\pi^2}\left(3 + \ln\frac{a^2}{\Lambda^2}\right)\right] \quad \text{for } |a| \to \infty \tag{253}$$

and renormalization group beta function

$$\beta(g(a)) \to -\frac{1}{4\pi^2} g_{\text{eff}}^3(a) \quad \text{for } |a| \to \infty \ . \tag{254}$$

This is the expected result of the beta function of SU(2) QCD with two Weyl fermions λ, ψ and a complex scalar ϕ in the adjoint representation of the gauge group, cf. [10], Chap. 10.1.

The expression for $h(a)$ is

$$\Phi_D \equiv h(\Phi) = \frac{i\Phi}{\pi} \left[1 + \ln \frac{\Phi^2}{\Lambda^2} \right] \ . \tag{255}$$

In the large $|\Phi|$ domain, Φ^2 is the gauge invariant quantity. Therefore

$$\Phi = \rho \exp(i\alpha) \qquad 0 \leq \alpha \leq \pi \qquad \rho \text{ fixed, large} \tag{256}$$

represents a closed path in moduli space. On such a path, using (255), the change will be

$$\Phi_D \to -\Phi_D - \frac{i\Phi}{\pi} i 2\pi = -\Phi_D + 2\Phi \tag{257}$$
$$\Phi \to -\Phi \ .$$

Therefore, there is a nontrivial monodromy around infinity represented by the matrix

$$M_\infty := \begin{pmatrix} -1 & 2 \\ 0 & -1 \end{pmatrix} = S^2 T^{-2} \ . \tag{258}$$

It is indeed a member of the duality group under which the theory is invariant.

Although we already know that in the classical limit only the particles without electric and magnetic charge are light and describe the long range behavior of the theory, it is instructive to re-derive this result: The electric and magnetic charge of the massless particles must be unaltered by the monodromy, so that

$$\mathbf{n}^T \stackrel{!}{=} \mathbf{n}^T M_\infty^{-1} \Longrightarrow \mathbf{n} = 0 \ , \tag{259}$$

as expected.

7.4 Branch Points at Finite u

Assuming that the moduli space has only isolated singularites, there are regions in moduli space free of monodromies. This requires that, besides infinity, there are other singular points with monodromies M_i in such a way that

$$\prod_{i=1}^{n} M_i = M_\infty \ . \tag{260}$$

The physical origin of a singularity at a finite value of the modulus is the appearence of another massless state. Therefore, one should study the influence on the effective action (233) of matter multiplets with various charges, when their mass approaches zero.

Their electric and magnetic charge $\mathbf{n}_i$ can again be determined by the monodromy,

$$\mathbf{n}_i^T \stackrel{!}{=} \mathbf{n}_i^T M_i^{-1} \; . \tag{261}$$

Furthermore, with u_i also $-u_i$ must be a branch point because of the residual $\mathbb{Z}_2$ symmetry of broken $U(1)_\mathcal{R}$.

Electric Charge Branch Points We start with a multiplet carrying electric charge 1. If the highest spin in the multiplet is not larger than 1/2, such a multiplet will necessarily be short, and its mass will vanish for $a \to 0$ as a consequence of (246). In this case, we are dealing with an $N = 2$ SUSY extension of QED. Since QED is infrared free and the vanishing mass represents exactly the infrared limit, a one loop calculation of the effective action is reliable. Through the Ward identity, the running of the effective coupling constant $\tau(a)$ is reduced to the one loop correction to the photon propagator due to the matter circulating in the loop. The result of the calculation is

$$\tau(a) = -\frac{i}{\pi} \ln \frac{m}{\Lambda} = -\frac{i}{\pi} \ln \frac{a}{\Lambda} \; . \tag{262}$$

Integrating once, we obtain

$$a_\mathrm{D} \equiv h(a) = -\frac{i}{\pi} a \ln \frac{a}{\Lambda} + \Phi \; . \tag{263}$$

If a circles the singular point $a = 0$, the transformation is

$$\begin{aligned} a_\mathrm{D} &\to a_\mathrm{D} + 2a \\ a &\to a \; . \end{aligned} \tag{264}$$

Therefore, the monodromy matrix corresponding to a multiplet with charges $\mathbf{n} = (n_\mathrm{m}, n_\mathrm{e}) = (0, 1)$ becoming massless is

$$M_{(0,1)} = \begin{pmatrix} 1 & 2 \\ 0 & 1 \end{pmatrix} = T^2 \; . \tag{265}$$

If instead of charge one, we have a field with an electric charge q, the off-diagonal element in (265) is multiplied by q^2. It is clear from (265) that the presence of only electrically charged multiplets of this type cannot satisfy (260). The above discussion does not include the possibility that *all* the gauge bosons become massless i.e. that the full SU(2) symmetry is restored at some point in the moduli space. The existence of such a "non-Abelian Coulomb phase" is highly implausible.

There is only one point, $a = u = 0$, in the classical moduli space at which the gauge symmetry is "restored" and all gauge bosons with their superpartners are massless, yielding a possible monodromy, but Seiberg and Witten give arguments that it will disappear in the quantum moduli space due to quantum corrections.

That $u = 0$ is the only branch point can be ruled out for mathematical reasons: In that case, u would be single valued everywhere, and $\tau(a^2)$ analytic; $\operatorname{Im} \tau$ had to be harmonic, reaching its minimum only at infinity and therefore changing sign, which clashes with the need for a real coupling (220).

Therefore, at least two more singularities at points $u_1 \neq 0, u_{-1} = -u_1$ are necessary for a consistent solution, and Seiberg and Witten show that two are also sufficient.

The Monopole Branch Point and Confinement Note first that for $a \neq 0$ the charged gauge bosons and their superpartners are massive (211). Even without above considerations, we expect therefore the collective excitations, which we found in the semiclassical analysis, to become massless: Monopoles for $a \neq 0, a_D = 0$, and dyons for $an_e + a_D n_m = 0$ for some $n_e, n_m \neq 0$.

One is thus led to the conclusion that at least some of the singular points are related to fields carrying magnetic charge becoming massless. Such objects, monopoles and dyons, exist in the theory for large a. We calculate now the monodromy matrix corresponding to a point in moduli space where the monopole becomes massless, i.e. $a_D \to 0$. Such a calculation can be easily done using duality: We first do an electromagnetic duality transformation (242) to an electric picture. Now we can use the result of the electric calculation (265) and after that we transform back to the original picture using the inverse of (242). As a result, the monodromy matrix for the monopole is becoming the conjugate of $M_{(0,1)}$ of (242):

$$M_{(1,0)} = \begin{pmatrix} 1 & 0 \\ -2 & 1 \end{pmatrix} = ST^2 S^{-1} = SM_\infty^{-1} S \ . \tag{266}$$

Under a small perturbation, a point where the magnetically charged object becomes massless leads to the condensation of these objects. When a magnetically charged field has an expectation value, the electrically charged fields are confined by a dual Meissner effect.

In type II superconductors, Cooper pairs (pairs of electrons) are the analogue of above magnetically charged objects. Cooper pairs condense, hence have a nonzero vacuum expectation value, and by that confine magnetic fields which try to penetrate the solid to a flux tube of very small diameter. The ends of this "string" lie on the surface of the superconductor and can be interpreted as magnetic charges. Since the magnetic photon has only a small penetration length, it can be said to have become very heavy and hence unobservable. G. 't Hooft and S. Mandelstam [18] suggested that the QCD vacuum confines the colour electric charges by a dual mechanism: Here, magnetic charges condense and confine the colour electric fields to small tubes between the quarks.

Since the existence of singular points corresponding to particles carrying magnetic charge becoming massless is a necessity in $N = 2$ SUSY gauge theories, Seiberg and Witten show the presence of this mechanism of confinement in four dimensions at least for this class of non-Abelian gauge theories.

The Dyon Branch Point The monodromy (266) does not represent by itself a full solution of the singularity structure since it does not satisfy (260). Postulating another singular point one obtains a plausible minimal solution. If the new singular point has monodromy M', the matrix can be calculated from (260),

$$M' = M_{(1,0)}^{-1} M_\infty = \begin{pmatrix} -1 & 2 \\ -2 & 3 \end{pmatrix} = ST^{-2}ST^{-2} . \qquad (267)$$

We can calculate the charges of the field whose vanishing mass produced M' by (261) or equivalently by calculating the matrix X which conjugates $M_{(0,1)}$ to M'. The result is

$$X = \begin{pmatrix} -1 & 1 \\ -2 & 1 \end{pmatrix} . \qquad (268)$$

The matrix X transforms the theory in the presence of the new object to an electric picture with a field with charge $(0,1)$ present. Using (248) we conclude that the charges of the new object are $(1,-1)$, a dyon, and therefore the new singular point in the moduli space is characterized by

$$-a + a_D = 0. \qquad (269)$$

The $\mathbb{Z}_2$ non-anomalous symmetry takes the point $a_D = 0$ and (269) into each other.

7.5 Solution and Conclusions

Once the singularity structure in the moduli space is guessed, the solution of the infrared behavior is reduced to a well defined mathematical problem. One has to calculate the function $\mathcal{F}$ or alternatively, since it contains the same information, the function $a_D = h(a)$. It is convenient to treat the problem in a symmetric fashion by parametrising a, a_D in terms of a variable $u, a(u)$ and $a_D(u)$. Eliminating u, a_D as a function of a can be recovered. The constraints on $a(u), a_D(u)$ we have derived are summarized as follows:

- The functions should have three singular points $a = \infty, a_D = 0$ and $-a + a_D = 0$ with monodromies $M_\infty, M_{(1,0)}$ and M' respectively.
- Following from (249),

$$\operatorname{Im} \frac{da_D}{da} = \operatorname{Im} \frac{\frac{da_D}{du}}{\frac{da}{du}} \geq 0. \qquad (270)$$

These requirements fix uniquely the functions $a(u), a_D(u)$ up to reparametrisation of u. The solution involves the mapping of the physical problem into the problem of calculating the periods of an elliptic curve. The solution is given in a parametrisation where the singular points are at $u = 1, -1, \infty$,

$$a = \frac{\sqrt{2}}{\pi} \int_{-1}^{1} dx \frac{\sqrt{x-u}}{\sqrt{x^2-1}} \quad , \quad a_D = \frac{\sqrt{2}}{\pi} \int_{1}^{u} dx \frac{\sqrt{x-u}}{\sqrt{x^2-1}} , \qquad (271)$$

and (212) yields the exact spectrum.

We discuss now further the implications of the results for the understanding of confinement. The exact result produces just two isolated points ($u = \pm 1$) where the monopoles and the dyons become massless, respectively. In a $N = 2$ SUSY framework we cannot go beyond these points. If we break $N = 2$ to $N = 1$, however, we can penetrate into the phase where the monopoles (or dyons) are condensed. Consider e.g. the perturbation $m \operatorname{tr} \Phi^2$ to the original Lagrangean, where Φ is the chiral multiplet in the adjoint representation. Near the point in the moduli space where there are massless monopoles for $m = 0$, the exact superpotential in the magnetic picture is

$$W = \sqrt{2}\Phi_D M \tilde{M} + m U(\Phi_D) , \qquad (272)$$

where $M, \tilde{M}$ is the monopole multiplet and the first term is required by $N = 2$ SUSY. The vacuum is obtained by minimizing (272) supplemented with the D-term condition

$$|M| = |\tilde{M}|. \qquad (273)$$

The equations are

$$\sqrt{2} M \tilde{M} + m \frac{dU}{d\Phi_D} = 0 \qquad (274)$$

$$\Phi_D M = \Phi_D \tilde{M} = 0 , \qquad (275)$$

which have as solution

$$M = \tilde{M} = \left(-\frac{m}{\sqrt{2}} \frac{dU}{d\Phi_D}\bigg|_{\Phi_D = 0} \right)^{1/2}. \qquad (276)$$

Therefore the monopole field has a vacuum expectation value, i.e. from the point of view of the effective theory the system is in a magnetic Higgs phase. Through the dual Meissner effect [18], electric lines are expelled from the system. If $\pm q$ external charges at a distance d are coupled to the system, the electric flux lines will be forced to join the charges on the shortest path i.e. the interaction energy will increase linearly with d, yielding confinement.

We remark that the monopoles which condense are U(1) monopoles, i.e. they will confine any particle having a non-vanishing Abelian charge. In this respect this mechanism seems different from the expected $\mathbb{Z}_N$ monopole confinement which would confine just particles with the nontrivial transformation properties under the center of the SU(N) group. However, it is clear that there is no phase boundary between regions where the two apparently different confinement mechanisms act. This is related to the fact that in four dimensions we cannot find a gauge invariant order parameter which would distinguish between the condensation of the two types of monopoles.

As a summary, the quantum moduli space of $N = 2$ SUSY SU(2) Yang Mills theory must have at least three branch points. The minimal solution consists of a moduli space in which three different kinds of particles become massless at certain points:

(i) At $|u| = \infty$, in the perturbative region, monopoles and dyons are heavy and become unobservable.
(ii) About some point u_1, monopoles are light and act as matter in dual $N = 2$ SUSY QED. They are local and stable. Electric charges are confined by the dual Meissner effect if monopoles are condensed, requiring an $N = 1$ SUSY.
(iii) At $u_{-1} = -u_1$, dyons become massless.

On the other hand, one should stress that a series of (very plausible) assumptions was made to arrive at this result:

- The Bogomolnyi bound (212) was assumed to hold non-perturbatively;
- Monopoles, dyons and the fundamental fields were assumed to be the only physical particles, and no new particles would arise in the quantum theory;
- The duality group is a invariance group of the quantum sector (This is proven if the path integral analogue of the dual transformation has no undiscovered pathologies.);
- Only massless particles can yield logarithmic behavior and hence branch points in the non-perturbative regime, at finite u.

In the end, we should keep in mind that the applicability of $N = 2$ SUSY pure SU(2) gauge theory to the real world is unfortunately limited, a fact acknowledged even by the most ardent of SUSY supporters:

- In the real world, supersymmetry (if it is a physical symmetry at all) is heavily broken, since we observe no supersymmetric partners to e.g. gluons. But then the Bogomolnyi bound (212) is invalidated, and no universal relation between masses and charges of the observed particles is known. Although e.g. all quarks have the same colour charge, their masses differ very much.
- Supersymmetry being broken, we lack the Higgs field ϕ in the adjoint representation of the gauge group, whose flat directions proved so vital and which "broke down" the gauge symmetry. The Higgs mechanism was essential to determine the effective low energy theory which was used to solve the model.
- A non-renormalisation theorem for the effective low energy theory, as well as for full QCD is not at hand.

The generalization to supersymmetric QCD with arbitrary number of colours and flavours and both massless and massive quarks has been performed as well [19]. Here, chiral symmetry breaking could be demonstrated besides the exact spectrum.

These notes follow loosely the lectures given by A. Schwimmer of the Weizmann Institute at Rehovot, Israel, at the workshop "Nonperturbative QCD", organised by the Graduiertenkolleg Erlangen-Regensburg on October 10-12, 1995 in Kloster Banz, Germany. We thank him for a critical reading of the manuscript and an extensive revision of the last two chapters.

A more thorough introduction into the work of Seiberg and Witten may be found in [20] [21] [22]. There is no need to remind the reader that the subject evolves rapidly.

References

[1] J. Wess, J. Bagger: *Supersymmetry and Supergravity*, Princeton, New Jersey: Princeton University Press, 1992
[2] D. Bailin, A. Love: *Supersymmetric Gauge Field Theory and String Theory*, Bristol, Philadelphia: Institute of Physics Publishing, 1994
[3] M.F. Sohnius: *Introducing Supersymmetry*, Physics Reports **128** (1985) 39-204
[4] R.U. Sexl, H.K. Urbantke: *Relativität, Gruppen, Teilchen*; Wien, New York: Springer 1992
[5] L.H. Ryder: *Quantum Field Theory*, Cambridge: Cambridge University Press, 1985
[6] L. O'Raifeartaigh: Phys. Rev. **139**B (1965) 1052
[7] Coleman, Mandula: Phys. Rev. **159** (1967) 1251
[8] R. Haag, M. F. Sohnius, J. T. Łopszuanski: Nucl. Phys. **B88** (1975) 257
[9] D. Olive, E. Witten, Phys. Lett. **B78**(1978), 97.
[10] T.-P. Cheng, L.-F. Li, *Gauge Theory of Elementary Particle Physics*, Claredon Press, Oxford 1989.
[11] G. 't Hooft, Nucl. Phys. **B79**(1974), 276; A. M. Polyakov, Sov. Phys. JETP Lett. **20**(1974), 194.
[12] R. Jackiw, *Topological Effects on the Physics of the Standard Model*, in this volume.
[13] M. Nakahara, *Geometry, Topology and Physics*, Graduate Student Series in Physics, Adam Hilger 1991.
[14] E. B. Bogomolnyi, Sov. J. Nucl. Phys. **24**(1976), 449; M. K. Prasad, C. M. Sommerfield, Phys. Rev. Lett. **35**(1975), 760.
[15] E. Witten, Phys. Lett. **B86**(1979), 283.
[16] N. Seiberg, E. Witten, Nucl. Phys. **B426**(1994), 19.
[17] M. T. Grisaru, P. C. West, Nucl. Phys. **B254**(1985), 249; N. Seiberg, Phys. Lett. **B206**(1988), 75.
[18] S. Mandelstam, Phys. Rep. **C23**(1976), 245; G. 't Hooft, Nucl. Phys. **B190**(1981), 455; G. 't Hooft, Phys. Scr. **25**(1982), 133.
[19] N. Seiberg, E. Witten, Nucl. Phys. **B431**(1995), 484.
[20] C. Gómez, R. Hernández, *Electric and Magnetic Duality and Effective Field Theories*, Lectures given at the Advanced School on Effective Theories, Almuñecar, Granada, 1995, preprint FTUAM 95/63, hep-th/9510023.
[21] E. Alvarez, L. Alvarez-Gaume, I. Bakas, *Supersymmetry and Dualities*, Lectures presented at the International Conference on Mirror Symmetry and S Duality, Trieste, Italy, Jun 5-9, 1995, preprint CERN-TH-95-258, hep-th/9510028.
[22] A. Bilal, *Duality in N=2 SUSY SU(2) Yang-Mills Theory: A Pedagogical Introduction to the Work of Seiberg and Witten*, Lectures presented at the Rencontre Theoristes et Mathematiciens, Strasbourg, France, Dec 1995, preprint LPTENS-95-53, hep-th/9601007.

Subject Index

Bold face refers to entire sections or subsections starting at the page indicated.

$1/N$-expansion **207**, see also Feynman diagram; Large N_c QCD
β-function 165, 256
η'-particle 155, 165, 176, see also U(1)-problem
σ model 3, **27**, **29**, **31**, 33, 173, 175, 176
θ-angle, -vacua **95**, **100**, 128, 137, 155, 164

Action
- classical, 3, 37, 130, 134, 136, 138, 141
- continuum, 22, 27, 36, 45, 62
- effective, 9, 13, 31, 120, 135, 136, 139, 146, 254, **260**, 262, 264, 266
- fixed point, 15, 19, 25, 30–33
- lattice, 24, 28, 29, 41, 45, 59, 65
- – perfect **27**
- Liouville, 150
- Wilson, **41**, 63, 80

Adiabatic approximation 75–77, 161, 163, 202
Aharonov–Bohm effect 141
Anomaly **139**
- axial, 62, 100, **102**, 112, 128, 141, **145**, 150, 154, 165, 254, 256, see also U(1)-problem
- chiral, **157**, 164, 165
- conformal, see Anomaly, Weyl
- physical consequences, 111, 148, **151**, 157, **164**, see also Decay: π^0, baryon, Higgs; U(1)-problem
- QCD, **164**, 202
- scale, 157, 165, 166, 168
- Standard Model, 151, 152, 217, 248
- trace, see Anomaly, scale
- U(1), see Anomaly, axial
- VAAA, 202
- Weyl, 143, **149**

Area law 47
Asymptotic freedom 21, 28, 32, 68, 71, 72, 251, 253, 264
Atiyah–Singer index theorem 141

B-meson 180, 181
Background field 99, 100, 103, 104, 109, 185, 186, 211
Baryon
- charmed, 215
- decay, see Decay
- number, 112, 154, 165, 188, 196, 212

Bethe–Salpeter amplitude 70
Bianchi identity 91, 96, 250, 251, 258
Bloch momentum 101, 102
Block spin transformation 12, 13, 26, 29, 31
Bogomolnyi bound 251–253, 255, 270
Borel transformation **176**, 179, 186
Born–Oppenheimer approximation 163, 164, see also Semiclassical approximation

Canonical quantization **91**, 122, 197, 212
Casimir operator 204, 219, 221
Central charge 229, 232, 233, **250**, 254, 259
Chern–Simons form 96, 108, see also Pontryagin density
Chern–Simons theory 113, 121, 122, **122**
Chiral perturbation theory 195, 210
Cluster decomposition 164

Subject Index

Coherent state 39
Coleman–Mandula theorem 220, 225
Coleman–Weinberg potential 138, 139
Collective
– behavior, 2, 4, 8
– coordinate, **207**, see also Rigid rotator
Condensate **176**
– fermion, 164, 179, 184, 186
– gluon, 155, 173, 184
– quark, 173
Confinement phase transition **51**
Continuum limit 21, 30, 41–43, 45, **55**, 60, 62, 63, 68
Cooling 79, 80, 82–85, 87
Correlation length **6**
Cosmological constant 111, 150, 151
Coupling, (ir)relevant, marginal see Operator, (ir)relevant, marginal
CP–invariance 111
Critical phenomena **5**
Critical surface 14, 16
Critical temperature 6

Debye screening 118
Decay
– α, 129
– η', 155
– τ, 71
– π^0, 112, 128, 148, **153**, 165, 168, 201
– B–meson, 180, 181
– baryon, 128, 137, **154**
– constants, 68, 191, 192
– Higgs, 157, **166, 167**
– – branching ratio 157, 168
Deformation effects 196, 199, 211
Density–density correlation 73, 76, 77, 85, 87
Dibaryon 214, 215
Dirac sea 141, 162
Dispersion relation 23, 24, 26, 177, 186
Double line notation 189, 191
Dual Meissner effect 267, 269, 270
Duality 73, 110, **256, 262**, 267, 270
Dyon **253, 254**, 267, **268**, 269, 270

Effective
– action, see Action
– Hamiltonian, 180
– interaction, 166, 168, 169, 201

– Lagrangean, 167, 202, 205, 212
– potential, 116, **138**, 155

Fermion
– doubling, **59**, 64
– quenched, 68, 70, 71, 77, 87
– staggered, **63**
– Wilson, **62**
Fermion generations **151**
Feynman diagram (planar) 67, 189, 191, 193
Fiducial domain **175**, 178, 179
Fixed point **13**, 25, 30, 79
– action, see Action
– Gaußian, 15, 16, **16**
Flux tube 47, 67, 76, 267
Fujikawa method 141, 144–146, 148
Functional
– Gaußian, 175
– generating, 9, 10, 12, 17, 18, 40, 113, **134**, 136, 138, 144, 258, 260
– integral, 108, 118, **131**

Gauß' law 44, 91–99, 122, 124, 254, 259, see also Canonical quantization; Schrödinger representation
Gauge
– Coulomb, 74, 75, 94, 159
– Fock–Schwinger, 168, 185, 186
– invariance, 3, 41, 107, 110, 120, 121, 162
– temporal, see Gauge, Weyl
– Weyl, 43, 44, 91–93, 97
Gauge transformation
– large, **95**, 99, 101, 102, 108
– small, 95, 99, 108
Gell-Mann–Oakes–Renner 176, 205
Gell-Mann–Okubo mass formula 210
General Relativity **142**
Georgi–Glashow-model 249–251
Glueball 176, 192, 193, 195
Goldstone boson 73, 138, 153–155, 165, 176, 200, 202, 245, see also Symmetry breakdown
Gravity 151
Green functions
– finite temperature, **113**
Group
– duality, see Duality

- gauge, 41, 45, 90, 103, 125, 189, 244, 248, 249, 265, 270
- Lie, 95, 124, 219, 221, 235
- Lorentz, 218, 220–224, 226, 237
- measure, 45
- Poincaré, 218, 219, 230, 233
- SU(N), **44**, 269
- U(1), 248

Hadron mass   70, 71
Hadron production   71
Hadronic observables   **68**
Hamilton–Jacobi formalism   121, 129, 155
Hard thermal loops   **116**, 118, **120**, 124
Hartree–Fock   74, 194, 199, 204
Heat bath method   43
Heavy quark theory   183
Hedgehog   196, 251
Higgs boson   166, 249–252, 255, 260, 270, see Decay
Higgs model   **247**
High temperature QCD   **113**
Higher loop effects   116, 139
Homotopy   62, 95, 196
Hopping parameter   **65**, 84
Hypercharge   152, 203
Hyperfine splittings   213

Imaginary–time formalism   114, 115, **115**, 116
Instanton   27, 71–73, **77**, **90**, **100**, 128, **136**, 154, 155, 176, 249
- action, 82, 112, 154
- background, 77, 154
- vacuum, 73, 176

Jacobian   40, 143–145

Kubo formula   **118**

Landau damping   113, 118
Landau levels   207
Laplace operator (perfect)   **22**, 32
Large N_c QCD   **189**
Lattice artifacts   23, 24, 28–30, 34
Level flow   163
Levinson theorem   141
Lie algebra   123, 124, 144, 220, 248, 249
- graded, 220, 224, 225, 235

- Lorentz group, **218**
Linear response theory   118
Liouville theorem   133
Locality   4, 29, 30
Lorentz invariance   92, 167

Magnetic moments   199, 214
Magnetic monopole   95, 112, **250**, 263, **267**, 269, 270
Mass splitting   182, 183, 194, 204–206, 213
Matsubara frequency   115
Mean field   194
Metropolis algorithm   43
Moduli space   249, 251, 255, 261, 263–269
Multiplet
- decuplet, 204, 205, 210
- exotic, 204, 208, 213
- octet, 165, 204, 205, 210
- supersymmetric, 233, 237, 246, 260

Nambu–Goldstone mode   165
Nielsen–Ninomiya theorem   62
Noether current   103, 148, 247
Noether theorem   209

Operator product expansion   71, 173, 176, 182, 184–186
- pragmatic, **173**, 176, 184
Operator: (ir)relevant, marginal   15, 18, 21, 30
Order prameter   65, 269
OZI-rule   192

Partition function   5, 32, 44, 66, 67, 133, 145
PCAC   112, 153, 176
Plasma waves   118, see also Landau damping
Polarization tensor   110, 118, 120, 171, 175, 178, 179, 185, 186
Polyakov line   **45**
Pontryagin density   97, 102, 107, 110, see also Chern–Simons form
Profile function   202, 212, 214

Quark model   67, 199, 204, 205, 214, 215
- constituent, 188

Subject Index   275

- non–relativistic, 188, 189, 199, 210, 213
Quark–antiquark potential   47
Quark–gluon plasma   113

Real–time formalism   115
Regularization
- ζ–function, 135, 138, 139, 144–146
- dimensional, 3
- lattice, **2**
- point splitting, 103, 104, 162
Renormalizability   112, 151, 152, 159, 260
Renormalization   **55**
Renormalization group   **3**
- coupling space, 14
- flow, **29**
- transformation, **8**, 17, 21
-- eigenvalue 15, 19
Renormalized coupling constant   **28**
Renormalized trajectory   30
Resonances
- J/ψ, 176, 178
- ρ, 178, 179
- charmonium, 171, 175
Response function   119
Rigid rotator   **207**, see also Collective coordinate
Running coupling constant   33, 189, 264

Scale invariance   164
Schrödinger representation   **93**, 95
Schwinger model   **98**, **109**, 111, **157**, 164
Semiclassical approximation   128, **129**, 139, 176, 199, 255, 260, see also WKB–approximation
Skyrme
- Lagrangean, see also Effective Lagrangean
Skyrme Lagrangean   211
Skyrme model
- three flavour, **200**
- two flavour, **195**
Solitons   128, **136**, 188, 194, 197, 198, 211, 258, see also Skyrme model
Spectral function   71, 72, 84, 119, 120
Stationary path approximation   132
Strangeness content of the nucleon   206
String tension   47, **55**, 83, 84

String theory   149
Strong coupling expansion   47, **49**, 66
Superfield   **234**, 260, 261
Superpotential   241, 245, 247, 269
Superspace   236
Supersymmetry Algebra   **224**, **230**
Symmetry
- $\mathcal{R}$, 229, 232, 233, 246, 249, 255, 261, 264, 266
- axial, 103, 110
- charge conjugation, 104
- chiral, 62, 63, 65, 102, 103, 107, 110, 141, 148, 153, 164, 176, 218, 229, 270
- crossing, 167
- gauge, 21, 103, 107, 110, 125, 248, 266, 270
- Poincaré, 141, **142**
- time reversal, 140
Symmetry breakdown   21, 155, 215, see also Goldstone boson
- spontaneous, 63, 111, 128, **138**, 139, 249, 255

Thermo field dynamics   115
Top quark   112, 152
Topological charge   79, 80, 82, 188, see also Winding number
Topological field theory   see Chern–Simons theory
Topological susceptibility   80, 84
Trace Class   135
Triangle anomaly   see Anomaly
Triangle graph   110, 148, 152, 153
Tunneling   101, 102, 112, 128, 129, 136, 137, 154, see also Instantons; WKB–approximation

U(1)–problem   111, **154**, see also η'–particle
Universality   4

Vlasov equation   120

Ward identity   107, 110, 266
Weak decay amplitudes   **180**, 214
Wess–Zumino–Witten term   201–203, 207, 212, 213, see also Skyrme model, three–flavour
Wigner–Weyl mode   165

Wilson loop 41, **45**, 67, 83
Winding number 78, 79, 96, 99, 101, 102, 108, 125, 141, 188, 196, 251, 252, *see also* Chern–Simons form; Pontryagin density
WKB–approximation 102, 123, 124, 128, **129**, 146

Young tableau 199, 204

Springer and the environment

At Springer we firmly believe that an international science publisher has a special obligation to the environment, and our corporate policies consistently reflect this conviction.

We also expect our business partners – paper mills, printers, packaging manufacturers, etc. – to commit themselves to using materials and production processes that do not harm the environment. The paper in this book is made from low- or no-chlorine pulp and is acid free, in conformance with international standards for paper permanency.

Lecture Notes in Physics

For information about Vols. 1–449
please contact your bookseller or Springer-Verlag

Vol. 450: M. F. Shlesinger, G. M. Zaslavsky, U. Frisch (Eds.), Lévy Flights and Related Topics in Physics. Proceedigs, 1994. XIV, 347 pages. 1995.

Vol. 451: P. Krée, W. Wedig (Eds.), Probabilistic Methods in Applied Physics. IX, 393 pages. 1995.

Vol. 452: A. M. Bernstein, B. R. Holstein (Eds.), Chiral Dynamics: Theory and Experiment. Proceedings, 1994. VIII, 351 pages. 1995.

Vol. 453: S. M. Deshpande, S. S. Desai, R. Narasimha (Eds.), Fourteenth International Conference on Numerical Methods in Fluid Dynamics. Proceedings, 1994. XIII, 589 pages. 1995.

Vol. 454: J. Greiner, H. W. Duerbeck, R. E. Gershberg (Eds.), Flares and Flashes, Germany 1994. XXII, 477 pages. 1995.

Vol. 455: F. Occhionero (Ed.), Birth of the Universe and Fundamental Physics. Proceedings, 1994. XV, 387 pages. 1995.

Vol. 456: H. B. Geyer (Ed.), Field Theory, Topology and Condensed Matter Physics. Proceedings, 1994. XII, 206 pages. 1995.

Vol. 457: P. Garbaczewski, M. Wolf, A. Weron (Eds.), Chaos – The Interplay Between Stochastic and Deterministic Behaviour. Proceedings, 1995. XII, 573 pages. 1995.

Vol. 458: I. W. Roxburgh, J.-L. Masnou (Eds.), Physical Processes in Astrophysics. Proceedings, 1993. XII, 249 pages. 1995.

Vol. 459: G. Winnewisser, G. C. Pelz (Eds.), The Physics and Chemistry of Interstellar Molecular Clouds. Proceedings, 1993. XV, 393 pages. 1995.

Vol. 460: S. Cotsakis, G. W. Gibbons (Eds.), Global Structure and Evolution in General Relativity. Proceedings, 1994. IX, 173 pages. 1996.

Vol. 461: R. López-Peña, R. Capovilla, R. García-Pelayo, H. Waelbroeck, F. Zertuche (Eds.), Complex Systems and Binary Networks. Lectures, México 1995. X, 223 pages. 1995.

Vol. 462: M. Meneguzzi, A. Pouquet, P.-L. Sulem (Eds.), Small-Scale Structures in Three-Dimensional Hydrodynamic and Magnetohydrodynamic Turbulence. Proceedings, 1995. IX, 421 pages. 1995.

Vol. 463: H. Hippelein, K. Meisenheimer, H.-J. Röser (Eds.), Galaxies in the Young Universe. Proceedings, 1994. XV, 314 pages. 1995.

Vol. 464: L. Ratke, H. U. Walter, B. Feuerbach (Eds.), Materials and Fluids Under Low Gravity. Proceedings, 1994. XVIII, 424 pages. 1996.

Vol. 465: S. Beckwith, J. Staude, A. Quetz, A. Natta (Eds.), Disks and Outflows Around Young Stars. Proceedings, 1994. XII, 361 pages, 1996.

Vol. 466: H. Ebert, G. Schütz (Eds.), Spin – Orbit-Influenced Spectroscopies of Magnetic Solids. Proceedings, 1995. VII, 287 pages, 1996.

Vol. 467: A. Steinchen (Ed.), Dynamics of Multiphase Flows Across Interfaces. 1994/1995. XII, 267 pages. 1996.

Vol. 468: C. Chiuderi, G. Einaudi (Eds.), Plasma Astrophysics. 1994. VII, 326 pages. 1996.

Vol. 469: H. Grosse, L. Pittner (Eds.), Low-Dimensional Models in Statistical Physics and Quantum Field Theory. Proceedings, 1995. XVII, 339 pages. 1996.

Vol. 470: E. Martínez-González, J. L. Sanz (Eds.), The Universe at High-z, Large-Scale Structure and the Cosmic Microwave Background. Proceedings, 1995. VIII, 254 pages. 1996.

Vol. 471: W. Kundt (Ed.), Jets from Stars and Galactic Nuclei. Proceedings, 1995. X, 290 pages. 1996.

Vol. 472: J. Greiner (Ed.), Supersoft X-Ray Sources. Proceedings, 1996. XIII, 350 pages. 1996.

Vol. 473: P. Weingartner, G. Schurz (Eds.), Law and Prediction in the Light of Chaos Research. X, 291 pages. 1996.

Vol. 474: Aa. Sandqvist, P. O. Lindblad (Eds.), Barred Galaxies and Circumnuclear Activity. Proceedings of the Nobel Symposium 98, 1995. XI, 306 pages. 1996.

Vol. 475: J. Klamut, B. W. Veal, B. M. Dabrowski, P. W. Klamut, M. Kazimierski (Eds.), Recent Developments in High Temperature Superconductivity. Proceedings, 1995. XIII, 362 pages. 1996.

Vol. 476: J. Parisi, S. C. Müller, W. Zimmermann (Eds.), Nonlinear Physics of Complex Systems. Current Status and Future Trends. XIII, 388 pages. 1996.

Vol. 477: Z. Petru, J. Przystawa, K. Rapcewicz (Eds.), From Quantum Mechanics to Technology. Proceedings, 1996. IX, 379 pages. 1996.

Vol. 479: H. Latal, W. Schweiger (Eds.), Perturbative and Nonperturbative Aspects of Quantum Field Theory. Proceedings, 1996. X, 430 pages. 1997.

Vol. 480: H. Flyvbjerg, J. Hertz, M. H. Jensen, O. G. Mouritsen, K. Sneppen (Eds.), Physics of Biological Systems. From Molecules to Species. X, 364 pages. 1997.

Vol. 481: F. Lenz, H. Grießhammer, D. Stoll (Eds.), Lectures on QCD. VII, 276 pages. 1997.

Vol. 482: X.-W. Pan, D. H. Feng, M. Vallières (Eds.), Contemporary Nuclear Shell Models. Proceedings, 1996. XII, 309 pages. 1997.

New Series m: Monographs

Vol. m 1: H. Hora, Plasmas at High Temperature and Density. VIII, 442 pages. 1991.

Vol. m 2: P. Busch, P. J. Lahti, P. Mittelstaedt, The Quantum Theory of Measurement. XIII, 165 pages. 1991. Second Revised Edition: XIII, 181 pages. 1996.

Vol. m 3: A. Heck, J. M. Perdang (Eds.), Applying Fractals in Astronomy. IX, 210 pages. 1991.

Vol. m 4: R. K. Zeytounian, Mécanique des fluides fondamentale. XV, 615 pages, 1991.

Vol. m 5: R. K. Zeytounian, Meteorological Fluid Dynamics. XI, 346 pages. 1991.

Vol. m 6: N. M. J. Woodhouse, Special Relativity. VIII, 86 pages. 1992.

Vol. m 7: G. Morandi, The Role of Topology in Classical and Quantum Physics. XIII, 239 pages. 1992.

Vol. m 8: D. Funaro, Polynomial Approximation of Differential Equations. X, 305 pages. 1992.

Vol. m 9: M. Namiki, Stochastic Quantization. X, 217 pages. 1992.

Vol. m 10: J. Hoppe, Lectures on Integrable Systems. VII, 111 pages. 1992.

Vol. m 11: A. D. Yaghjian, Relativistic Dynamics of a Charged Sphere. XII, 115 pages. 1992.

Vol. m 12: G. Esposito, Quantum Gravity, Quantum Cosmology and Lorentzian Geometries. Second Corrected and Enlarged Edition. XVIII, 349 pages. 1994.

Vol. m 13: M. Klein, A. Knauf, Classical Planar Scattering by Coulombic Potentials. V, 142 pages. 1992.

Vol. m 14: A. Lerda, Anyons. XI, 138 pages. 1992.

Vol. m 15: N. Peters, B. Rogg (Eds.), Reduced Kinetic Mechanisms for Applications in Combustion Systems. X, 360 pages. 1993.

Vol. m 16: P. Christe, M. Henkel, Introduction to Conformal Invariance and Its Applications to Critical Phenomena. XV, 260 pages. 1993.

Vol. m 17: M. Schoen, Computer Simulation of Condensed Phases in Complex Geometries. X, 136 pages. 1993.

Vol. m 18: H. Carmichael, An Open Systems Approach to Quantum Optics. X, 179 pages. 1993.

Vol. m 19: S. D. Bogan, M. K. Hinders, Interface Effects in Elastic Wave Scattering. XII, 182 pages. 1994.

Vol. m 20: E. Abdalla, M. C. B. Abdalla, D. Dalmazi, A. Zadra, 2D-Gravity in Non-Critical Strings. IX, 319 pages. 1994.

Vol. m 21: G. P. Berman, E. N. Bulgakov, D. D. Holm, Crossover-Time in Quantum Boson and Spin Systems. XI, 268 pages. 1994.

Vol. m 22: M.-O. Hongler, Chaotic and Stochastic Behaviour in Automatic Production Lines. V, 85 pages. 1994.

Vol. m 23: V. S. Viswanath, G. Müller, The Recursion Method. X, 259 pages. 1994.

Vol. m 24: A. Ern, V. Giovangigli, Multicomponent Transport Algorithms. XIV, 427 pages. 1994.

Vol. m 25: A. V. Bogdanov, G. V. Dubrovskiy, M. P. Krutikov, D. V. Kulginov, V. M. Strelchenya, Interaction of Gases with Surfaces. XIV, 132 pages. 1995.

Vol. m 26: M. Dineykhan, G. V. Efimov, G. Ganbold, S. N. Nedelko, Oscillator Representation in Quantum Physics. IX, 279 pages. 1995.

Vol. m 27: J. T. Ottesen, Infinite Dimensional Groups and Algebras in Quantum Physics. IX, 218 pages. 1995.

Vol. m 28: O. Piguet, S. P. Sorella, Algebraic Renormalization. IX, 134 pages. 1995.

Vol. m 29: C. Bendjaballah, Introduction to Photon Communication. VII, 193 pages. 1995.

Vol. m 30: A. J. Greer, W. J. Kossler, Low Magnetic Fields in Anisotropic Superconductors. VII, 161 pages. 1995.

Vol. m 31: P. Busch, M. Grabowski, P. J. Lahti, Operational Quantum Physics. XI, 230 pages. 1995.

Vol. m 32: L. de Broglie, Diverses questions de mécanique et de thermodynamique classiques et relativistes. XII, 198 pages. 1995.

Vol. m 33: R. Alkofer, H. Reinhardt, Chiral Quark Dynamics. VIII, 115 pages. 1995.

Vol. m 34: R. Jost, Das Märchen vom Elfenbeinernen Turm. VIII, 286 pages. 1995.

Vol. m 35: E. Elizalde, Ten Physical Applications of Spectral Zeta Functions. XIV, 228 pages. 1995.

Vol. m 36: G. Dunne, Self-Dual Chern-Simons Theories. X, 217 pages. 1995.

Vol. m 37: S. Childress, A.D. Gilbert, Stretch, Twist, Fold: The Fast Dynamo. XI, 410 pages. 1995.

Vol. m 38: J. González, M. A. Martín-Delgado, G. Sierra, A. H. Vozmediano, Quantum Electron Liquids and High-T_c Superconductivity. X, 299 pages. 1995.

Vol. m 39: L. Pittner, Algebraic Foundations of Non-Commutative Differential Geometry and Quantum Groups. XII, 469 pages. 1996.

Vol. m 40: H.-J. Borchers, Translation Group and Particle Representations in Quantum Field Theory. VII, 131 pages. 1996.

Vol. m 41: B. K. Chakrabarti, A. Dutta, P. Sen, Quantum Ising Phases and Transitions in Transverse Ising Models. X, 204 pages. 1996.

Vol. m 42: P. Bouwknegt, J. McCarthy, K. Pilch, The W_3 Algebra. Modules, Semi-infinite Cohomology and BV Algebras. XI, 204 pages. 1996.

Vol. m 43: M. Schottenloher, A Mathematical Introduction to Conformal Field Theory. VIII, 142 pages. 1997.

Vol. m 44: A. Bach, Indistinguishable Classical Particles. VIII, 157 pages. 1997.

Vol. m 45: M. Ferrari, V. T. Granik, A. Imam, J. C. Nadeau (Eds.), Advances in Doublet Mechanics. XVI, 214 pages. 1997.

Vol. m 48: P. Kopietz, Bosonization of Interacting Fermions in Arbitrary Dimensions. XII, 257 pages. 1997.